PRAISE FOR THE MANGA GUIDE SE

"Highly recommended."
—CHOICE MAGAZINE ON *THE MANGA GUIDE TO DATABASES*

"Stimulus for the next generation of scientists."
—SCIENTIFIC COMPUTING ON *THE MANGA GUIDE TO MOLECULAR BIOLOGY*

"A great fit of form and subject. Recommended."
—OTAKU USA MAGAZINE ON *THE MANGA GUIDE TO PHYSICS*

"The art is charming and the humor engaging. A fun and fairly painless lesson on what many consider to be a less-than-thrilling subject."
—SCHOOL LIBRARY JOURNAL ON *THE MANGA GUIDE TO STATISTICS*

"This is really what a good math text should be like. Unlike the majority of books on subjects like statistics, it doesn't just present the material as a dry series of pointless-seeming formulas. It presents statistics as something *fun*, and something enlightening."
—GOOD MATH, BAD MATH ON *THE MANGA GUIDE TO STATISTICS*

"I found the cartoon approach of this book so compelling and its story so endearing that I recommend that every teacher of introductory physics, in both high school and college, consider using it."
—AMERICAN JOURNAL OF PHYSICS ON *THE MANGA GUIDE TO PHYSICS*

"A single tortured cry will escape the lips of every thirty-something biochem major who sees *The Manga Guide to Molecular Biology*: 'Why, oh why couldn't this have been written when I was in college?'"
—THE SAN FRANCISCO EXAMINER

"A lot of fun to read. The interactions between the characters are lighthearted, and the whole setting has a sort of quirkiness about it that makes you keep reading just for the joy of it."
—HACK A DAY ON *THE MANGA GUIDE TO ELECTRICITY*

"*The Manga Guide to Databases* was the most enjoyable tech book I've ever read."
—RIKKI KITE, LINUX PRO MAGAZINE

"For parents trying to give their kids an edge or just for kids with a curiosity about their electronics, *The Manga Guide to Electricity* should definitely be on their bookshelves."
—SACRAMENTO BOOK REVIEW

"This is a solid book and I wish there were more like it in the IT world."
—SLASHDOT ON *THE MANGA GUIDE TO DATABASES*

"*The Manga Guide to Electricity* makes accessible a very intimidating subject, letting the reader have fun while still delivering the goods."
—GEEKDAD BLOG, WIRED.COM

"If you want to introduce a subject that kids wouldn't normally be very interested in, give it an amusing storyline and wrap it in cartoons."
—MAKE ON *THE MANGA GUIDE TO STATISTICS*

"This book does exactly what it is supposed to: offer a fun, interesting way to learn calculus concepts that would otherwise be extremely bland to memorize."
—DAILY TECH ON *THE MANGA GUIDE TO CALCULUS*

"The art is fantastic, and the teaching method is both fun and educational."
—ACTIVE ANIME ON *THE MANGA GUIDE TO PHYSICS*

"An awfully fun, highly educational read."
—FRAZZLEDDAD ON *THE MANGA GUIDE TO PHYSICS*

"Makes it possible for a 10-year-old to develop a decent working knowledge of a subject that sends most college students running for the hills."
—SKEPTICBLOG ON *THE MANGA GUIDE TO MOLECULAR BIOLOGY*

"This book is by far the best book I have read on the subject. I think this book absolutely rocks and recommend it to anyone working with or just interested in databases."
—GEEK AT LARGE ON *THE MANGA GUIDE TO DATABASES*

"The book purposefully departs from a traditional physics textbook and it does it very well."
—DR. MARINA MILNER-BOLOTIN, RYERSON UNIVERSITY ON *THE MANGA GUIDE TO PHYSICS*

"Kids would be, I think, much more likely to actually pick this up and find out if they are interested in statistics as opposed to a regular textbook."
—GEEK BOOK ON *THE MANGA GUIDE TO STATISTICS*

"*The Manga Guide to Statistics* offers a visualization of statistics that can't be found in any mere textbook."
—ANIME 3000

"A great introduction for readers of any age, and an exemplar of technical communication."
—LINUX USERS OF VICTORIA ON *THE MANGA GUIDE TO ELECTRICITY*

THE MANGA GUIDE™ TO THE UNIVERSE

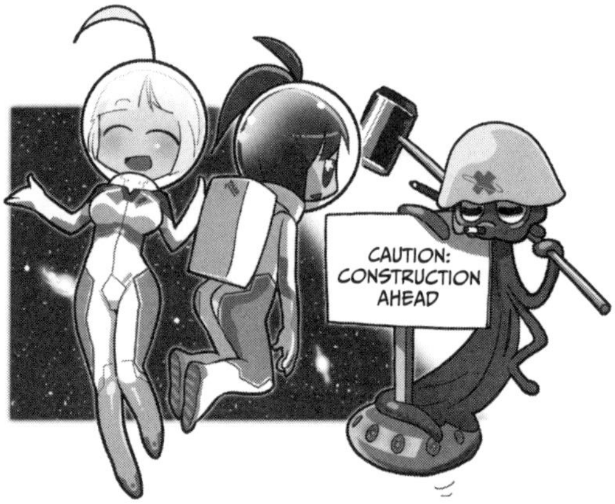
CAUTION:
CONSTRUCTION
AHEAD

THE MANGA GUIDE™ TO THE
UNIVERSE

KENJI ISHIKAWA,
KIYOSHI KAWABATA,
AND VERTE CORP.

THE MANGA GUIDE TO THE UNIVERSE. Copyright © 2011 by Kenji Ishikawa, Kiyoshi Kawabata, and Verte Corp.

The Manga Guide to the Universe is a translation of the Japanese original, *Manga de wakaru uchu*, published by Ohmsha, Ltd. of Tokyo, Japan, © 2008 by Kenji Ishikawa, Kiyoshi Kawabata, and Verte Corp.

This English edition is co-published by No Starch Press, Inc. and Ohmsha, Ltd.

All rights reserved. No part of this work may be reproduced or transmitted in any form or by any means, electronic or mechanical, including photocopying, recording, or by any information storage or retrieval system, without the prior written permission of the copyright owner and the publisher.

Printed in Canada

15 14 13 12 11   1 2 3 4 5 6 7 8 9

ISBN-10: 1-59327-267-7
ISBN-13: 978-1-59327-267-8

Publisher: William Pollock
Supervising Editor: Kiyoshi Kawabata
Author: Kenji Ishikawa
Illustrator: Yutaka Hiiragi
Producer: Verte Corp.
Production Editors: Ansel Staton and Serena Yang
Developmental Editor: Tyler Ortman
Translator: Arnie Rusoff
Technical Reviewers: Adam Blythe Smith and Kebra Ward
Compositor: Riley Hoffman
Copyeditor: Paula L. Fleming
Indexer: BIM Indexing & Proofreading Services

For information on book distributors or translations, please contact No Starch Press, Inc. directly:
No Starch Press, Inc.
38 Ringold Street, San Francisco, CA 94103
phone: 415.863.9900; fax: 415.863.9950; info@nostarch.com; http://www.nostarch.com/

Library of Congress Cataloging-in-Publication Data

Kawabata, Kiyoshi, 1940-
  [Manga de wakaru uchu. English]
  The manga guide to the universe / by Kiyoshi Kawabata, Kenji Ishikawa, and Verte Corp.
       p. cm.
  Includes index.
  ISBN-13: 978-1-59327-267-8
  ISBN-10: 1-59327-267-7
  1. Cosmology--Comic books, strips, etc. 2. Graphic novels. I. Ishikawa, Kenji, 1958- II. Verte Corp. III. Title.
  QB982.K3913 2010
  523.1--dc22
                                                         2009023969

No Starch Press and the No Starch Press logo are registered trademarks of No Starch Press, Inc. Other product and company names mentioned herein may be the trademarks of their respective owners. Rather than use a trademark symbol with every occurrence of a trademarked name, we are using the names only in an editorial fashion and to the benefit of the trademark owner, with no intention of infringement of the trademark.

The information in this book is distributed on an "As Is" basis, without warranty. While every precaution has been taken in the preparation of this work, neither the author nor No Starch Press, Inc. shall have any liability to any person or entity with respect to any loss or damage caused or alleged to be caused directly or indirectly by the information contained in it.

All characters in this publication are fictitious, and any resemblance to real persons, living or dead, is purely coincidental.

CONTENTS

FOREWORD... xi

PREFACE.. xiii

PROLOGUE
A TALE THAT BEGINS ON THE MOON.................................. 1
The Story of Kaguya-hime.. 10
Cosmic Myths... 18
    Ancient India's View of the Universe.............................. 18
    Ancient Egypt's View of the Universe.............................. 18
    Ancient Babylonia's View of the Universe.......................... 19
In China, Where Astronomy Was Originally Developed................... 19
In Ancient Greece, Where the Size of Earth Was Calculated............ 20
    Eratosthenes' Calculation Method................................. 20
If Earth Is Round, the Moon Must Be Round Too....................... 21

1
IS EARTH THE CENTER OF THE UNIVERSE?............................ 23
A Mysterious Light Appeared in the Sky.............................. 24
Close Encounters... 27
Does the Sun Revolve Around Earth?................................. 34
A Heliocentric Model Was Proposed 2,300 Years Ago................... 40
From the Geocentric Theory to the Heliocentric Theory............... 50
Galileo's Discoveries—and Trial..................................... 56
Putting Things in Perspective....................................... 59
What Is the Approximate Distance to the Horizon?.................... 66
Measuring the Size of the Universe: How Far to the Moon?............ 67
    Corner Cube Mirrors.. 67
    How a Corner Cube Mirror Works................................... 67
    Before the Corner Cube Prism..................................... 68
Geocentric Theory vs. Heliocentric Theory—the Outcome of a Battle Royale 69
    What Kind of Orbit Did a Planet Trace in the Geocentric Theory?... 70
    The Tychonic System That Embellished the Geocentric Theory....... 70
    Just How Progressive Was Copernicus?............................. 71
    Kepler Completed the Heliocentric Theory......................... 72
    What Did Galileo Do?... 72
    What Has the Heliocentric Theory Taught Us?...................... 73
A Somewhat Complicated Explanation of Kepler's Laws................. 73
    First Law: The Orbit of Every Planet Is an Ellipse with the Sun at the Focus 73
    Second Law: A Line Joining a Planet and the Sun Sweeps Out Equal Areas
        During Equal Intervals of Time................................ 75
    Third Law: The Square of the Orbital Period of a Planet Is Directly Proportional
        to the Cube of the Semimajor Axis of Its Orbit................ 77

2
FROM THE SOLAR SYSTEM TO THE MILKY WAY 79

What If Kaguya-hime Came from a Planet in Our Solar System?..................... 80
Kaguya-hime and the Solar System.. 82
    Mercury.. 83
    Venus 84
    Mars 85
    Jupiter.. 86
    Saturn.. 87
    Uranus 88
    Neptune 89
    Pluto 90
    Earth.. 91
    The Moon 92
    The Sun 95
The Size of the Milky Way Galaxy.. 104
What's in the Middle of the Galaxy?.. 106
Top Five Mysteries of the Galaxy That Have Not Yet Been Explained! 108
    What Is the Galaxy's Shape, and How Did It Form? 108
    What's at the Center?... 108
    How Are Supermassive Black Holes Formed? 109
    What Is the Galaxy Made Of?.. 109
    What Will Happen When We Collide with the Andromeda Galaxy? 109
The Milky Way Galaxy Is One of Many Galaxies 110
The Universe Is Steadily Getting Larger... 116
    Why Can We See the Milky Way?... 116
    A Disc-Shaped Galactic Model Is the Easiest to Understand 117
    Results of Scientific Observation Also Prove a Disc-Shaped Universe... 118
    An Idea from Kant Enlarged the Perceived Universe in a Flash............. 119
    How Did Technology for Observing the Universe Progress?.................. 120
    Famous Telescopes 122
    What Can a Radio Telescope Observe?.. 124
Another Way to Measure the Size of the Universe: A Triangulation Trick ... 125
    Triangulation Can Give Us the Distance to Stars Beyond the Solar System 126
How Big Is the Solar System?... 127

3
THE UNIVERSE WAS BORN WITH A BIG BANG 129

Galaxies Are Islands of Light in the Void of Space 130
The Winning Team Learns a Lesson... 133
What Is the Large-Scale Structure of the Cosmos?..................................... 140
    Planetary System 140
    Galaxy.. 140
    Group of Galaxies or Cluster of Galaxies... 140
    Supercluster of Galaxies... 141

Hubble's Great Discovery 142
    The Origins of the Universe: "Hubble's Great Discovery—Act I" 143
    Back to the Play: "Hubble's Great Discovery—Act II". 146
If the Universe Is Expanding... 151
Everything Started with the Big Bang 161
Hubble's Theory of the Expansion of the Universe Was Imperfect. 162
Three Pieces of Evidence for the Big Bang Theory 166
Do Aliens Exist? 180
    Calculating the Number of Extraterrestrial Civilizations 180
    Extraterrestrial Life and a World-Renowned Physicist. 181
    Has Life Been Created Often? 182
    Which Is the Closest Star System That Could Support Extraterrestrial Life?. 183
    Can We Contact an Extraterrestrial Civilization?. 184
    Tardigrades (Water Bears) Are the Toughest Astronauts 185
A Third Method of Measuring the Size of the Universe: If You Know the Properties of a Star,
    Can You Figure Out How Far Away It Is? 186
    Stars with Varying Brightness Are "Lighthouses of the Universe" 188
    Methods of Measuring Even Greater Distances 189

4
WHAT IS IT LIKE AT THE EDGE OF THE UNIVERSE? 191

Where Is the Universe Going? 192
The Closest Earthlike Planet. 203
The Kaguya-go Journey Board Game 206
Arrival at the "Edge" of the Universe 208
Professor Sanuki's Soliloquy. 209

5
OUR EVER-EXPANDING UNIVERSE. 213

The Big Show. 215
The Multiverse Contains Numerous Universes 219
The Edge, Birth, and End of the Universe.... 219
    Why Might Space Be Curved? 219
    Will You Return to the Same Location in a Plane, a Cylinder, and a Sphere? 220
    Negative Curvature. 221
    Friedmann's Dynamic Universe. 222
What Will Ultimately Become of the Universe? 227
WMAP and Our Flat Universe. 229
The True Age of the Universe 232

INDEX 235

GALLERY OF ASTRONOMICAL MARVELS . . . . following page 242

FOREWORD

It was a great pleasure to have been able to provide assistance to Kenji Ishikawa in creating this book. I was very thankful for the help I had received when I wrote my own book, *A Distant 14.6 Billion Light Year Journey*, and felt that I wanted to return the favor now. Be that as it may, I inspected the manuscript as carefully as possible, and I aimed for complete accuracy by requesting corrections or revisions with little thought of any nuisance it might be to the author or publisher.

Research related to the universe is progressing steadily now, and even researchers find it difficult to sufficiently understand the leading edge of their own fields. Even more, it's next to impossible for anyone to gain a complete understanding of the universe. From that perspective, when I first read this book, I was surprised that it discussed major observational and theoretical results, starting from a view of the solar system and finally arriving at cosmology. I was also pleased to find diligent explanations of the basics of astrophysics and astronomy. This book, which overflows with the author's extraordinary care and enthusiasm for solving the riddles of the universe, makes an extremely lively, unique, and wholesome handbook.

Furthermore, the power of manga as a means of communication is enormous, and it goes without saying that it is much more effective than any mere accumulation of words. If a new perspective on the universe can be gained with this book, new interest in the universe can be generated, or readers can be encouraged to aspire to gaining a better understanding of its mysteries, it will be an unexpected delight to me, since I have been involved in cosmology for a great many years.

KIYOSHI KAWABATA
SUPERVISING EDITOR
NOVEMBER 2008

PREFACE

While I was writing the scenario for this book, a cameraman (whom I was working with on an entirely different project) unexpectedly said to me, "Recently, I've been enjoying thinking about the universe."

I had no idea why he brought up the subject—it just popped up in the middle of an ordinary conversation. When I asked him why he mentioned it, the cameraman said, "Well, imagining what's happening in the universe uses my brain in a completely different way than my work does, so it's really refreshing to think about."

Of course! In our work, we're constantly worrying about random details so that we won't make mistakes. Our minds get fatigued, just like how certain muscles ache if we do a repetitive task for a long time. And just like how we can relieve muscle fatigue by doing some light exercise, it's always good to take a break from work to think about something else for a while. A topic such as "What is happening in the universe?" fits the bill perfectly. Since I also like thinking about the universe, I came up with a few pearls of wisdom to share with the cameraman:

- "Since the entire universe is moving and space itself is expanding, there's no way to indicate a specific place using coordinates."
- "We still don't seem to know what most of the matter and energy that make up the galaxy consists of."
- "There may be universes other than our own universe."

Even though these thoughts are more like hints at mysteries than true knowledge, the cameraman was intrigued, so the two of us batted around ideas for a little while. It was just a short conversation, but I have extremely fond memories of it.

So why is the universe so interesting? Perhaps it's because we may not arrive at any answers, no matter how much we think about it.

Of course, mankind has compiled a great deal of knowledge about the universe so far. The Big Bang theory, which sheds light on the secrets of the creation of matter as well as the beginning of the universe, the discovery of the large-scale structure of space, and other great discoveries have provided valuable answers in our quest for the full picture of the cosmos. However, whenever we gain new knowledge, we only seem to discover more mysteries. The history of research about the universe is like climbing up a mountain to see what's beyond it, only to see another mountain, and then another, and another . . .

For example, consider the Moon. The question of whether there is water on the Moon has been debated for quite a long time. If there is a large amount of water there, oxygen could be created by decomposing it, and since it could also be used as drinking water, it may make it possible to construct a lunar outpost on the surface of the Moon. This is an important subject for mankind, and opinions about it have been tossed back and forth for some time.

Since the matter constituting the Moon is similar to that of Earth, simple consideration would lead us to believe that there should have been water there at some point. However, the Moon hardly has any atmosphere, and it seems more like a desert landscape—what is left behind after moisture evaporates and is dispersed into space. But if we knew that there were craters that were always in a shadow near the north and south poles, then there would be a greater possibility that water may be retained in ice in those locations. What would the result be?

Unfortunately, according to the reports from the Japanese spacecraft Kaguya, the existence of ice near the south pole could not be confirmed. The most recent conclusion is that although there is a possibility that water may be hidden in the ground, even if there were water or ice, it would be an extremely small amount. However, if we are able to explore under the surface of the Moon, that "answer" may change again.

Since this kind of mystery still remains about the Moon—our closest neighbor in the universe!—when we expand our target to the solar system, galaxy, group of galaxies, and so on, we'll surely be inundated with things we do not understand. Nevertheless, any thought experiments we conduct in order to try to make conjectures, while paying our respect to the efforts of our predecessors who devoted their utmost efforts to get closer to the truth, will not be just mental calisthenics, but they could lead to Nobel Prize–grade discoveries.

Kanna, Gloria, and Yamane, the three high school girls who are the heroines of this book, show a light-hearted interest in the universe at first. But as their knowledge grows, they gradually become fascinated with its allure. By the end of the story, the girls are learning about the outer limits of astronomy and astrophysics.

To encourage you to enjoy their story, I wanted to avoid difficult discussions in the comics and text as best as I could. Since I am the type who wants to throw away science books the moment numeric formulas appear, I tried my best to keep this in mind.

The universe is all around us, and we are but a small part of it. It's only natural that we ponder our planet's place in the universe and dream of exploring our universe's farthest edges. "I've been enjoying thinking about the universe." I will take great pleasure as an author if this is your thought after reading this book.

KENJI ISHIKAWA
OCTOBER 2008

PROLOGUE
A TALE THAT BEGINS ON THE MOON

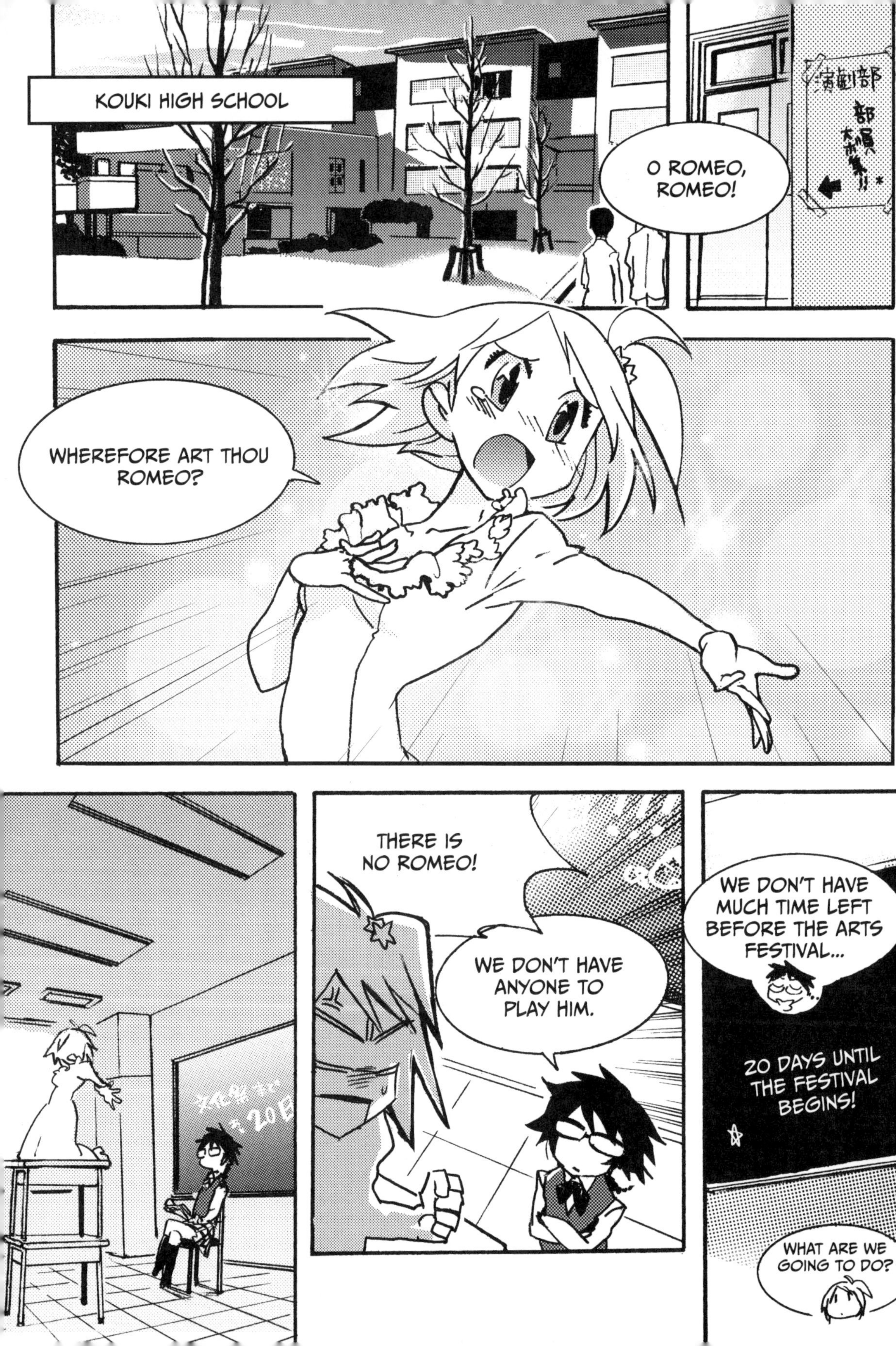
KOUKI HIGH SCHOOL
O ROMEO, ROMEO!
WHEREFORE ART THOU ROMEO?
THERE IS NO ROMEO!
WE DON'T HAVE ANYONE TO PLAY HIM.
WE DON'T HAVE MUCH TIME LEFT BEFORE THE ARTS FESTIVAL...
20 DAYS UNTIL THE FESTIVAL BEGINS!
WHAT ARE WE GOING TO DO?

* DRAMA CLUB! WANTED: CLUB MEMBERS!

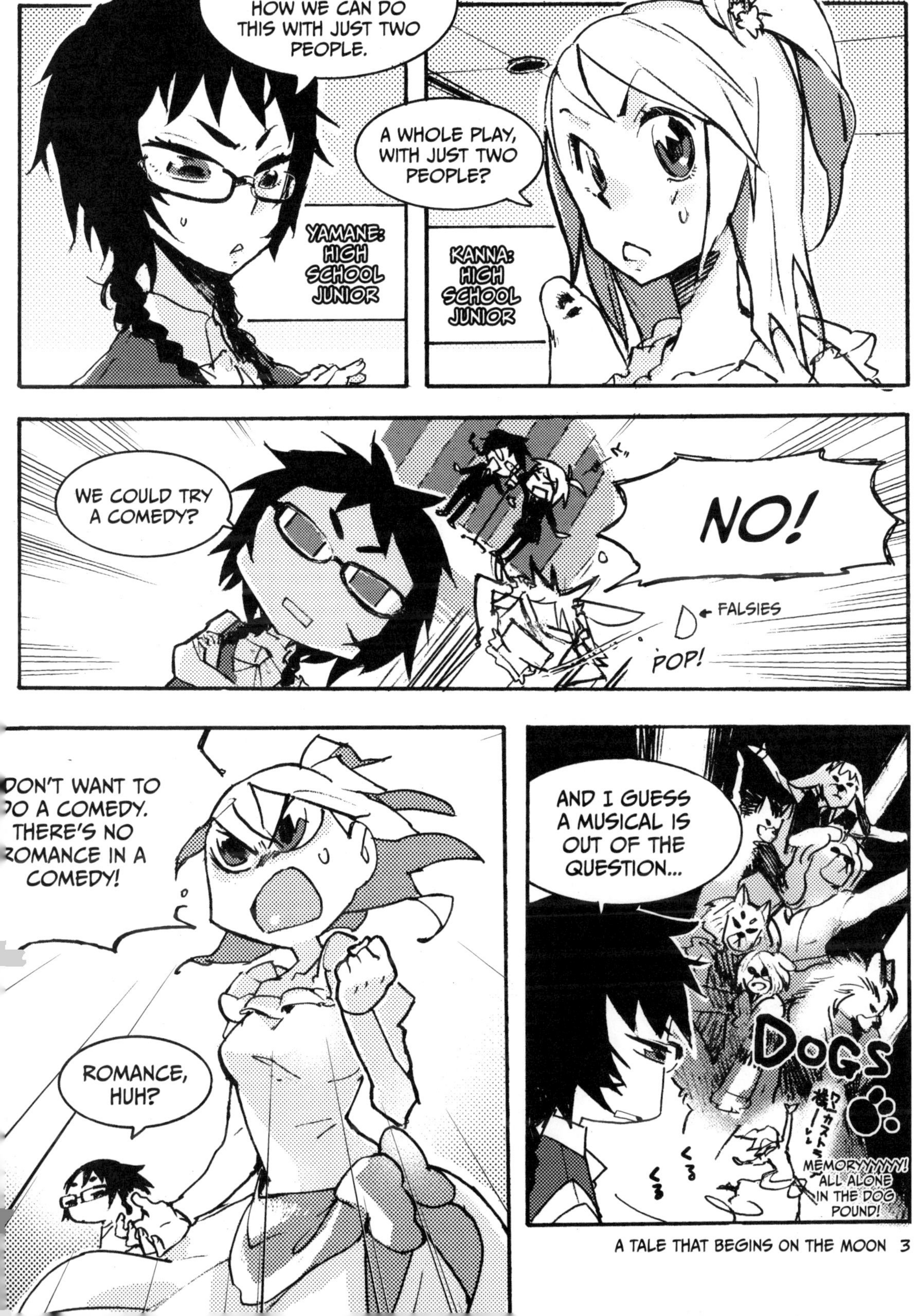
WELL, WE'D BETTER START THINKING ABOUT HOW WE CAN DO THIS WITH JUST TWO PEOPLE.
YAMANE: HIGH SCHOOL JUNIOR
A WHOLE PLAY, WITH JUST TWO PEOPLE?
KANNA: HIGH SCHOOL JUNIOR
WE COULD TRY A COMEDY?
NO!
FALSIES
POP!
DON'T WANT TO DO A COMEDY. THERE'S NO ROMANCE IN A COMEDY!
ROMANCE, HUH?
AND I GUESS A MUSICAL IS OUT OF THE QUESTION...
DOGS
MEMORYYYYYY! ALL ALONE IN THE DOG POUND!

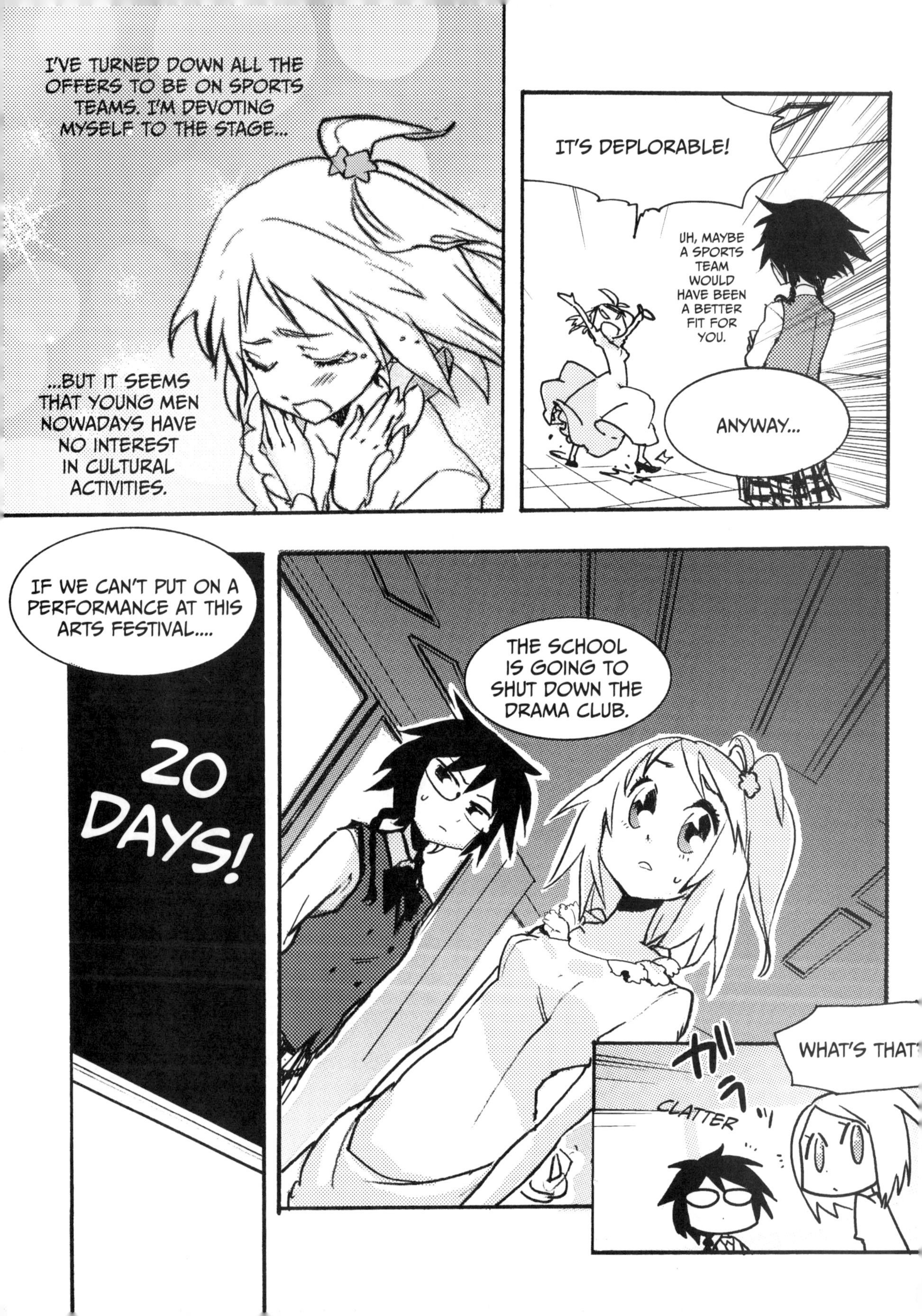
I'VE TURNED DOWN ALL THE OFFERS TO BE ON SPORTS TEAMS. I'M DEVOTING MYSELF TO THE STAGE...
...BUT IT SEEMS THAT YOUNG MEN NOWADAYS HAVE NO INTEREST IN CULTURAL ACTIVITIES.
IT'S DEPLORABLE!
UH, MAYBE A SPORTS TEAM WOULD HAVE BEEN A BETTER FIT FOR YOU.
ANYWAY...
IF WE CAN'T PUT ON A PERFORMANCE AT THIS ARTS FESTIVAL....
20 DAYS!
THE SCHOOL IS GOING TO SHUT DOWN THE DRAMA CLUB.
WHAT'S THAT?
CLATTER

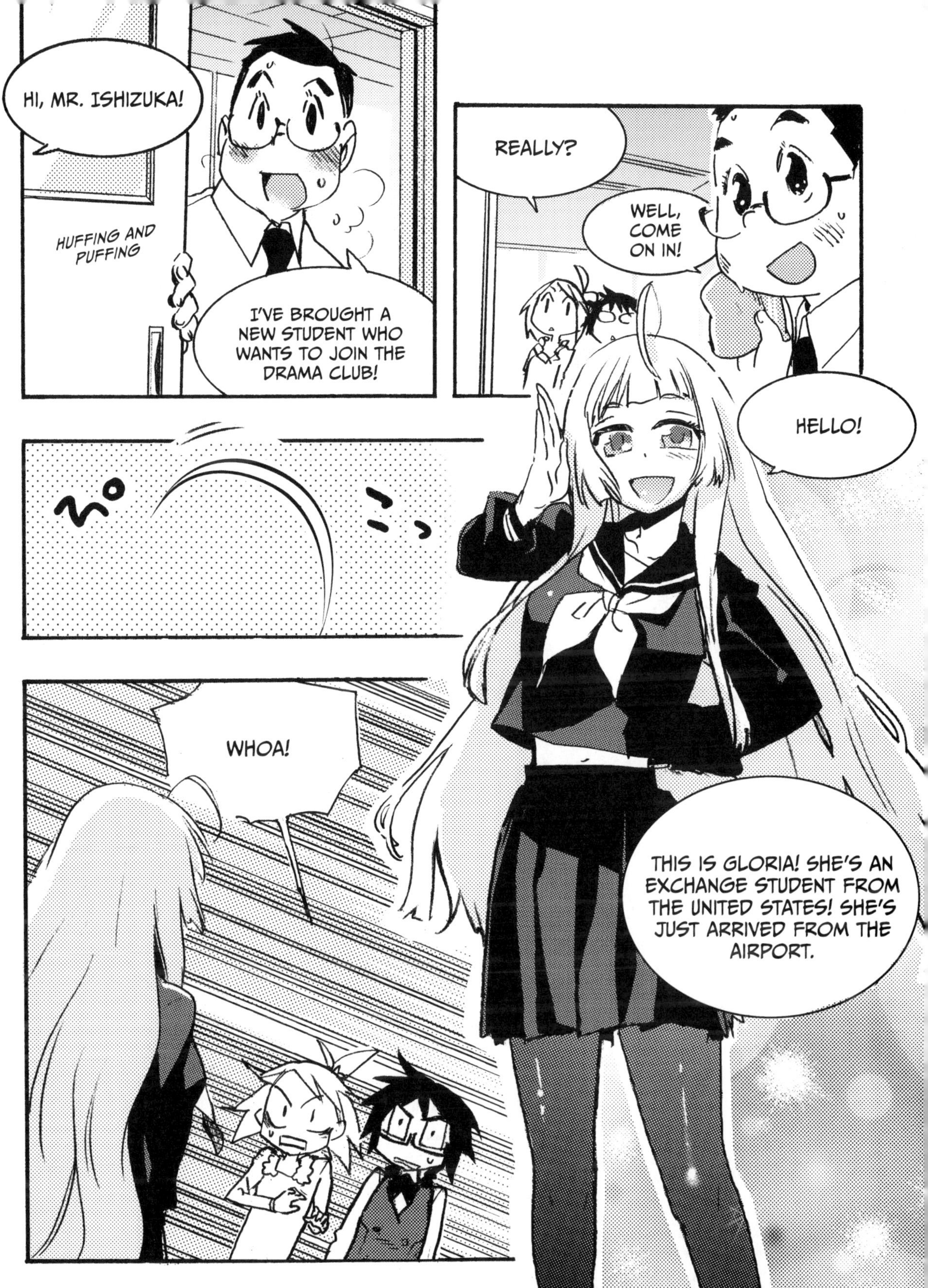

HI, MR. ISHIZUKA!
HUFFING AND PUFFING
I'VE BROUGHT A NEW STUDENT WHO WANTS TO JOIN THE DRAMA CLUB!
REALLY?
WELL, COME ON IN!
WHOA!
HELLO!
THIS IS GLORIA! SHE'S AN EXCHANGE STUDENT FROM THE UNITED STATES! SHE'S JUST ARRIVED FROM THE AIRPORT.

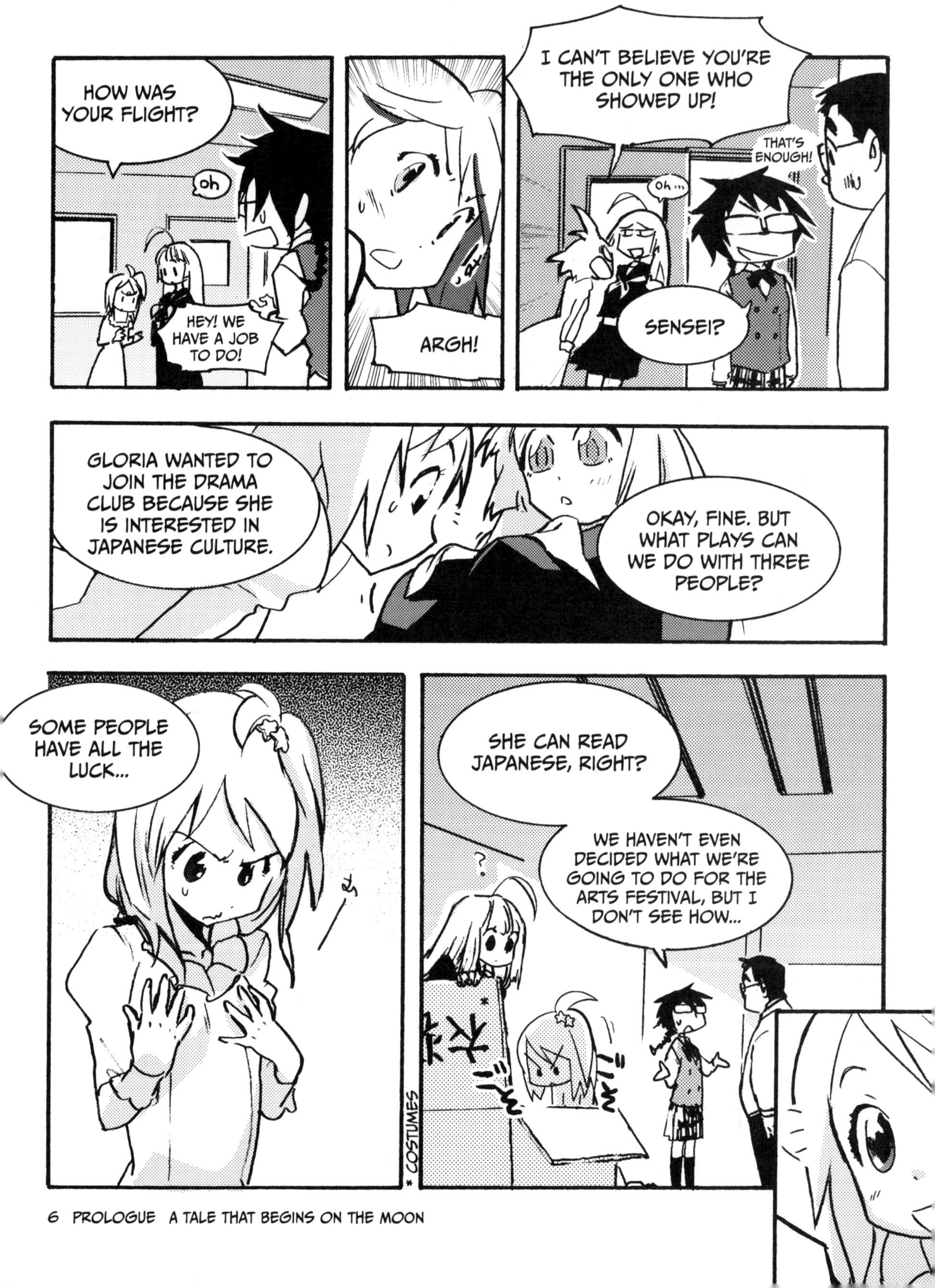
HOW WAS YOUR FLIGHT?
oh
HEY! WE HAVE A JOB TO DO!
ARGH!
I CAN'T BELIEVE YOU'RE THE ONLY ONE WHO SHOWED UP!
THAT'S ENOUGH!
Oh...
SENSEI?
GLORIA WANTED TO JOIN THE DRAMA CLUB BECAUSE SHE IS INTERESTED IN JAPANESE CULTURE.
OKAY, FINE. BUT WHAT PLAYS CAN WE DO WITH THREE PEOPLE?
SOME PEOPLE HAVE ALL THE LUCK...
SHE CAN READ JAPANESE, RIGHT?
WE HAVEN'T EVEN DECIDED WHAT WE'RE GOING TO DO FOR THE ARTS FESTIVAL, BUT I DON'T SEE HOW...
?
* COSTUMES

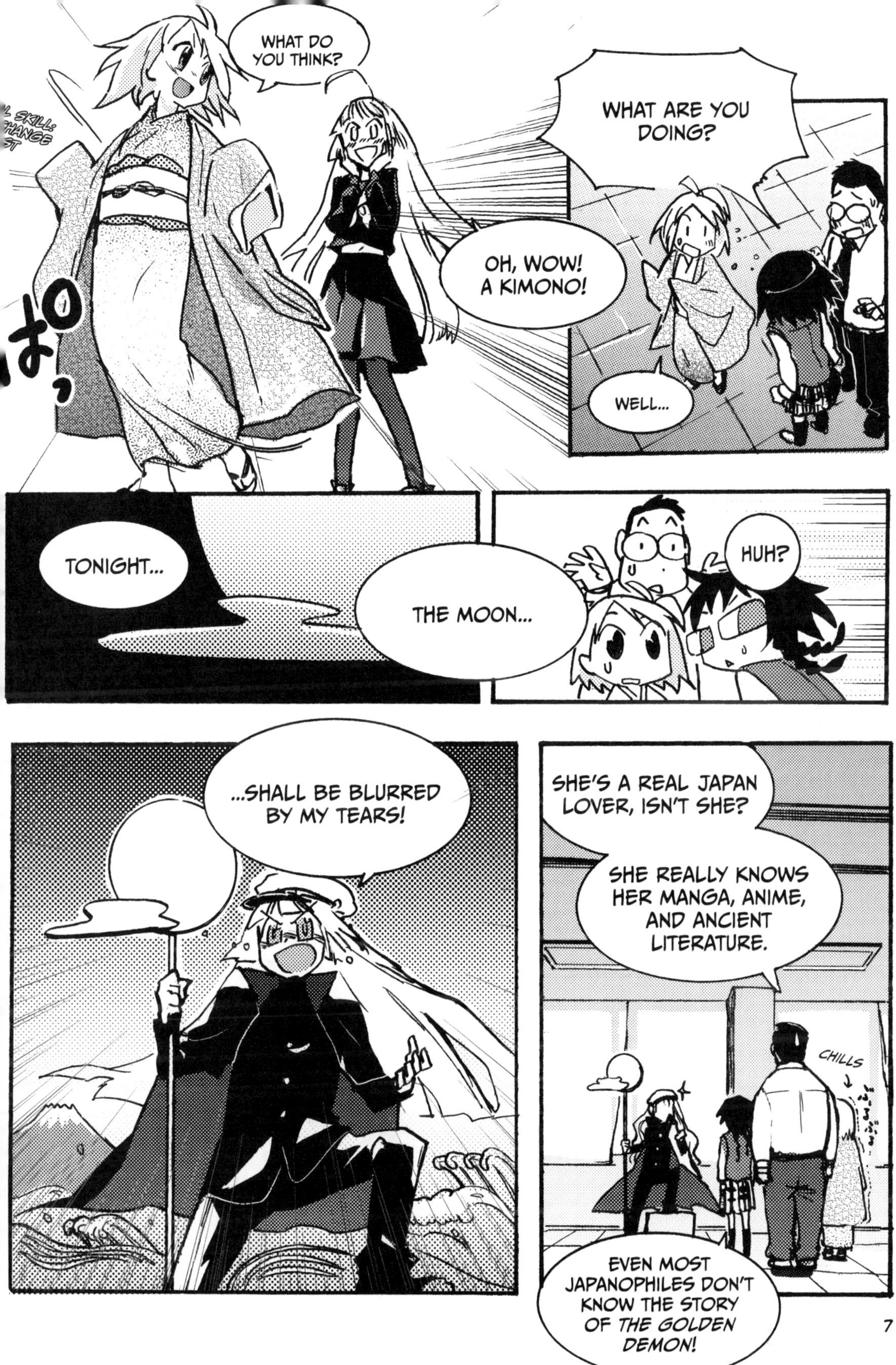
WHAT DO YOU THINK?
OH, WOW! A KIMONO!
WHAT ARE YOU DOING?
WELL...
TONIGHT...
THE MOON...
HUH?
...SHALL BE BLURRED BY MY TEARS!
SHE'S A REAL JAPAN LOVER, ISN'T SHE?
SHE REALLY KNOWS HER MANGA, ANIME, AND ANCIENT LITERATURE.
CHILLS
EVEN MOST JAPANOPHILES DON'T KNOW THE STORY OF *THE GOLDEN DEMON*!

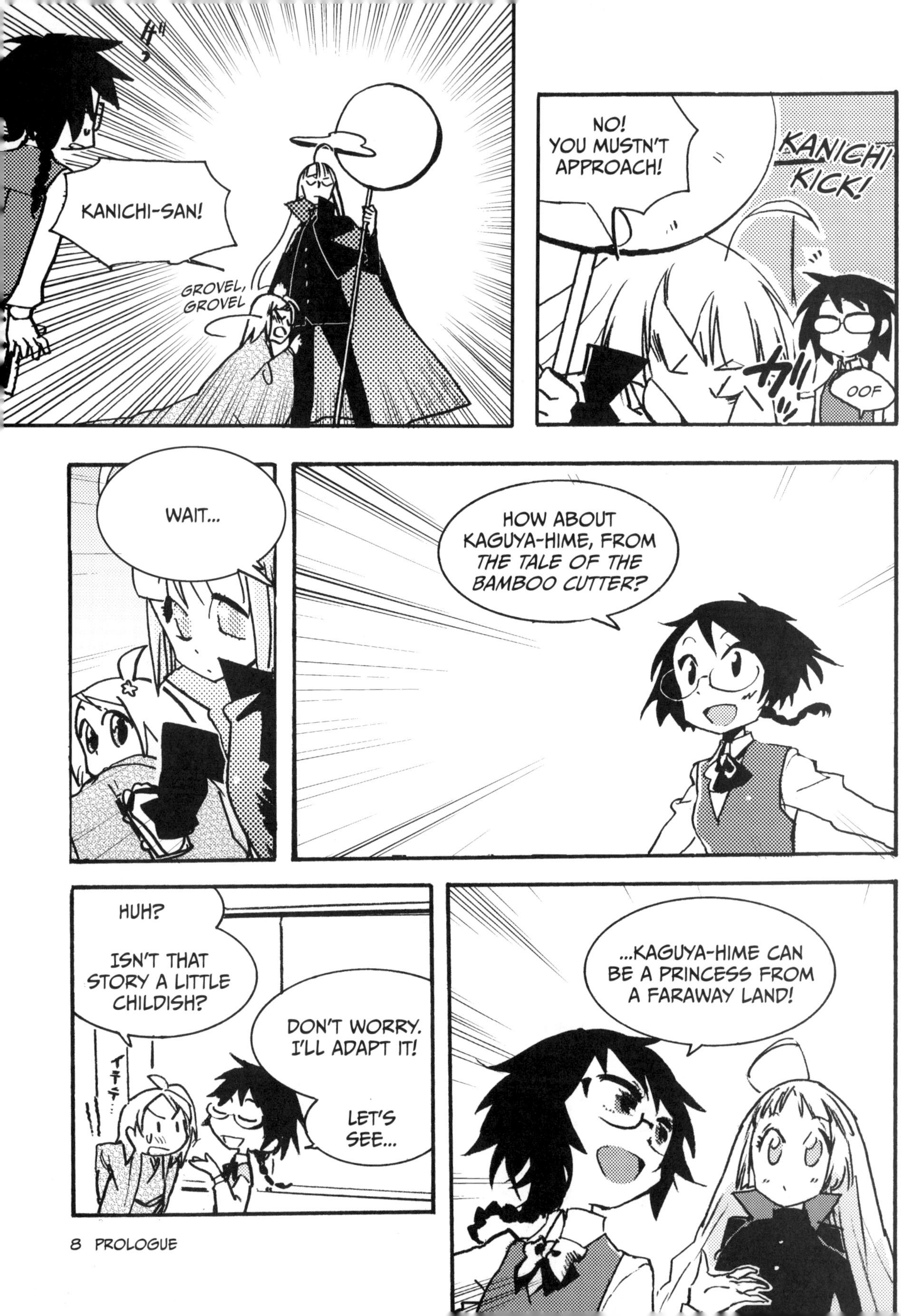
KANICHI-SAN!
GROVEL, GROVEL
NO! YOU MUSTN'T APPROACH!
KANICHI KICK!
OOF
WAIT...
HOW ABOUT KAGUYA-HIME, FROM *THE TALE OF THE BAMBOO CUTTER*?
HUH?
ISN'T THAT STORY A LITTLE CHILDISH?
DON'T WORRY. I'LL ADAPT IT!
LET'S SEE...
...KAGUYA-HIME CAN BE A PRINCESS FROM A FARAWAY LAND!

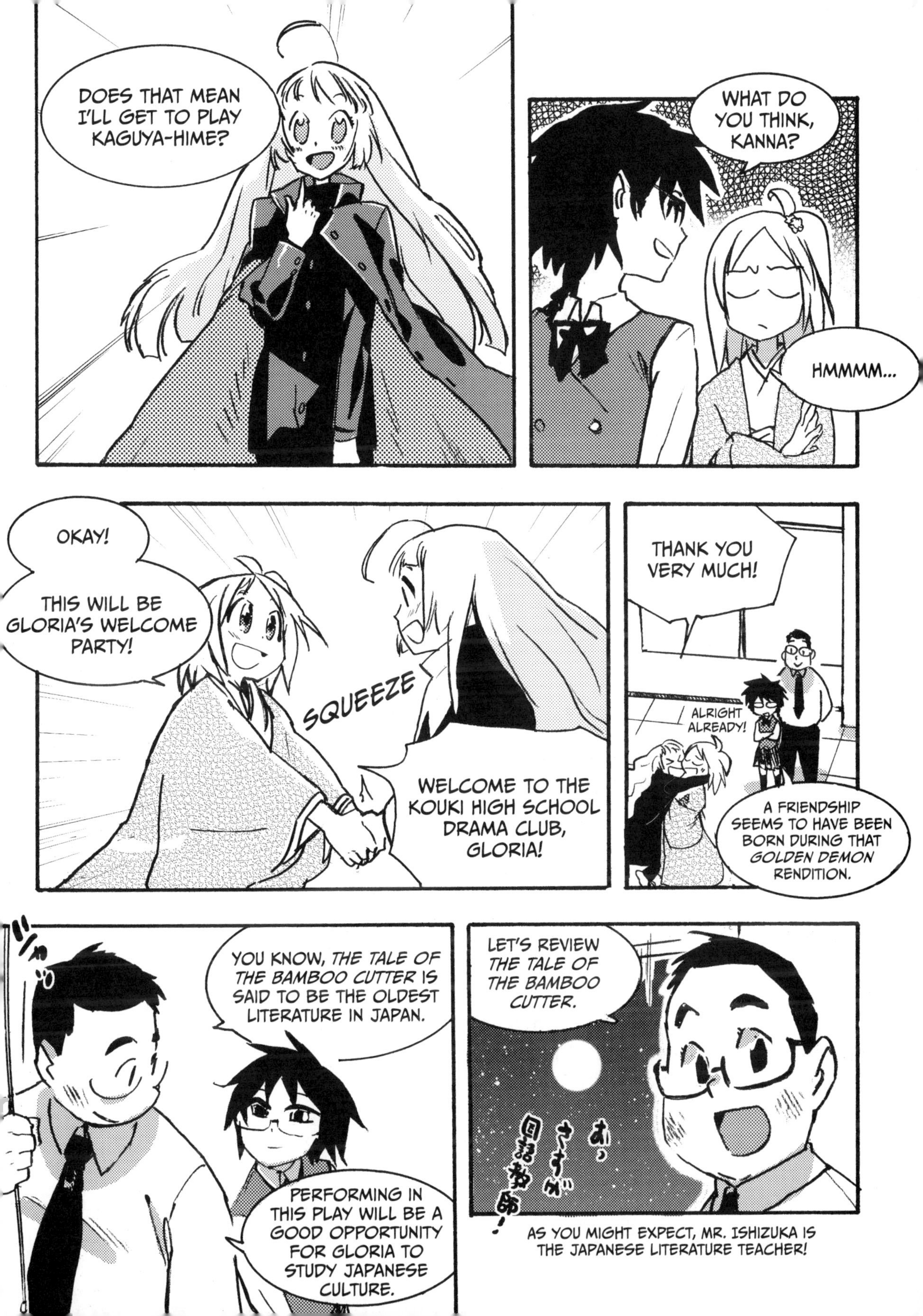

DOES THAT MEAN I'LL GET TO PLAY KAGUYA-HIME?
WHAT DO YOU THINK, KANNA?
HMMMM...
OKAY!
THIS WILL BE GLORIA'S WELCOME PARTY!
SQUEEZE
WELCOME TO THE KOUKI HIGH SCHOOL DRAMA CLUB, GLORIA!
THANK YOU VERY MUCH!
ALRIGHT ALREADY!
A FRIENDSHIP SEEMS TO HAVE BEEN BORN DURING THAT *GOLDEN DEMON* RENDITION.
YOU KNOW, *THE TALE OF THE BAMBOO CUTTER* IS SAID TO BE THE OLDEST LITERATURE IN JAPAN.
PERFORMING IN THIS PLAY WILL BE A GOOD OPPORTUNITY FOR GLORIA TO STUDY JAPANESE CULTURE.
LET'S REVIEW *THE TALE OF THE BAMBOO CUTTER.*
あっ さすが 国語教師！
AS YOU MIGHT EXPECT, MR. ISHIZUKA IS THE JAPANESE LITERATURE TEACHER!

THE STORY OF KAGUYA-HIME

Long, long ago, an elderly bamboo cutter was walking through a grove when he came upon a mysterious glowing stalk of bamboo. When he cut it open, he found a tiny girl inside—so tiny that she fit in the palm of his hand. Thinking that the gods had taken pity on him and his wife, an old childless couple, he decided to bring her home so that he and his wife could raise her as their own child.

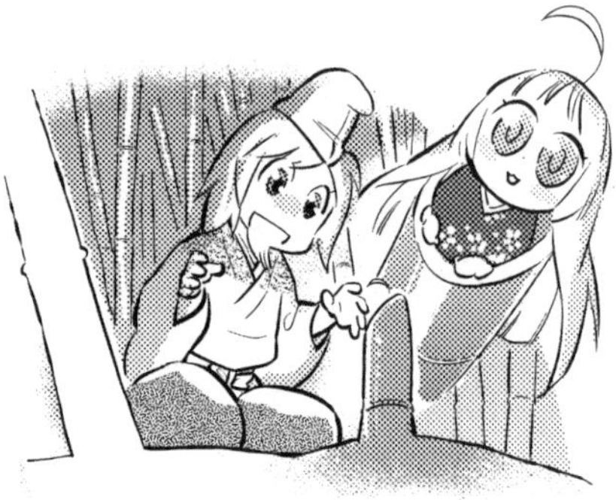

From that day forward, whenever the old man cut down a stalk of bamboo, he found a piece of gold inside. Little by little, he became very wealthy. The girl grew up quickly, and in just three months she grew into a kind and loving daughter.

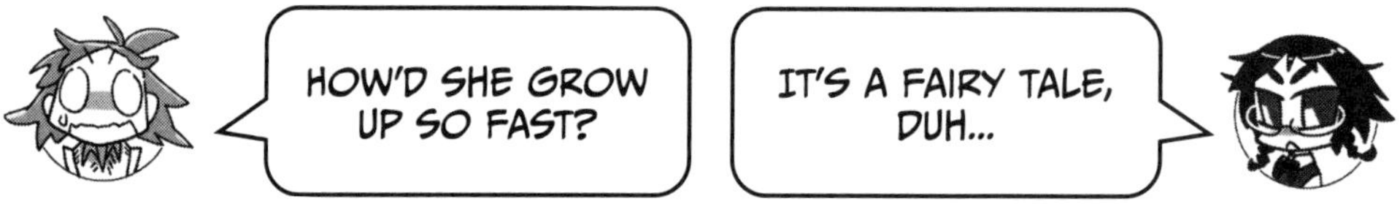
HOW'D SHE GROW UP SO FAST?
IT'S A FAIRY TALE, DUH...

The girl, who was named Kaguya-hime, was so exceptionally beautiful that word of her beauty reached even the faraway capital. Many suitors called on her, but she wasn't interested in any of them.

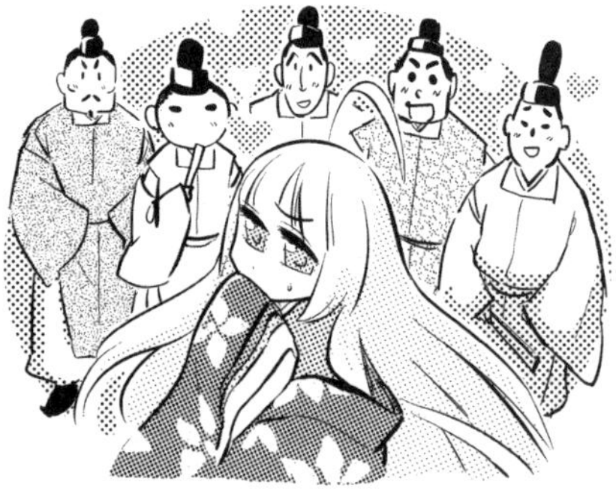

However, five of these men were unable to ignore her beauty, and they asked for her hand in marriage.

As the condition for accepting a marriage proposal, Kaguya-hime asked each of her suitors to bring back a rare treasure that was impossible to find. Naturally, no one succeeded.

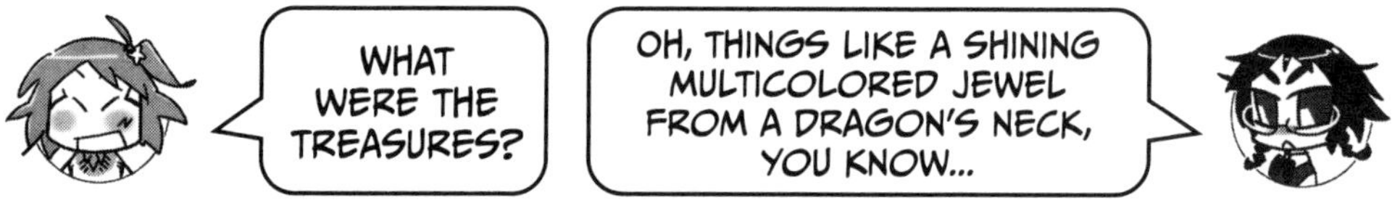
WHAT WERE THE TREASURES?
OH, THINGS LIKE A SHINING MULTICOLORED JEWEL FROM A DRAGON'S NECK, YOU KNOW...

PRINCE OTOMO WAS ASKED TO FIND THE DRAGON'S JEWEL, BUT HE KNEW IF HE ENTRUSTED THE TASK TO HIS SAMURAI, ONE OF THEM WOULD STEAL IT. SO HE SET SAIL HIMSELF. BUT ALONG THE WAY, HE ENCOUNTERED A TERRIBLE STORM. THIS KIND OF ADVENTURE CERTAINLY MAKES AN INTERESTING STORY, BUT LET'S MOVE ALONG...

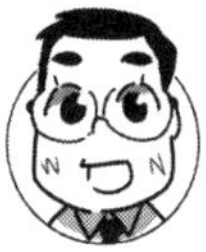

Kaguya-hime also caught the eye of the emperor—but he too was rejected. As the years passed, Kaguya-hime became more and more pensive whenever she gazed at the Moon, and as the autumnal full Moon approached, she would often burst into tears. The old bamboo cutter was very worried. When he asked her what was wrong, she replied, "I am not from this world! I come from the capital of the Moon, and I must return there when the Moon is full."

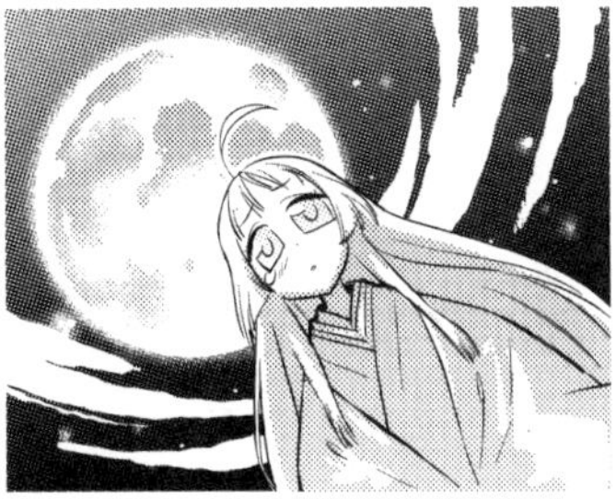

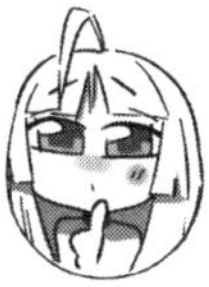

WHEN WOULD THE MOON BE FULL?

BY THE OLD CALENDAR, IT WAS THE 15TH NIGHT OF THE 8TH MONTH. NOWADAYS IT WOULD BE THE FULL MOON THAT OCCURS SOMETIME IN SEPTEMBER—THE HARVEST MOON.

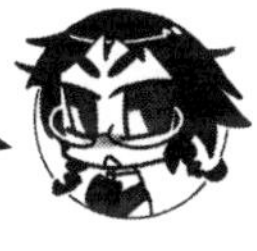

Hearing of this, the emperor tried to capture Kaguya-hime for himself before she could return to the Moon. He surrounded her house with soldiers, but then soldiers from the Moon came down and defeated them.

Before leaving for the Moon, Kaguya-hime gave the old bamboo cutter a letter and an elixir of immortality to give to the Emperor. Then the Moon's emissaries put the celestial maiden's feathered robe on her shoulders, and all of her memories of Earth disappeared. She returned to the Moon, pulled upward by an invisible force.

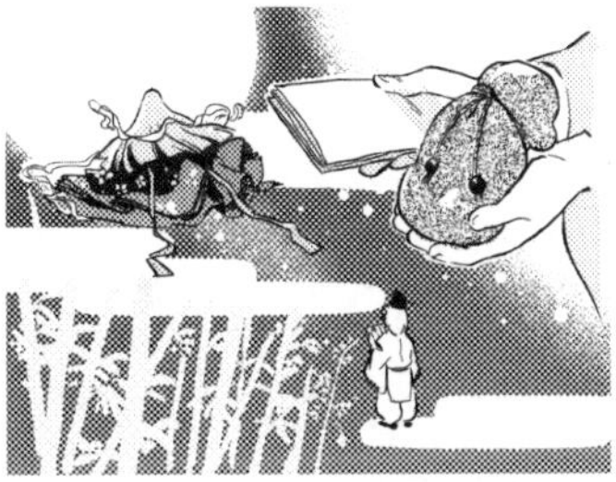

The Emperor read her letter but decided that he didn't want to live forever if he couldn't see her again. So he ordered his men to burn the elixir on top of the highest mountain in the country—the one that was closest to the Moon.

From then on, the mountain where the elixir was burned became known as Mt. Fuji, from the Japanese word for *immortality* (*fushi*).

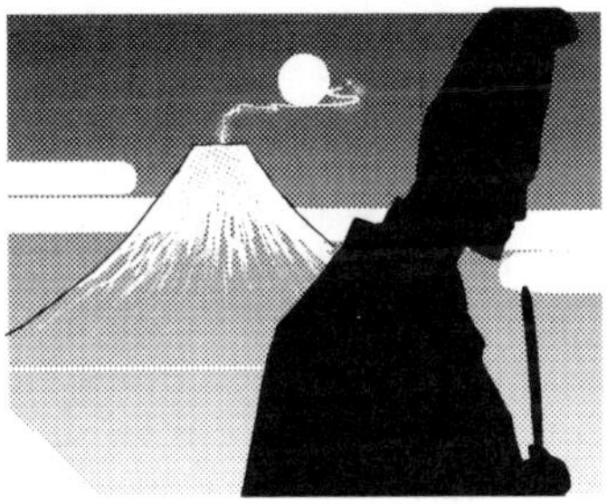

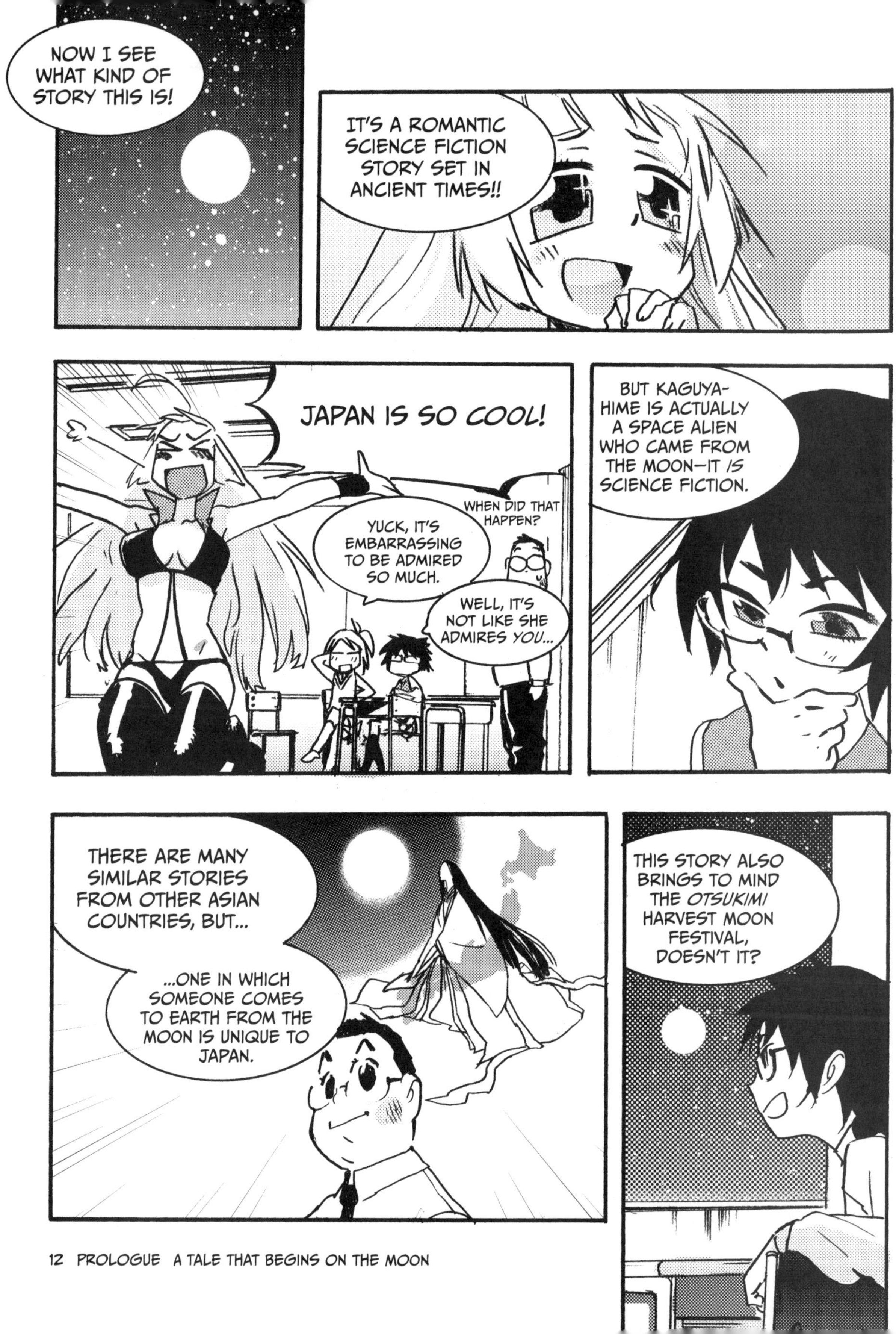
NOW I SEE WHAT KIND OF STORY THIS IS!
IT'S A ROMANTIC SCIENCE FICTION STORY SET IN ANCIENT TIMES!!
JAPAN IS SO COOL!
YUCK, IT'S EMBARRASSING TO BE ADMIRED SO MUCH.
WHEN DID THAT HAPPEN?
WELL, IT'S NOT LIKE SHE ADMIRES *YOU*...
BUT KAGUYA-HIME IS ACTUALLY A SPACE ALIEN WHO CAME FROM THE MOON—IT *IS* SCIENCE FICTION.
THERE ARE MANY SIMILAR STORIES FROM OTHER ASIAN COUNTRIES, BUT...
...ONE IN WHICH SOMEONE COMES TO EARTH FROM THE MOON IS UNIQUE TO JAPAN.
THIS STORY ALSO BRINGS TO MIND THE *OTSUKIMI* HARVEST MOON FESTIVAL, DOESN'T IT?

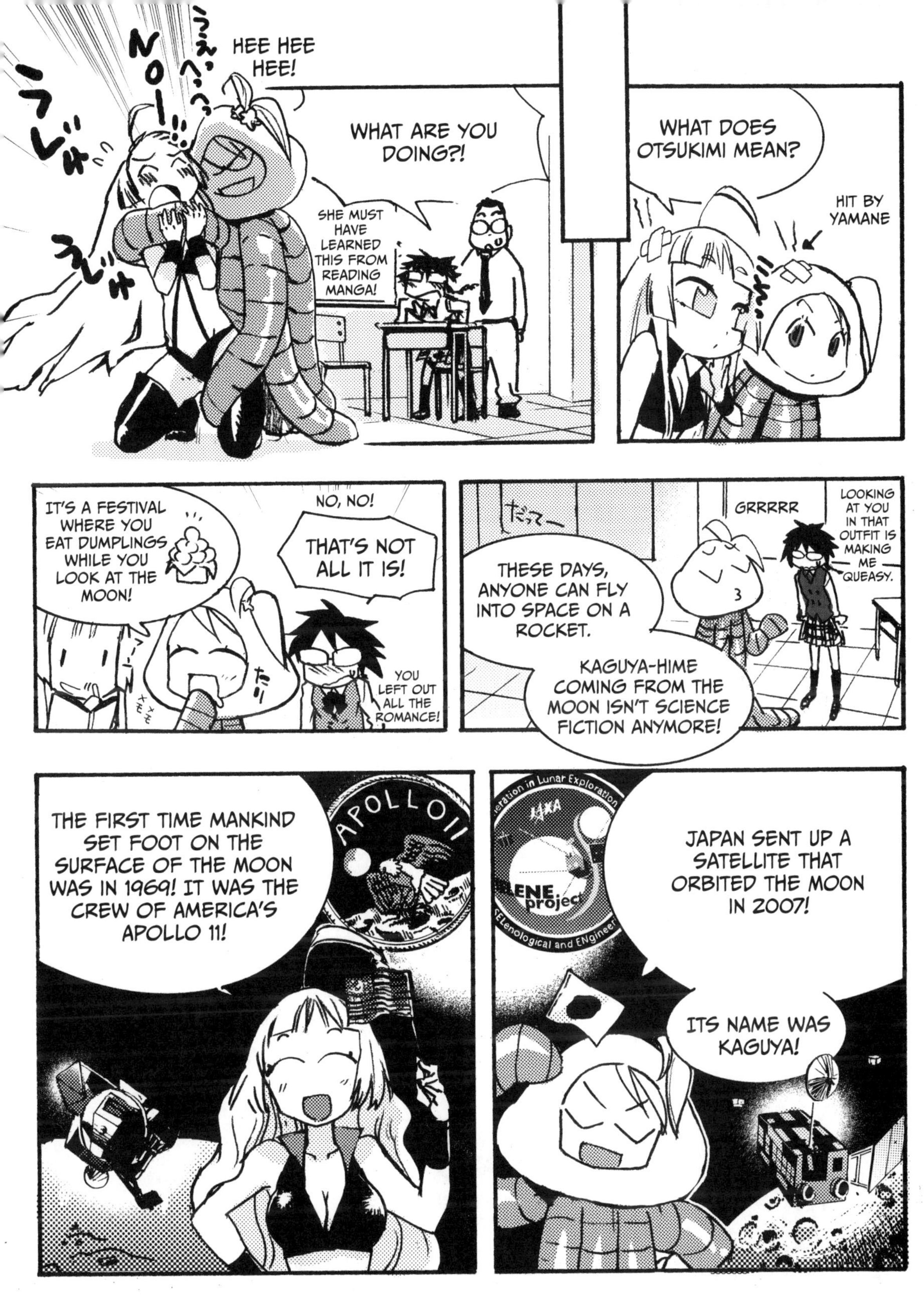
NO—!!
HEE HEE HEE!
WHAT ARE YOU DOING?!
SHE MUST HAVE LEARNED THIS FROM READING MANGA!
WHAT DOES OTSUKIMI MEAN?
HIT BY YAMANE
IT'S A FESTIVAL WHERE YOU EAT DUMPLINGS WHILE YOU LOOK AT THE MOON!
NO, NO!
THAT'S NOT ALL IT IS!
YOU LEFT OUT ALL THE ROMANCE!
GRRRRR
LOOKING AT YOU IN THAT OUTFIT IS MAKING ME QUEASY.
THESE DAYS, ANYONE CAN FLY INTO SPACE ON A ROCKET.
KAGUYA-HIME COMING FROM THE MOON ISN'T SCIENCE FICTION ANYMORE!
THE FIRST TIME MANKIND SET FOOT ON THE SURFACE OF THE MOON WAS IN 1969! IT WAS THE CREW OF AMERICA'S APOLLO 11!
APOLLO 11
JAPAN SENT UP A SATELLITE THAT ORBITED THE MOON IN 2007!
ITS NAME WAS KAGUYA!

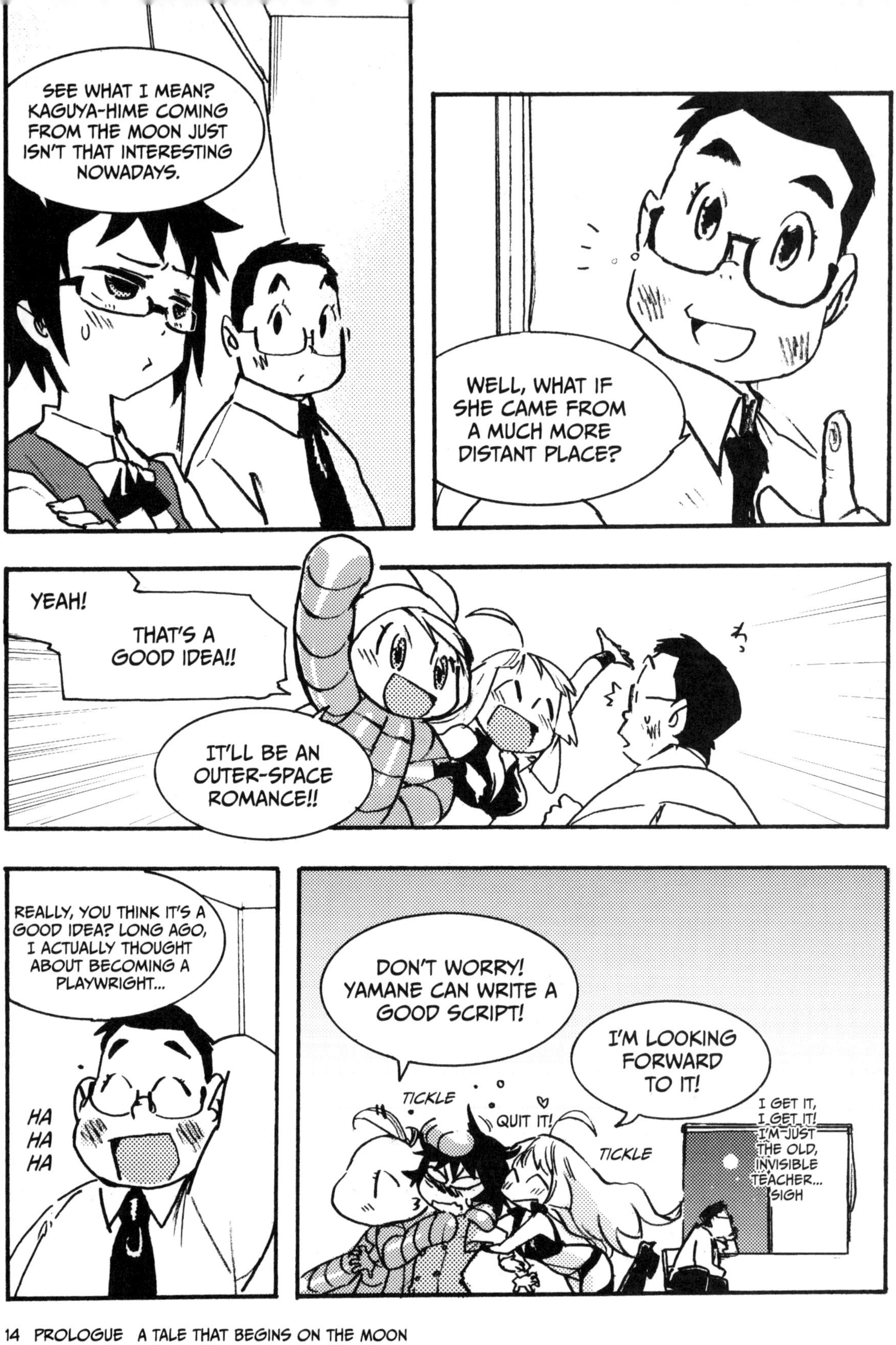
SEE WHAT I MEAN? KAGUYA-HIME COMING FROM THE MOON JUST ISN'T THAT INTERESTING NOWADAYS.
WELL, WHAT IF SHE CAME FROM A MUCH MORE DISTANT PLACE?
YEAH!
THAT'S A GOOD IDEA!!
IT'LL BE AN OUTER-SPACE ROMANCE!!
REALLY, YOU THINK IT'S A GOOD IDEA? LONG AGO, I ACTUALLY THOUGHT ABOUT BECOMING A PLAYWRIGHT...
HA HA HA
DON'T WORRY! YAMANE CAN WRITE A GOOD SCRIPT!
I'M LOOKING FORWARD TO IT!
TICKLE
QUIT IT!
TICKLE
I GET IT, I GET IT! I'M JUST THE OLD, INVISIBLE TEACHER... SIGH

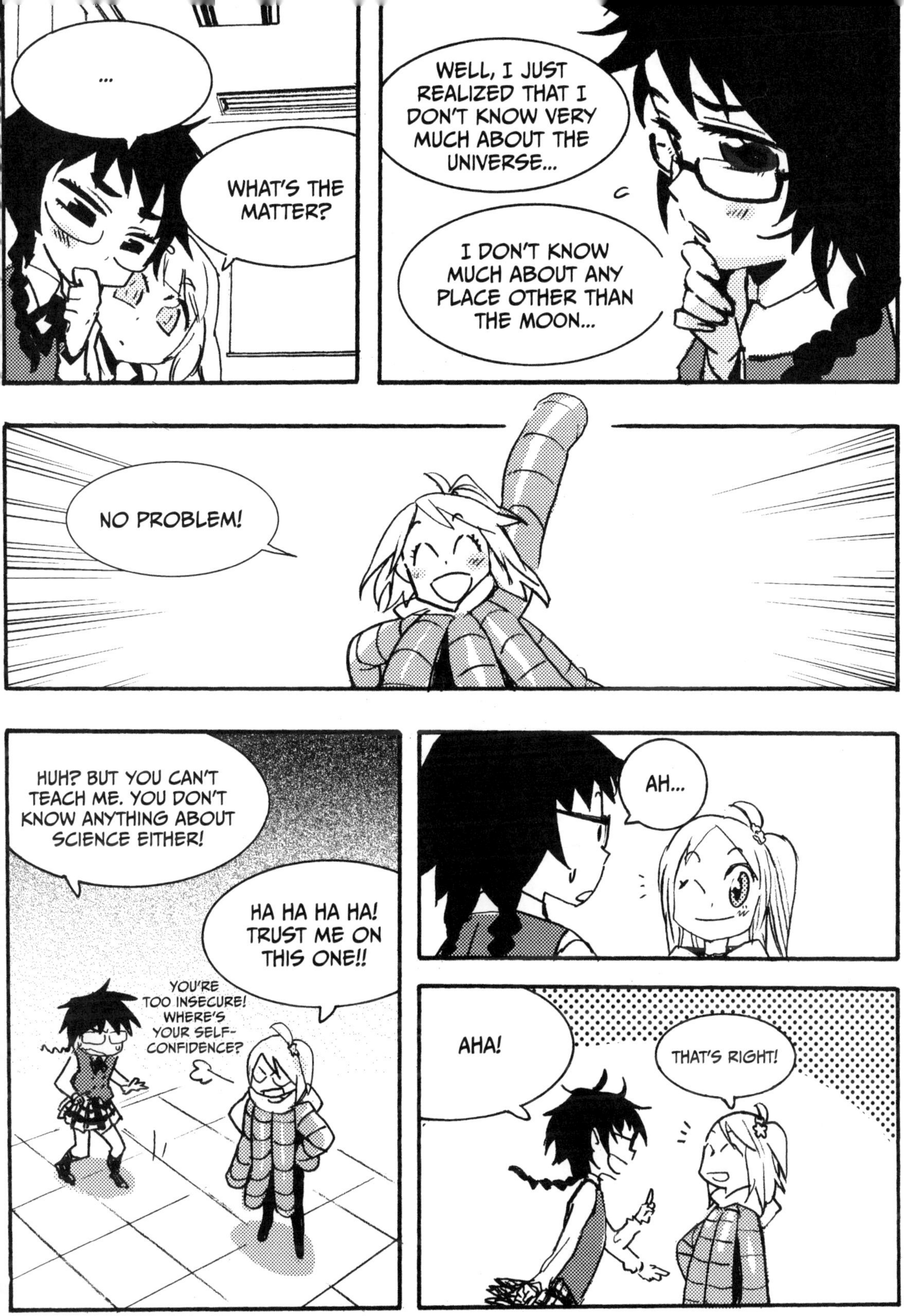
...
WHAT'S THE MATTER?
WELL, I JUST REALIZED THAT I DON'T KNOW VERY MUCH ABOUT THE UNIVERSE...
I DON'T KNOW MUCH ABOUT ANY PLACE OTHER THAN THE MOON...
NO PROBLEM!
HUH? BUT YOU CAN'T TEACH ME. YOU DON'T KNOW ANYTHING ABOUT SCIENCE EITHER!
HA HA HA HA! TRUST ME ON THIS ONE!!
YOU'RE TOO INSECURE! WHERE'S YOUR SELF-CONFIDENCE?
AH...
AHA!
THAT'S RIGHT!

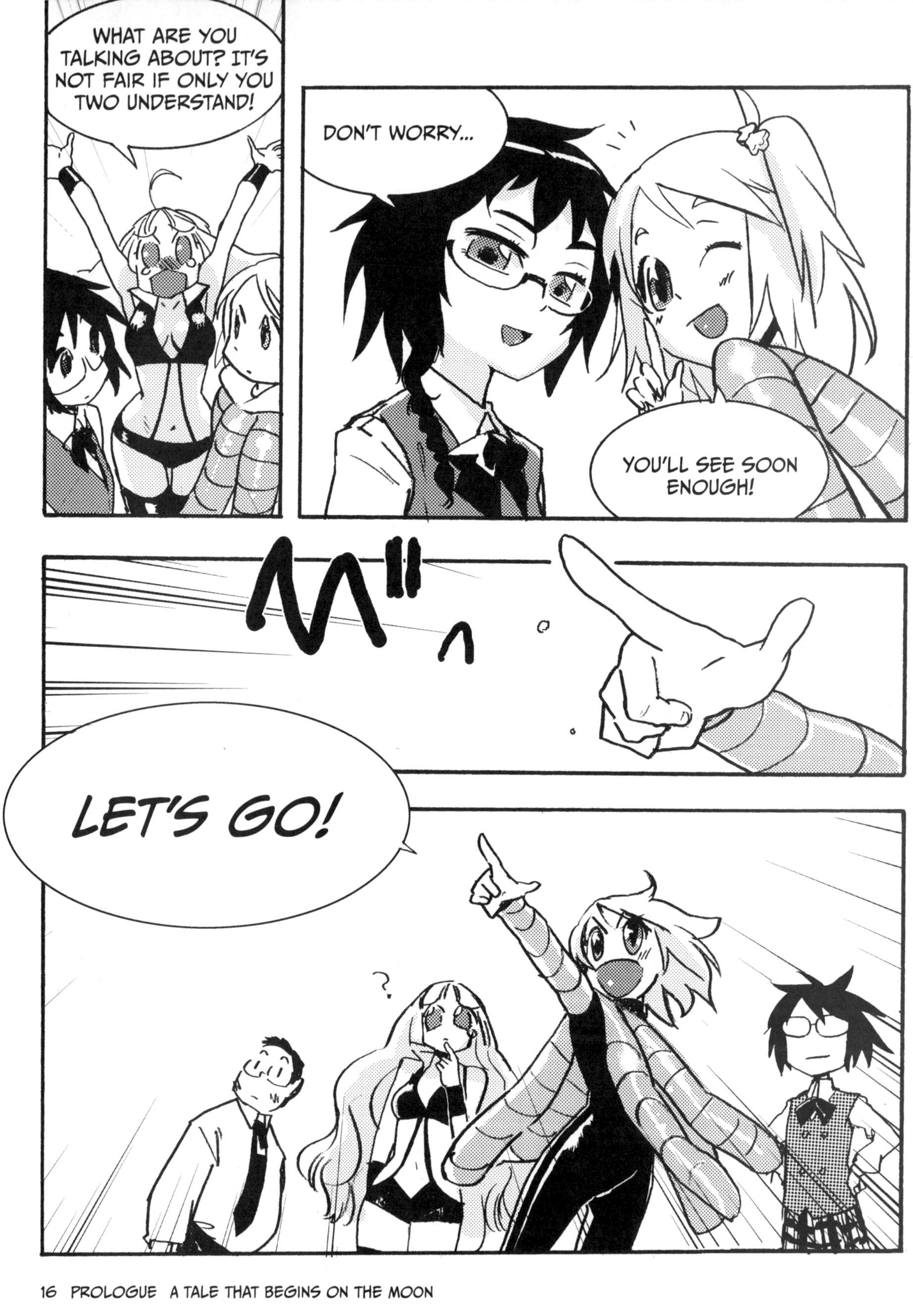
WHAT ARE YOU TALKING ABOUT? IT'S NOT FAIR IF ONLY YOU TWO UNDERSTAND!
DON'T WORRY...
YOU'LL SEE SOON ENOUGH!
LET'S GO!

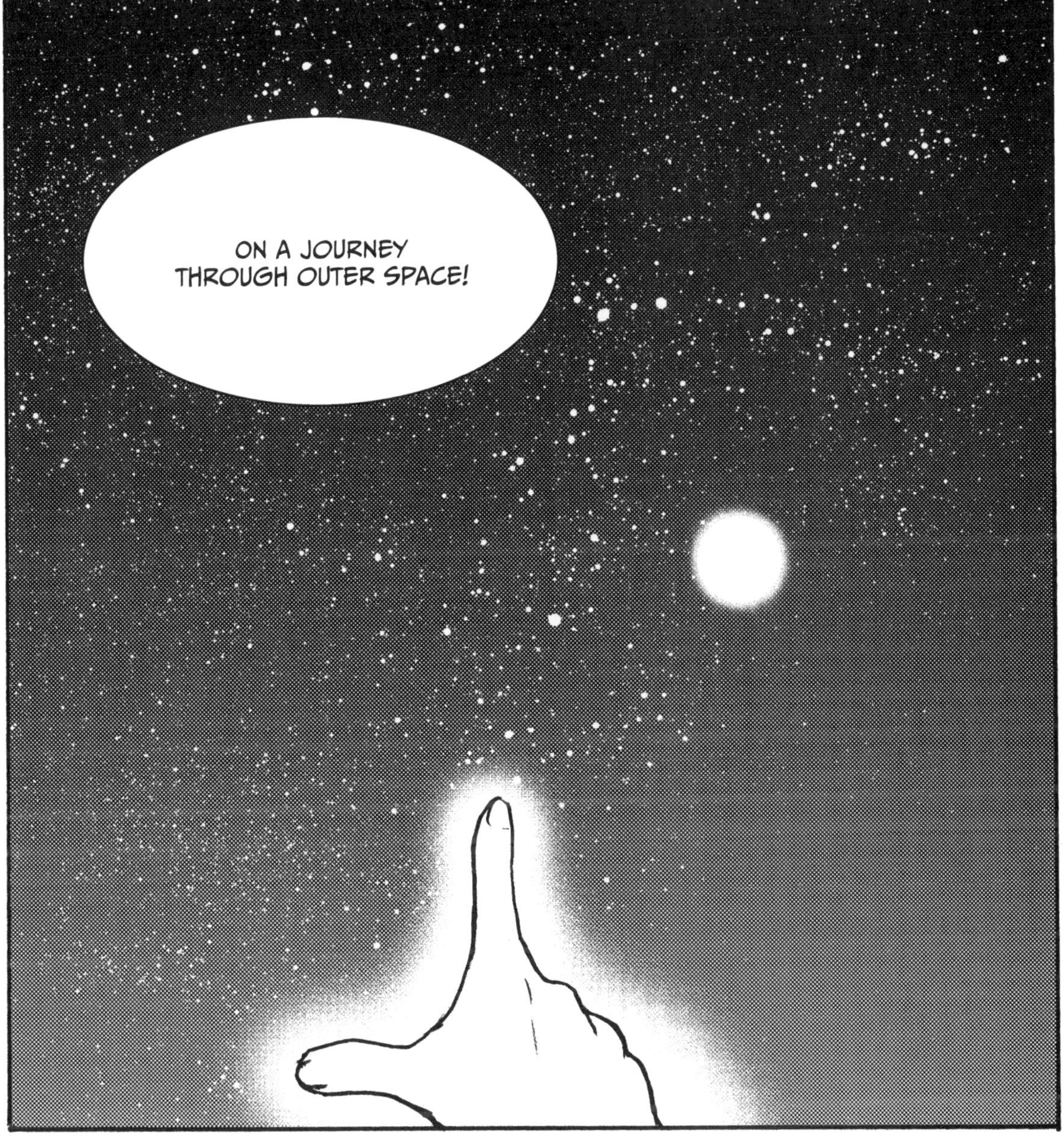
ON A JOURNEY
THROUGH OUTER SPACE!

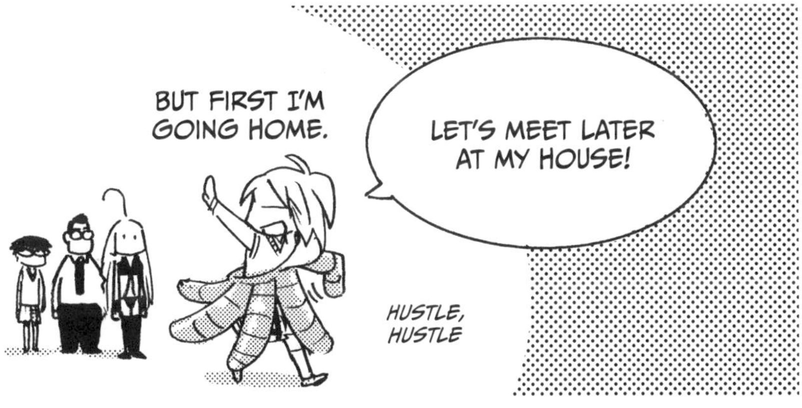
BUT FIRST I'M
GOING HOME.
LET'S MEET LATER
AT MY HOUSE!
HUSTLE,
HUSTLE

COSMIC MYTHS

How did ancient Japanese people know that the Moon was a celestial body like Earth?

The Tale of the Bamboo Cutter is an ancient Japanese fairy tale known to almost everyone in Japan. *The Tale of Genji*, written approximately 1,000 years ago, mentioned that the first fairy tale ever told was about an old bamboo cutter. However, it is rather surprising that the ancient Japanese believed that there was a city on the Moon where people lived.

For a long time, mankind believed that the universe was a small amount of space that wrapped around the world in which they lived. Maps from ancient times showed celestial bodies such as the Sun, Moon, and stars as tiny entities affixed to the surface of a shell that surrounded Earth. But in a universe like that, Kaguya-hime's story doesn't make sense. The people who created her story had a different view of the universe, in which the Moon was another world where people lived. Let's look at some other views of the universe from ancient times.

ANCIENT INDIA'S VIEW OF THE UNIVERSE

In ancient India, people believed in a turtle that rode on top of an enormous coiled snake and that elephants stood on the turtle's back to support a hemispherical Earth. The Sun was thought to appear and disappear as it revolved around the highest mountain, which stood at the center of the world. (This is Mt. Sumeru, which likely represented the Himalayas.) The Moon, which was the lamp belonging to the night watchman on this mountain, was thought to wax and wane depending on the direction the watchman was facing.

Ancient India's view of the universe

ANCIENT EGYPT'S VIEW OF THE UNIVERSE

In ancient Egypt, people believed that Nut, goddess of the sky, was supported by Shu, god of the air. Nut was said to be a symbol of the Nile River, and daytime and nighttime occurred when the Sun god Ra went back and forth across the river by boat every day. The Moon and stars were though to be suspended from Nut's body.

Ancient Egypt's view of the universe

ANCIENT BABYLONIA'S VIEW OF THE UNIVERSE

The Babylonians thought that the Moon and stars were affixed to an enormous arched ceiling called the *celestial sphere*. The celestial sphere was supported by Mt. Ararat, and the Sun moved from east to west across its inner surface.

Ancient Babylonia's view of the universe

IN CHINA, WHERE ASTRONOMY WAS ORIGINALLY DEVELOPED

In contrast to these imaginary universes, people in ancient China and Greece tried to develop models of the universe scientifically. It was in China that astronomy was first developed.

In China, several cosmologies were conceived approximately 2,000 to 2,400 years ago, based on observations of the heavens. Two such cosmologies were called Gai Tian and Hun Tian.

Gai Tian described a dome-shaped sky, like a cap, over a hemispherical Earth. This was surrounded by water (the ocean) and rotated once a day from east to west around the north pole. The Sun also traced a circle in the sky, and the size of that circle varied with the seasons.

Hun Tian, whose name means *the entire sky*, expanded upon the concept of Gai Tian to try to more accurately represent the movements of the celestial bodies. The celestial sphere enveloped everything like an eggshell rather than just capping it like a dome, and the variation in the constellations according to the seasons was explained by the notion that the north pole shifted, rather than always remaining directly overhead.

Gai Tian:
A cosmology positing a hemispherical dome over Earth

Hun Tian:
A spherical cosmology

In Ancient Greece, Where the Size of Earth Was Calculated

The ancient Greeks tried to explain the shape of the universe by using the logical thinking that permeates modern mathematics and physics. One of their greatest achievements was the discovery that Earth is a spherical body floating in space. The ancient Greeks were also the first to calculate the size of Earth.

Eratosthenes (who lived from about 276 BC to 195 BC) was a Greek scholar who was active in Egypt during the Hellenistic period. He calculated the size of Earth using the following method.

Eratosthenes' Calculation Method

Eratosthenes discovered an account that said that a rod standing vertically at midday on the summer solstice in Syene (in the southern part of Egypt) did not cast a shadow. It seemed that this phenomenon could only occur south of the Tropic of Cancer, when the Sun appeared at the zenith (directly overhead).

The astonished scholar wondered what would happen in Alexandria, in the northern part of Egypt, and he immediately performed the experiment under the same conditions. The result was that the shadow of the rod remained visible. Eratosthenes concluded from this evidence that Earth is a sphere, a theory that was being discussed among some scholars at the time.

Alexandria
Tropic of Cancer
Equator
Tropic of Capricorn
Syene

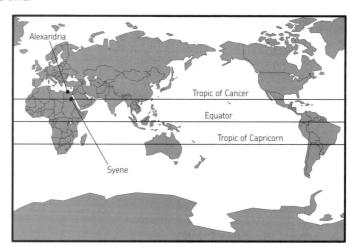

The Tropics and the Equator

Eratosthenes also used his observations to try to measure the size of Earth. First, he measured the length of the rod's shadow. He calculated that in Alexandria at the same time on the same day, the Sun's rays arrived from a direction that was offset from the vertical by 7.2 degrees.

Next, he had a man walk from Alexandria to Syene and determined from the man's stride that the distance was 5,000 stadia (an ancient unit of measurement), or approximately 925 km (575 miles). Then he used the following formula to determine the circumference of Earth.

$$575 \text{ miles } (925 \text{ km}) \times \frac{360°}{7.2°} = 28{,}750 \text{ miles } (46{,}250 \text{ km})$$

Although we now know that the circumference of Earth is 40,000 km (24,855 miles), Eratosthenes' calculation is remarkably close.

Sun
Perpendicular
to the surface
of Earth
Sun's rays
L
7.2°
North pole
Alexandria
Syene
7.2°
Earth
South pole

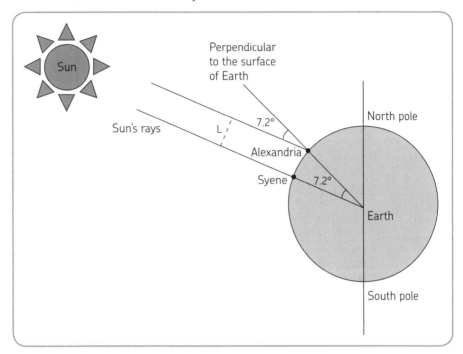

Eratosthenes' calculation method

Another version of the story says that Eratosthenes got his idea by seeing the Sun's rays reach the bottom of a well rather than observing the shadow cast by a rod. Nevertheless, it is generally accepted that he calculated Earth's circumference as being approximately 25,000 miles, which is roughly accurate.

IF EARTH IS ROUND, THE MOON MUST BE ROUND TOO

It is quite reasonable to suppose that scholars like Eratosthenes were not the only ones to realize that Earth was round. Certain phenomena—such as the fact that you cannot see beyond the horizon, or that the top of the sail always appears first when a ship is approaching—were obvious to people whose lives were closely related to the sea, and these occurrences are inconceivable on a flat surface.

Ancient Greece, where Eratosthenes lived, was a maritime nation bordered by the Ionian and Aegean Seas and located not far from the Mediterranean Sea. For that reason alone, many seafaring Greeks may have sensed that the world might be round.

On the other hand, when people with good eyesight observe light striking the Moon they should easily see that its surface is spherical rather than flat. For example, if you look at an enlarged photograph, there are clearly gradations at the outer edge and the waxing or waning border line. This wouldn't happen if the Moon were flat.

Now, let's return to the story of Kaguya-hime.

The round Earth

Japan is an island country, surrounded by the sea. This means that even in ancient times, some Japanese people probably recognized the existence of the curved horizon and from that concluded that Earth was round.

For example, when European missionaries traveled to Japan in the 16th century, they tried to introduce their scientific knowledge to the feudal lords they found there. One item they presented as a display of their knowledge was a globe. However, contrary to the expectations of the Europeans, most Japanese people did not show surprise at the suggestion that the world was a sphere.

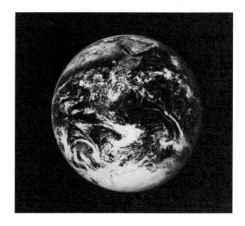

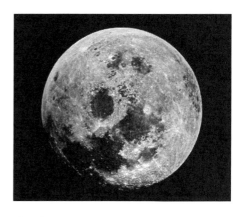

The Moon has been appreciated by the Japanese since ancient times.

The fact that Japanese people have gazed at and felt affection for the Moon since ancient times is also apparent from folklore, such as the story of the Moon Rabbit. And although Otsukimi (moon-viewing) festivals seem to have originated in China, the custom of appreciating the Moon is said to have existed in Japan since the *Jōmon* period (approximately 14,000 BC to 400 BC). Most likely, it would have been recognized then that the Moon was a sphere.

If Earth—like the Moon—is round and floats in space, then the idea that people should live on both Earth and the Moon is a natural conclusion. Therefore, it's not surprising that this idea appears in the tale of Kaguya-hime.

1
IS EARTH THE CENTER OF THE UNIVERSE?

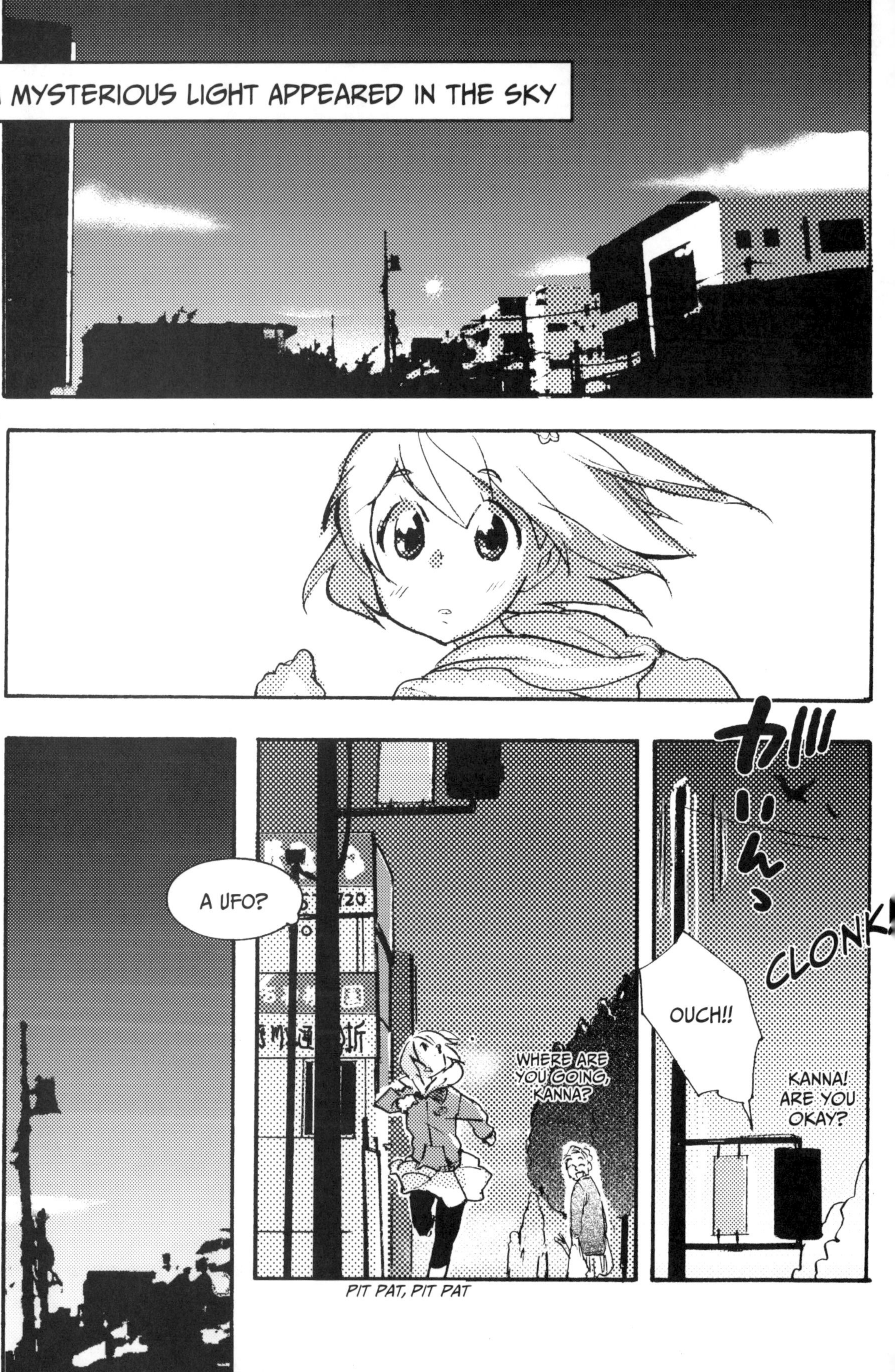
A MYSTERIOUS LIGHT APPEARED IN THE SKY
A UFO?
WHERE ARE YOU GOING, KANNA?
PIT PAT, PIT PAT
CLONK!
OUCH!!
KANNA! ARE YOU OKAY?

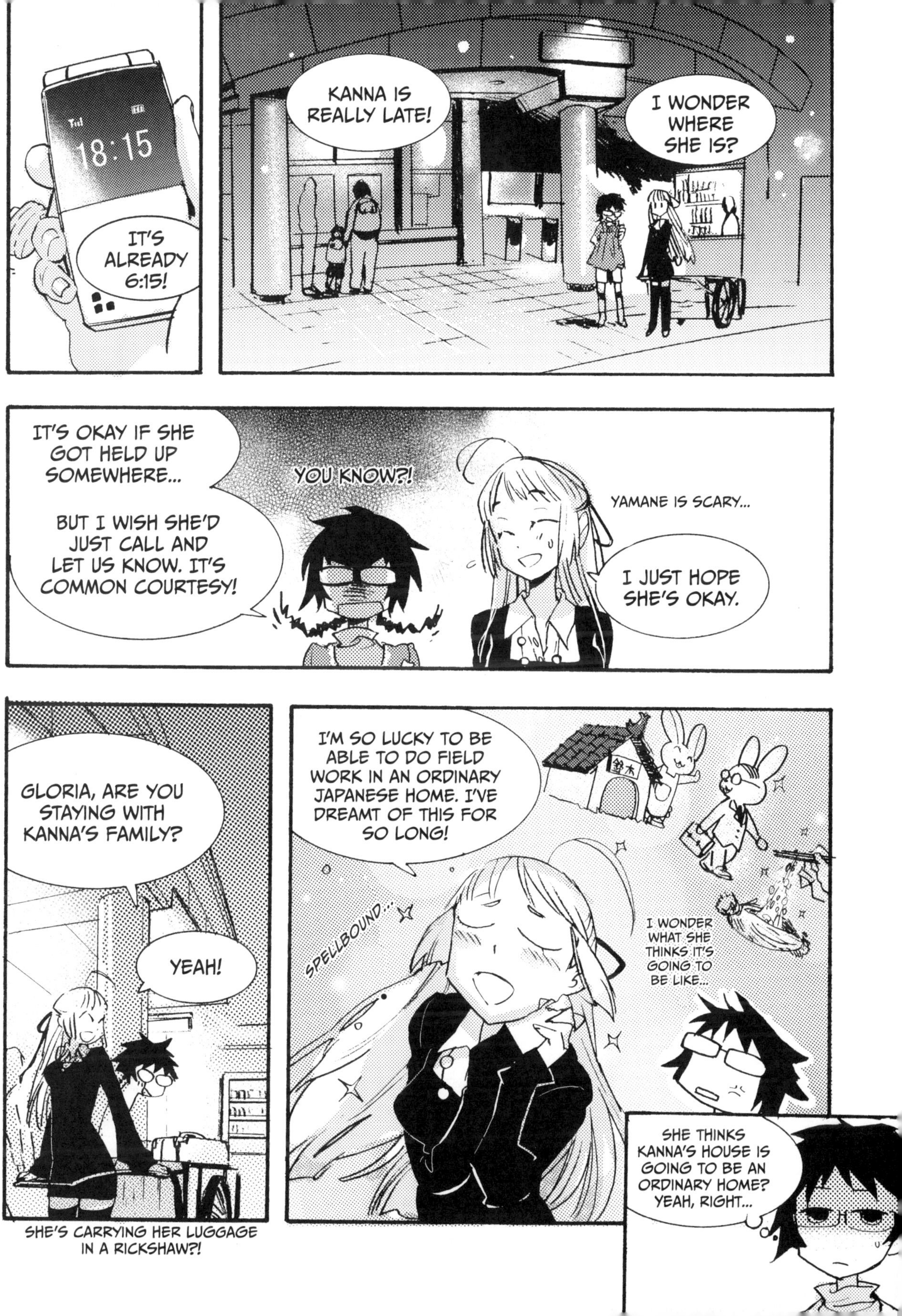
18:15
IT'S ALREADY 6:15!
KANNA IS REALLY LATE!
I WONDER WHERE SHE IS?
IT'S OKAY IF SHE GOT HELD UP SOMEWHERE...
BUT I WISH SHE'D JUST CALL AND LET US KNOW. IT'S COMMON COURTESY!
YOU KNOW?!
YAMANE IS SCARY...
I JUST HOPE SHE'S OKAY.
GLORIA, ARE YOU STAYING WITH KANNA'S FAMILY?
YEAH!
SHE'S CARRYING HER LUGGAGE IN A RICKSHAW?!
I'M SO LUCKY TO BE ABLE TO DO FIELD WORK IN AN ORDINARY JAPANESE HOME. I'VE DREAMT OF THIS FOR SO LONG!
SPELLBOUND...
I WONDER WHAT SHE THINKS IT'S GOING TO BE LIKE...
SHE THINKS KANNA'S HOUSE IS GOING TO BE AN ORDINARY HOME? YEAH, RIGHT...

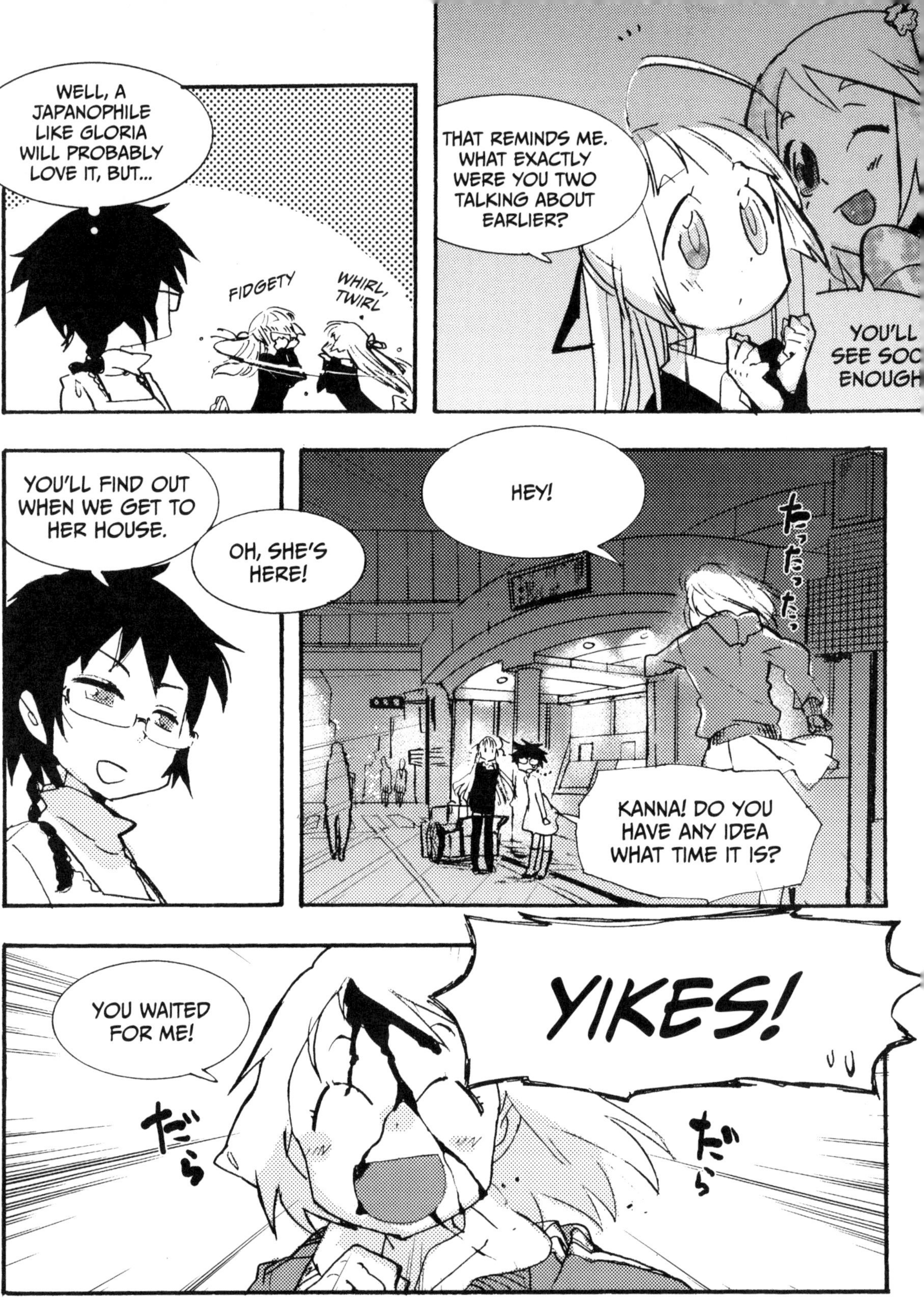
WELL, A JAPANOPHILE LIKE GLORIA WILL PROBABLY LOVE IT, BUT...
FIDGETY
WHIRL, TWIRL
THAT REMINDS ME. WHAT EXACTLY WERE YOU TWO TALKING ABOUT EARLIER?
YOU'LL SEE SOO ENOUGH
YOU'LL FIND OUT WHEN WE GET TO HER HOUSE.
OH, SHE'S HERE!
HEY!
たったった
KANNA! DO YOU HAVE ANY IDEA WHAT TIME IT IS?
YOU WAITED FOR ME!
YIKES!
だら
だら

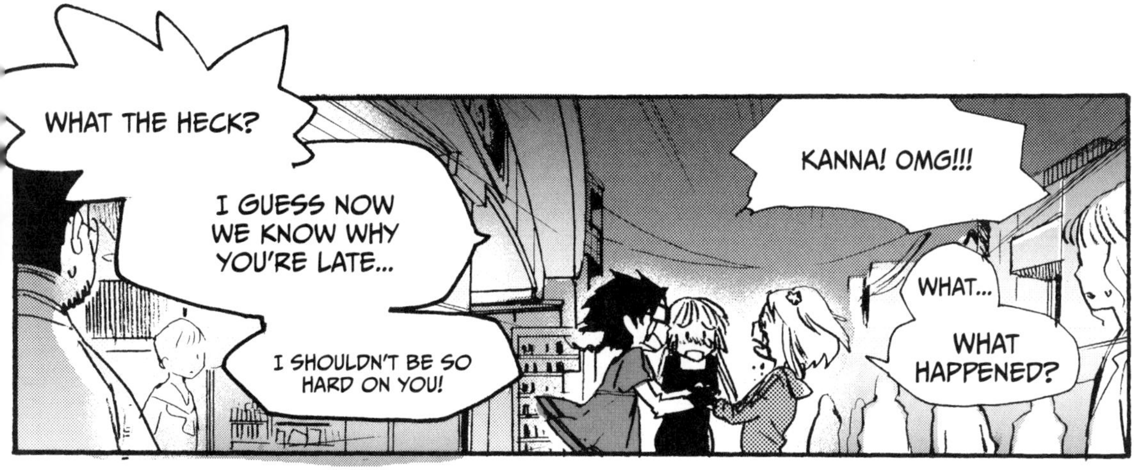
WHAT THE HECK?
I GUESS NOW WE KNOW WHY YOU'RE LATE...
I SHOULDN'T BE SO HARD ON YOU!
KANNA! OMG!!!
WHAT...
WHAT HAPPENED?

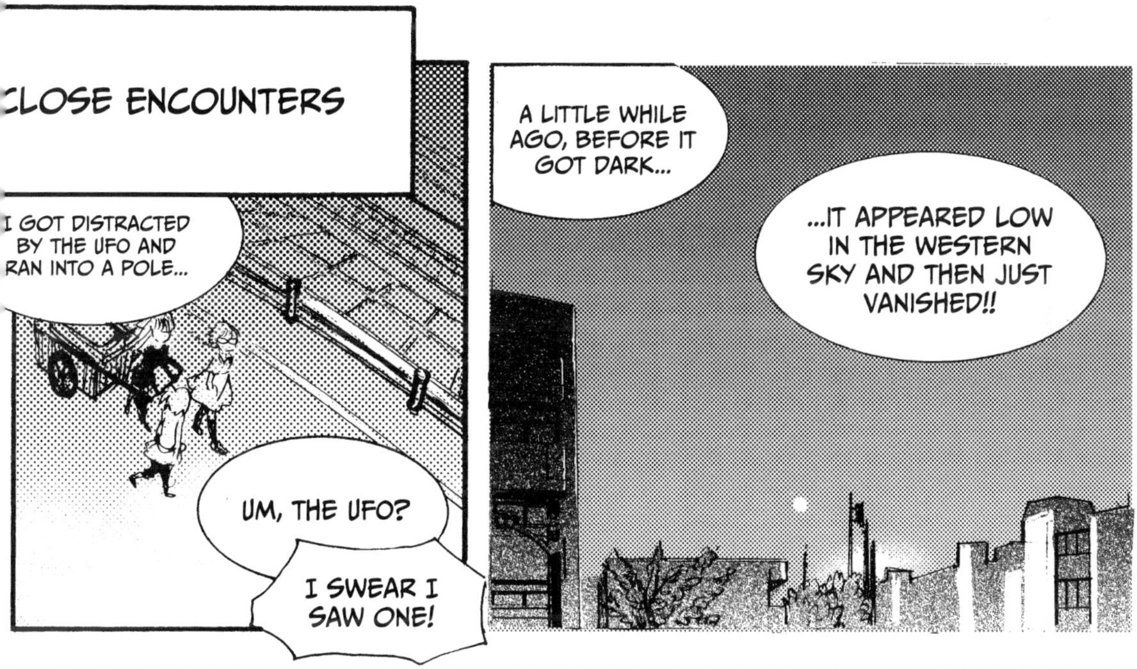
CLOSE ENCOUNTERS
I GOT DISTRACTED BY THE UFO AND RAN INTO A POLE...
UM, THE UFO?
I SWEAR I SAW ONE!
A LITTLE WHILE AGO, BEFORE IT GOT DARK...
...IT APPEARED LOW IN THE WESTERN SKY AND THEN JUST VANISHED!!

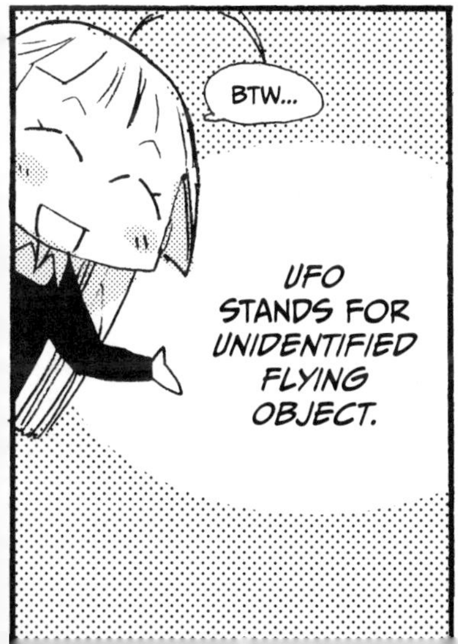
BTW...
UFO STANDS FOR UNIDENTIFIED FLYING OBJECT.

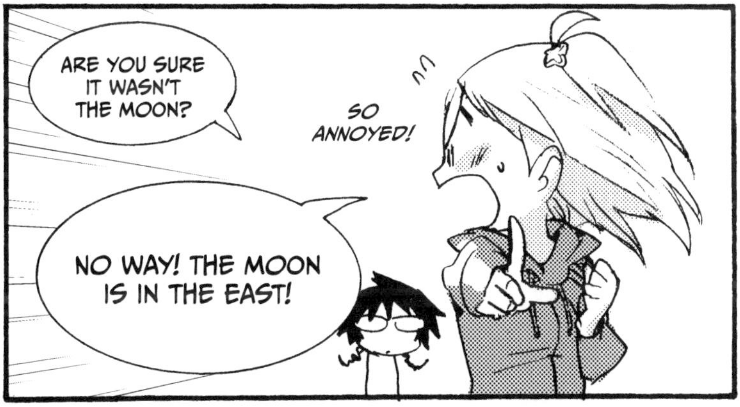
ARE YOU SURE IT WASN'T THE MOON?
SO ANNOYED!
NO WAY! THE MOON IS IN THE EAST!

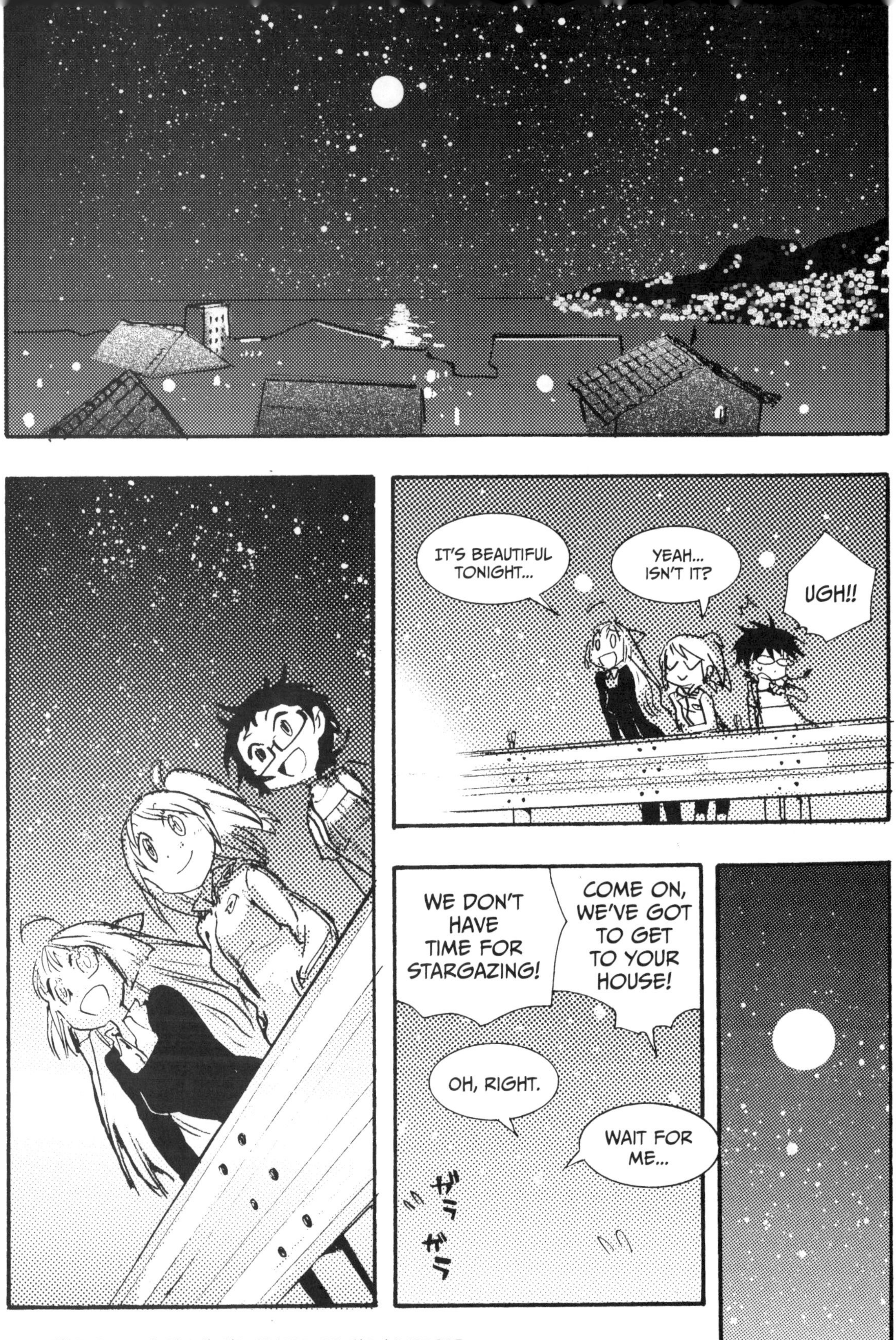
IT'S BEAUTIFUL TONIGHT...
YEAH... ISN'T IT?
UGH!!
WE DON'T HAVE TIME FOR STARGAZING!
COME ON, WE'VE GOT TO GET TO YOUR HOUSE!
OH, RIGHT.
WAIT FOR ME...
ガラ ガラ

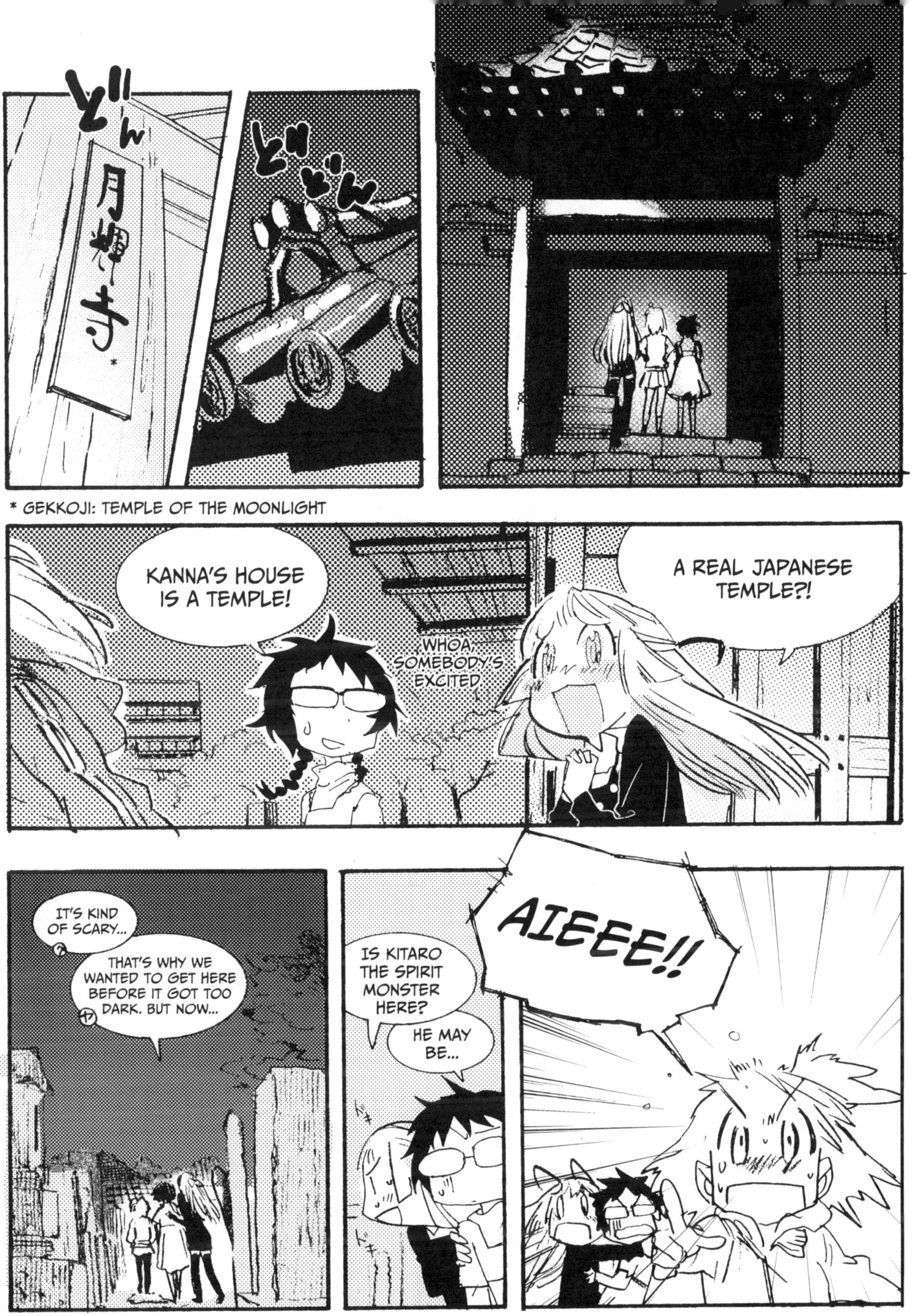
* GEKKOJI: TEMPLE OF THE MOONLIGHT
KANNA'S HOUSE IS A TEMPLE!
WHOA, SOMEBODY'S EXCITED
A REAL JAPANESE TEMPLE?!
IT'S KIND OF SCARY...
THAT'S WHY WE WANTED TO GET HERE BEFORE IT GOT TOO DARK. BUT NOW...
IS KITARO THE SPIRIT MONSTER HERE?
HE MAY BE...
AIEEE!!

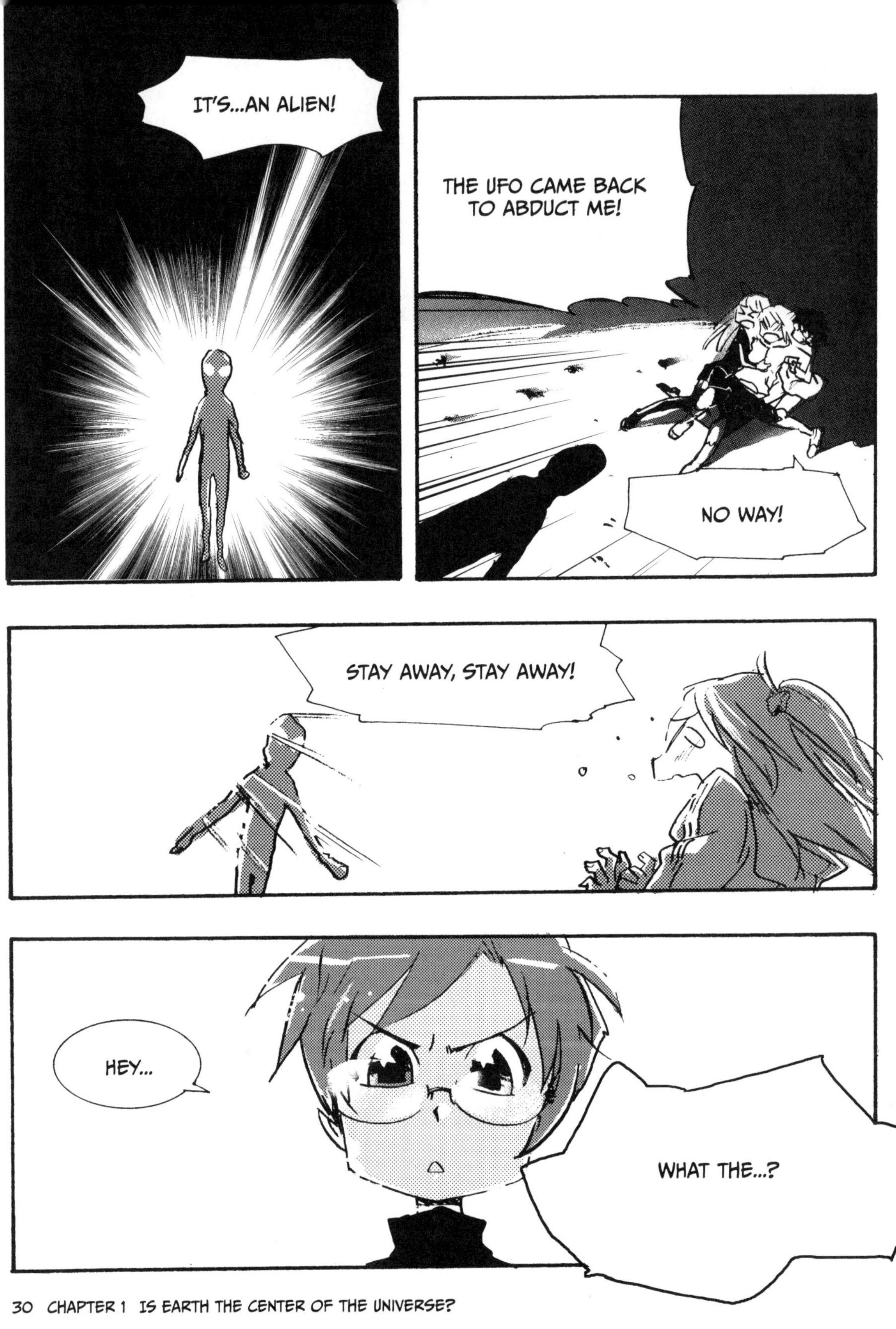
IT'S...AN ALIEN!
THE UFO CAME BACK TO ABDUCT ME!
NO WAY!
STAY AWAY, STAY AWAY!
HEY...
WHAT THE...?

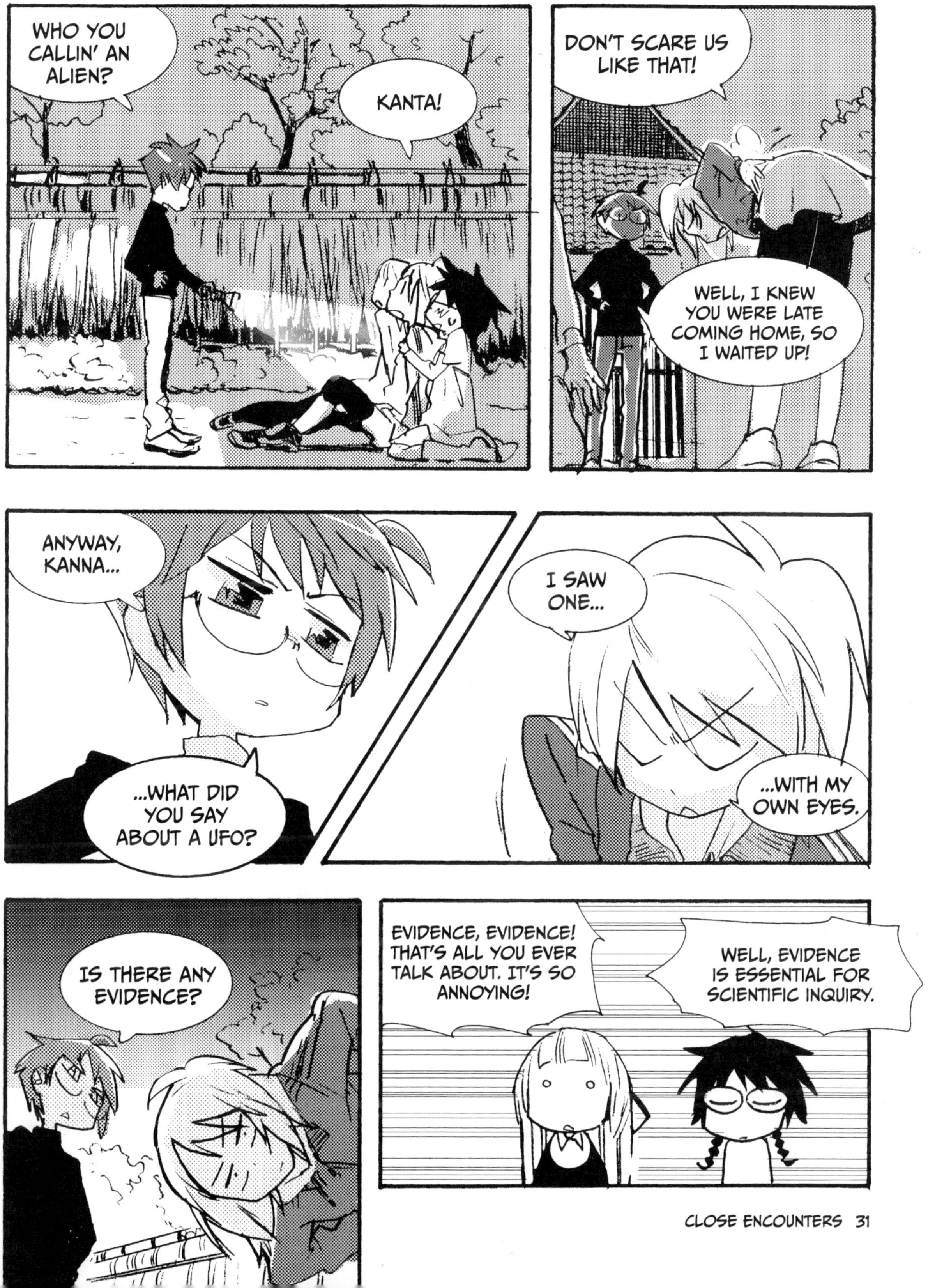
WHO YOU CALLIN' AN ALIEN?
KANTA!
DON'T SCARE US LIKE THAT!
WELL, I KNEW YOU WERE LATE COMING HOME, SO I WAITED UP!
ANYWAY, KANNA...
...WHAT DID YOU SAY ABOUT A UFO?
I SAW ONE...
...WITH MY OWN EYES.
IS THERE ANY EVIDENCE?
EVIDENCE, EVIDENCE! THAT'S ALL YOU EVER TALK ABOUT. IT'S SO ANNOYING!
WELL, EVIDENCE IS ESSENTIAL FOR SCIENTIFIC INQUIRY.

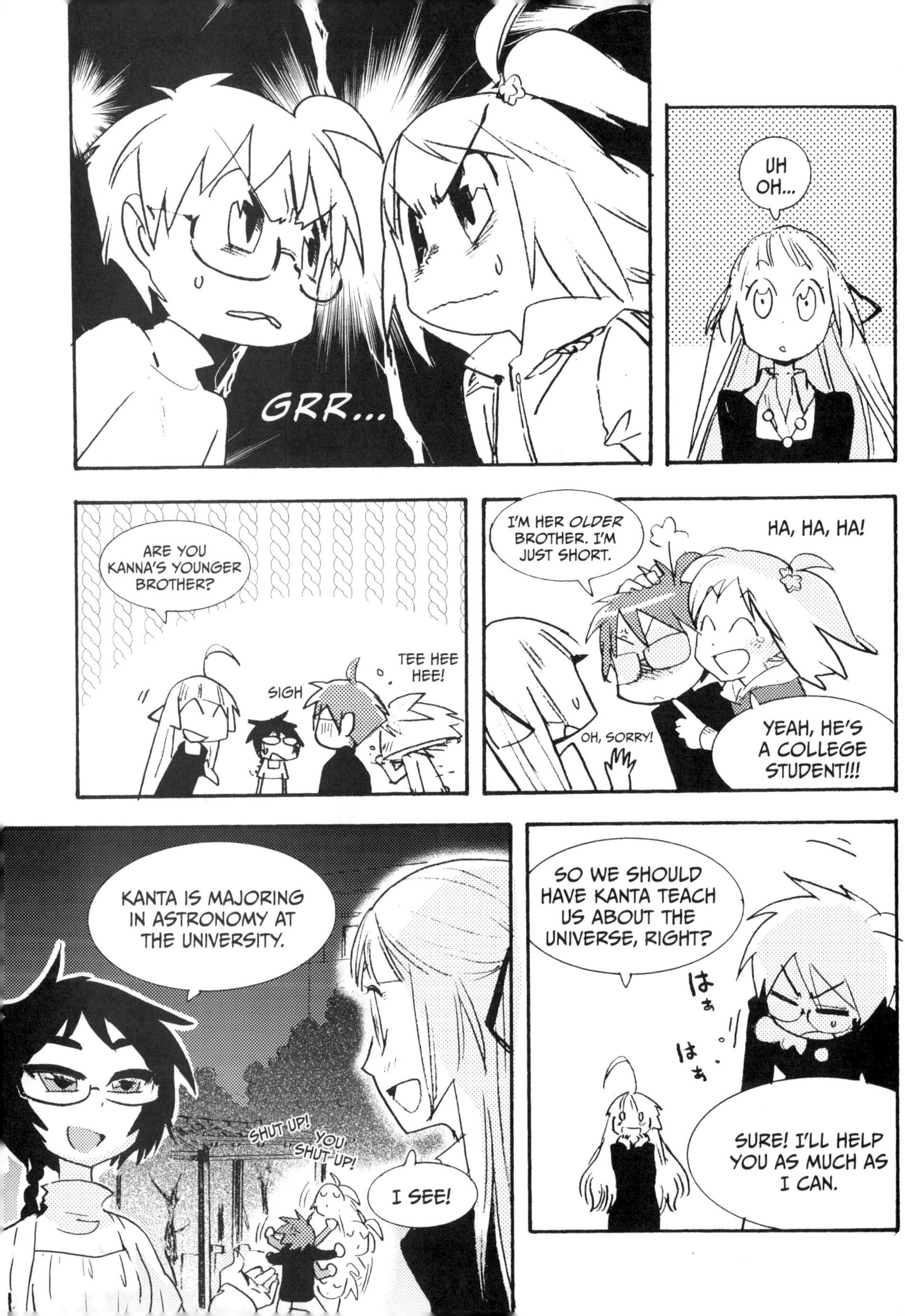
GRR...
UH OH...
ARE YOU KANNA'S YOUNGER BROTHER?
SIGH
TEE HEE HEE!
I'M HER *OLDER* BROTHER. I'M JUST SHORT.
HA, HA, HA!
OH, SORRY!
YEAH, HE'S A COLLEGE STUDENT!!!
KANTA IS MAJORING IN ASTRONOMY AT THE UNIVERSITY.
SHUT UP!
YOU SHUT UP!
I SEE!
SO WE SHOULD HAVE KANTA TEACH US ABOUT THE UNIVERSE, RIGHT?
はぁ
はぁ
SURE! I'LL HELP YOU AS MUCH AS I CAN.

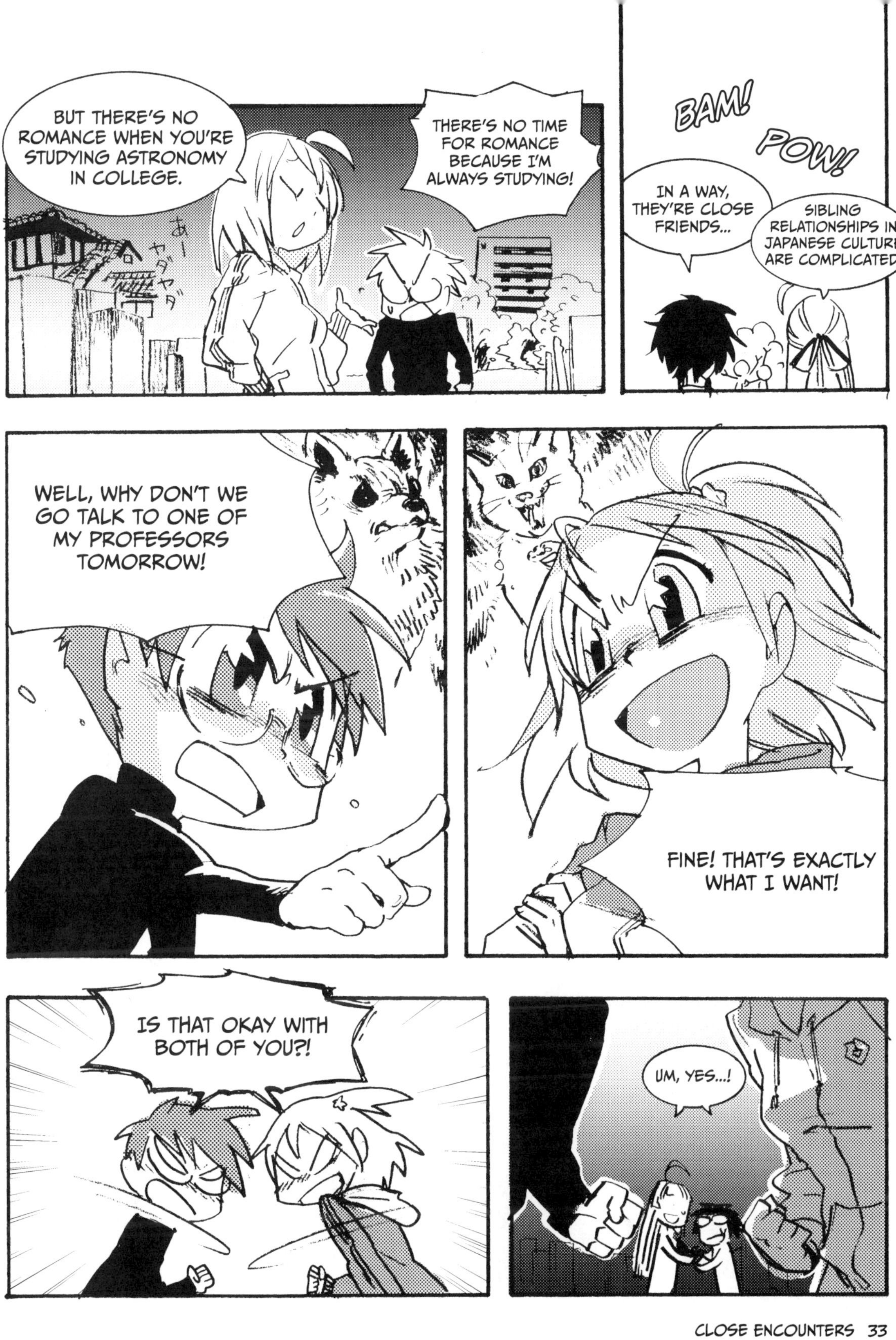
BUT THERE'S NO ROMANCE WHEN YOU'RE STUDYING ASTRONOMY IN COLLEGE.
あー ヤダヤダ
THERE'S NO TIME FOR ROMANCE BECAUSE I'M ALWAYS STUDYING!
BAM!
POW!
IN A WAY, THEY'RE CLOSE FRIENDS...
SIBLING RELATIONSHIPS IN JAPANESE CULTURE ARE COMPLICATED.
WELL, WHY DON'T WE GO TALK TO ONE OF MY PROFESSORS TOMORROW!
FINE! THAT'S EXACTLY WHAT I WANT!
IS THAT OKAY WITH BOTH OF YOU?!
UM, YES...!

DOES THE SUN REVOLVE AROUND EARTH?

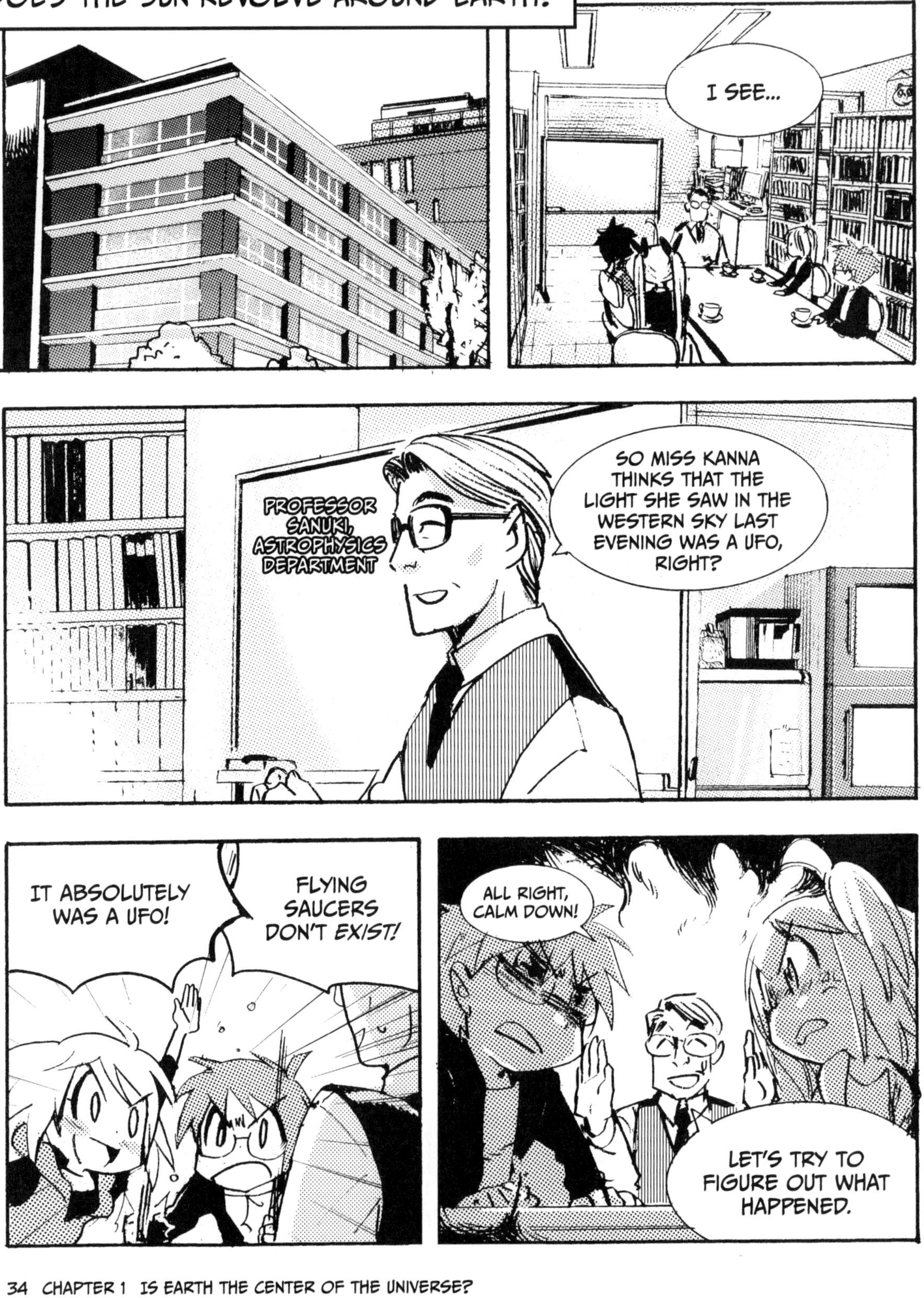

I SEE...
PROFESSOR SANUKI, ASTROPHYSICS DEPARTMENT
SO MISS KANNA THINKS THAT THE LIGHT SHE SAW IN THE WESTERN SKY LAST EVENING WAS A UFO, RIGHT?
IT ABSOLUTELY WAS A UFO!
FLYING SAUCERS DON'T *EXIST!*
ALL RIGHT, CALM DOWN!
LET'S TRY TO FIGURE OUT WHAT HAPPENED.

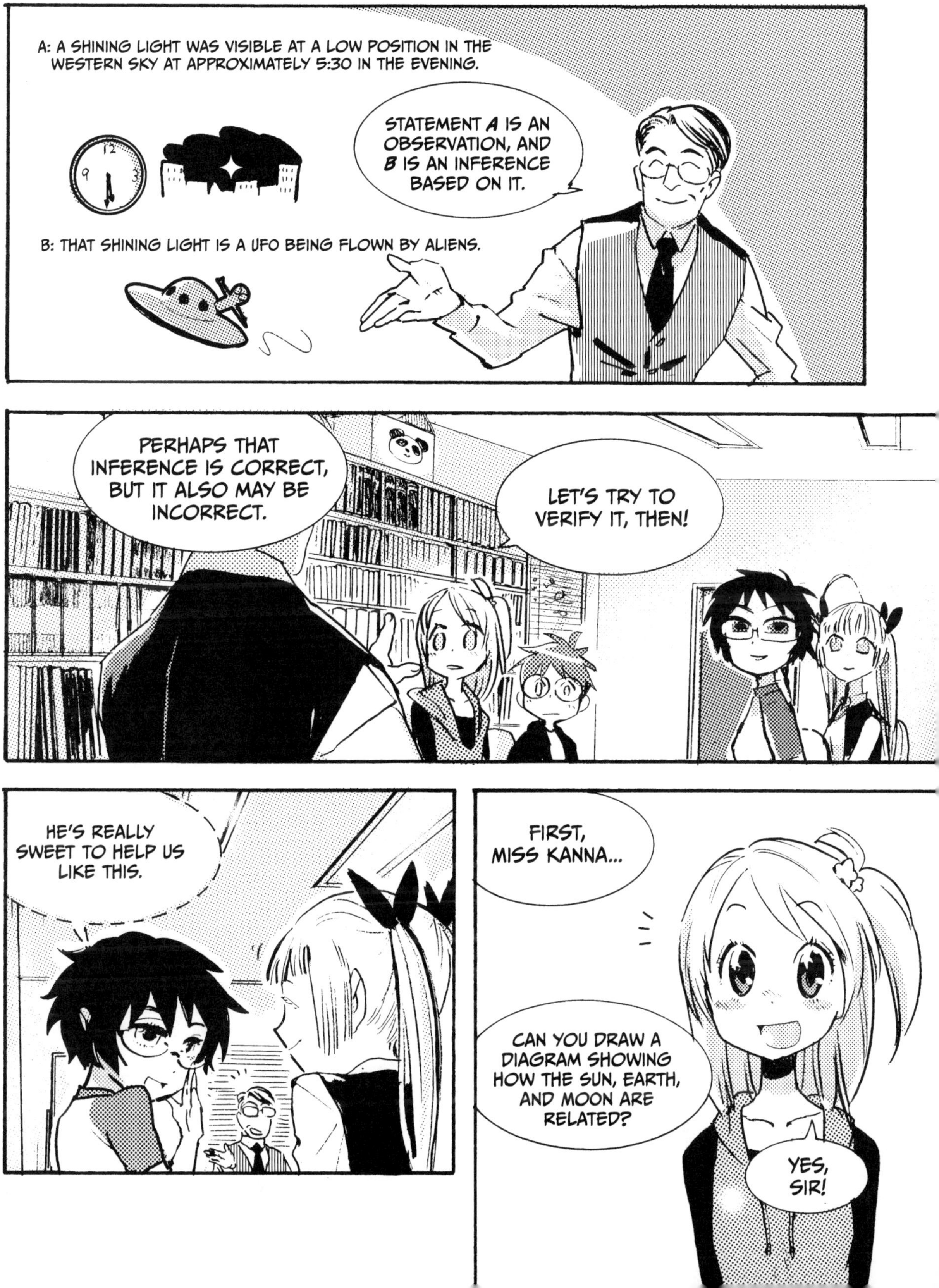
A: A SHINING LIGHT WAS VISIBLE AT A LOW POSITION IN THE WESTERN SKY AT APPROXIMATELY 5:30 IN THE EVENING.
STATEMENT *A* IS AN OBSERVATION, AND *B* IS AN INFERENCE BASED ON IT.
B: THAT SHINING LIGHT IS A UFO BEING FLOWN BY ALIENS.
PERHAPS THAT INFERENCE IS CORRECT, BUT IT ALSO MAY BE INCORRECT.
LET'S TRY TO VERIFY IT, THEN!
HE'S REALLY SWEET TO HELP US LIKE THIS.
FIRST, MISS KANNA...
CAN YOU DRAW A DIAGRAM SHOWING HOW THE SUN, EARTH, AND MOON ARE RELATED?
YES, SIR!

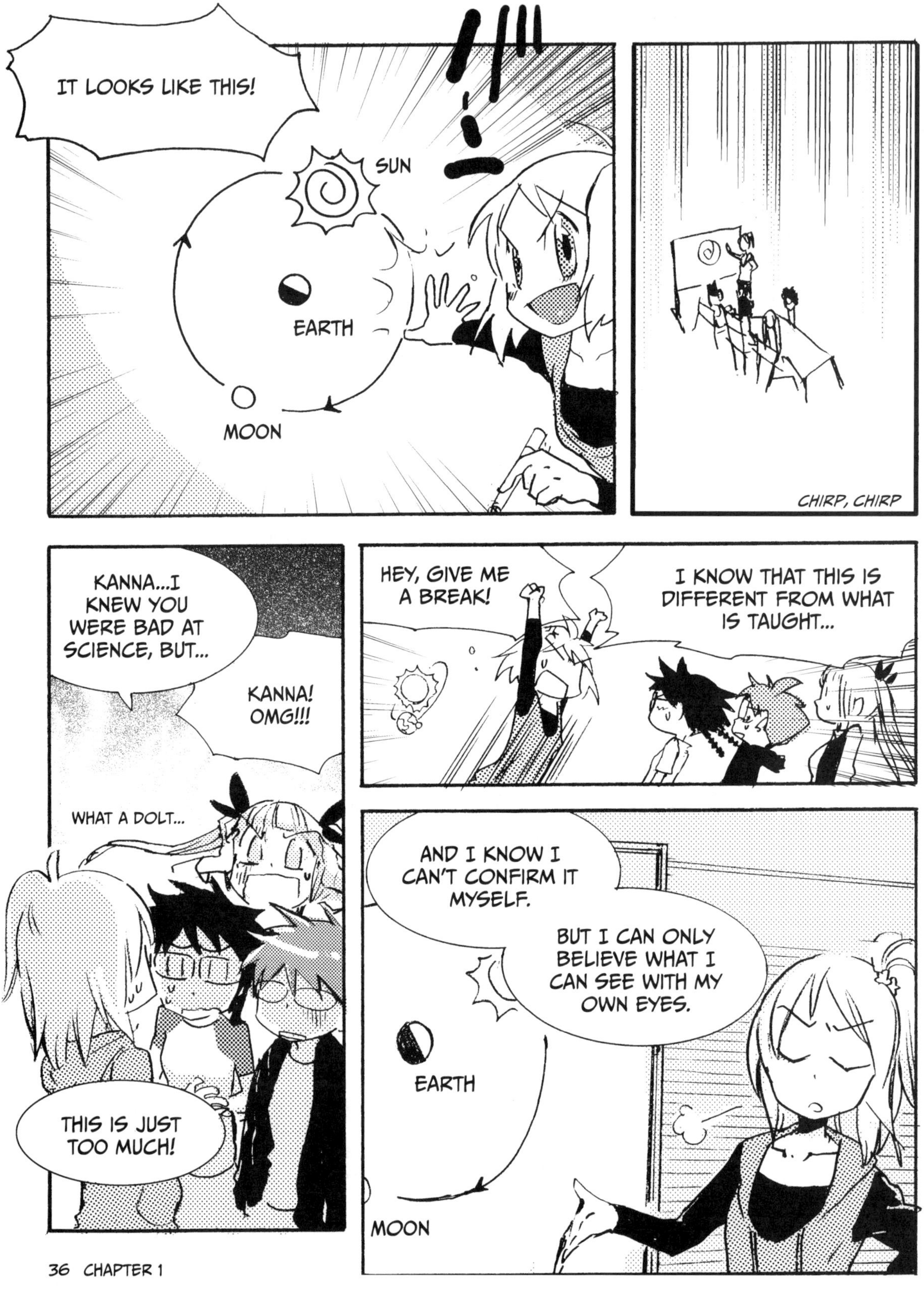
IT LOOKS LIKE THIS!
SUN
EARTH
MOON
CHIRP, CHIRP
KANNA...I KNEW YOU WERE BAD AT SCIENCE, BUT...
KANNA! OMG!!!
WHAT A DOLT...
THIS IS JUST TOO MUCH!
HEY, GIVE ME A BREAK!
I KNOW THAT THIS IS DIFFERENT FROM WHAT IS TAUGHT...
AND I KNOW I CAN'T CONFIRM IT MYSELF.
BUT I CAN ONLY BELIEVE WHAT I CAN SEE WITH MY OWN EYES.
EARTH
MOON

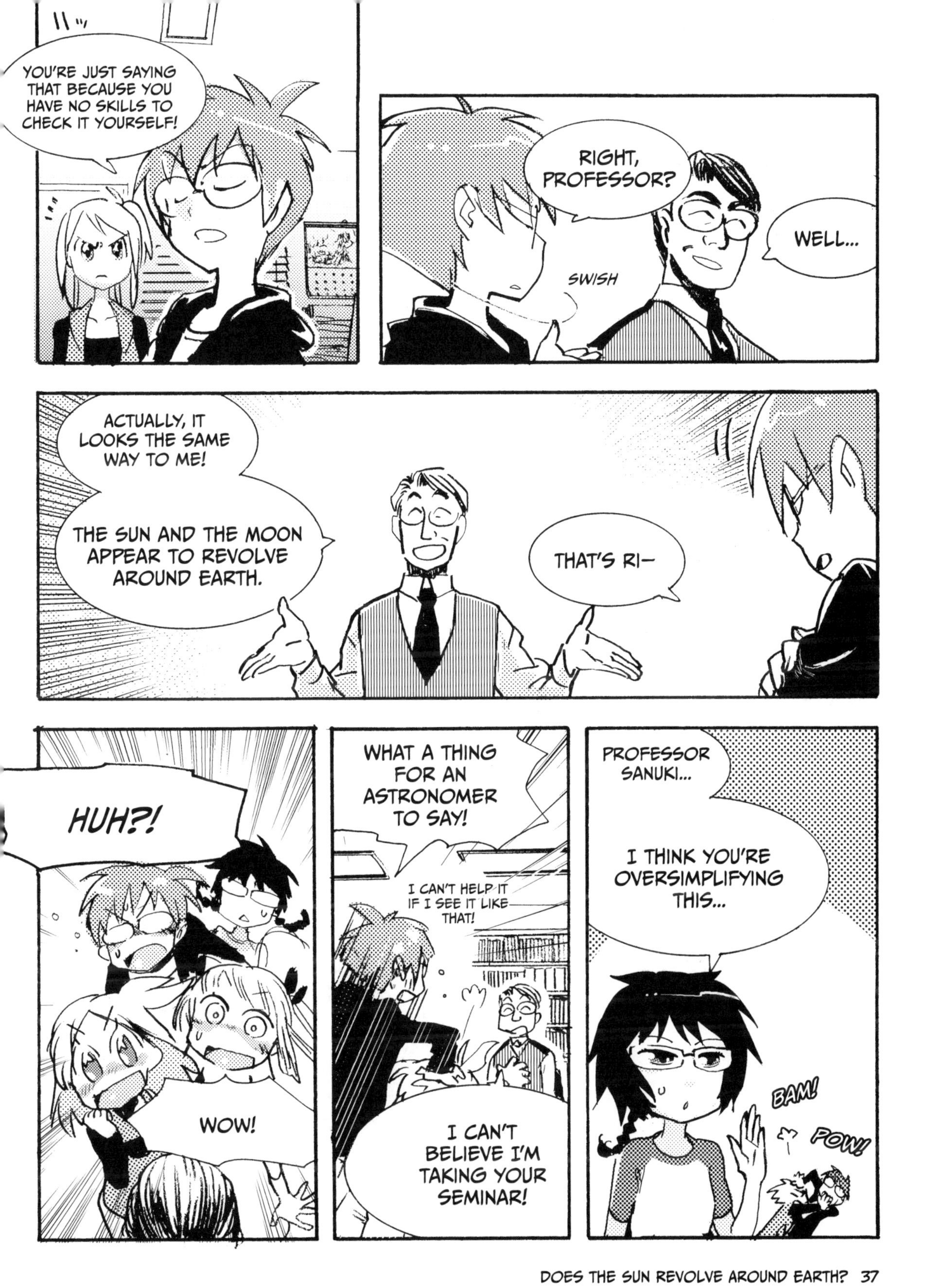
YOU'RE JUST SAYING THAT BECAUSE YOU HAVE NO SKILLS TO CHECK IT YOURSELF!
RIGHT, PROFESSOR?
SWISH
WELL...
ACTUALLY, IT LOOKS THE SAME WAY TO ME!
THE SUN AND THE MOON APPEAR TO REVOLVE AROUND EARTH.
THAT'S RI–
HUH?!
WOW!
WHAT A THING FOR AN ASTRONOMER TO SAY!
I CAN'T HELP IT IF I SEE IT LIKE THAT!
I CAN'T BELIEVE I'M TAKING YOUR SEMINAR!
PROFESSOR SANUKI...
I THINK YOU'RE OVERSIMPLIFYING THIS...
BAM!
POW!

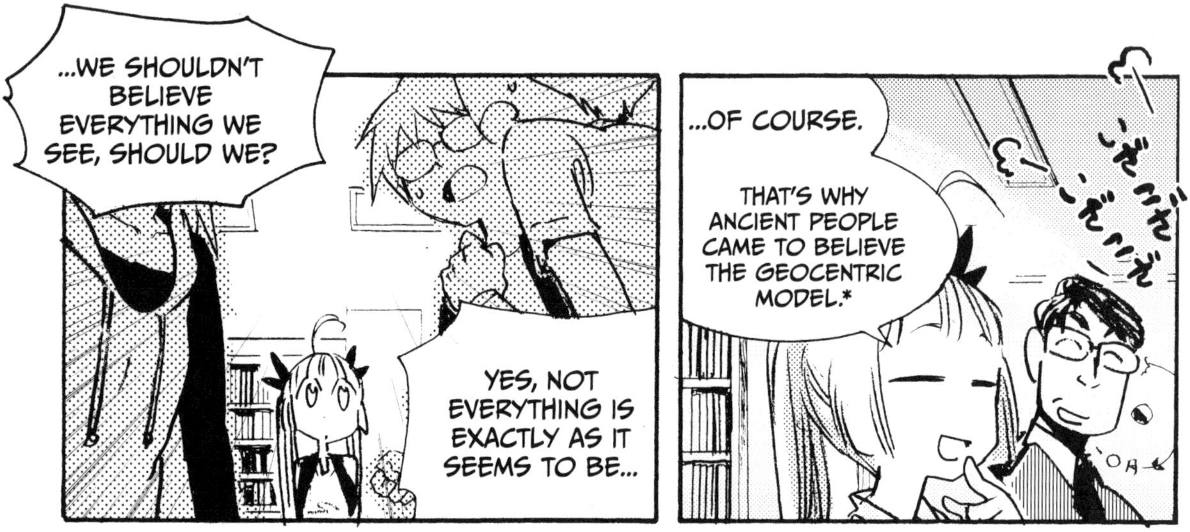
...WE SHOULDN'T BELIEVE EVERYTHING WE SEE, SHOULD WE?
YES, NOT EVERYTHING IS EXACTLY AS IT SEEMS TO BE...
...OF COURSE.
THAT'S WHY ANCIENT PEOPLE CAME TO BELIEVE THE GEOCENTRIC MODEL.*

* A MODEL OF OUR PLANETARY SYSTEM WITH EARTH AT ITS CENTER, LIKE THE ONE KANNA DREW, IS CALLED *GEOCENTRIC*.

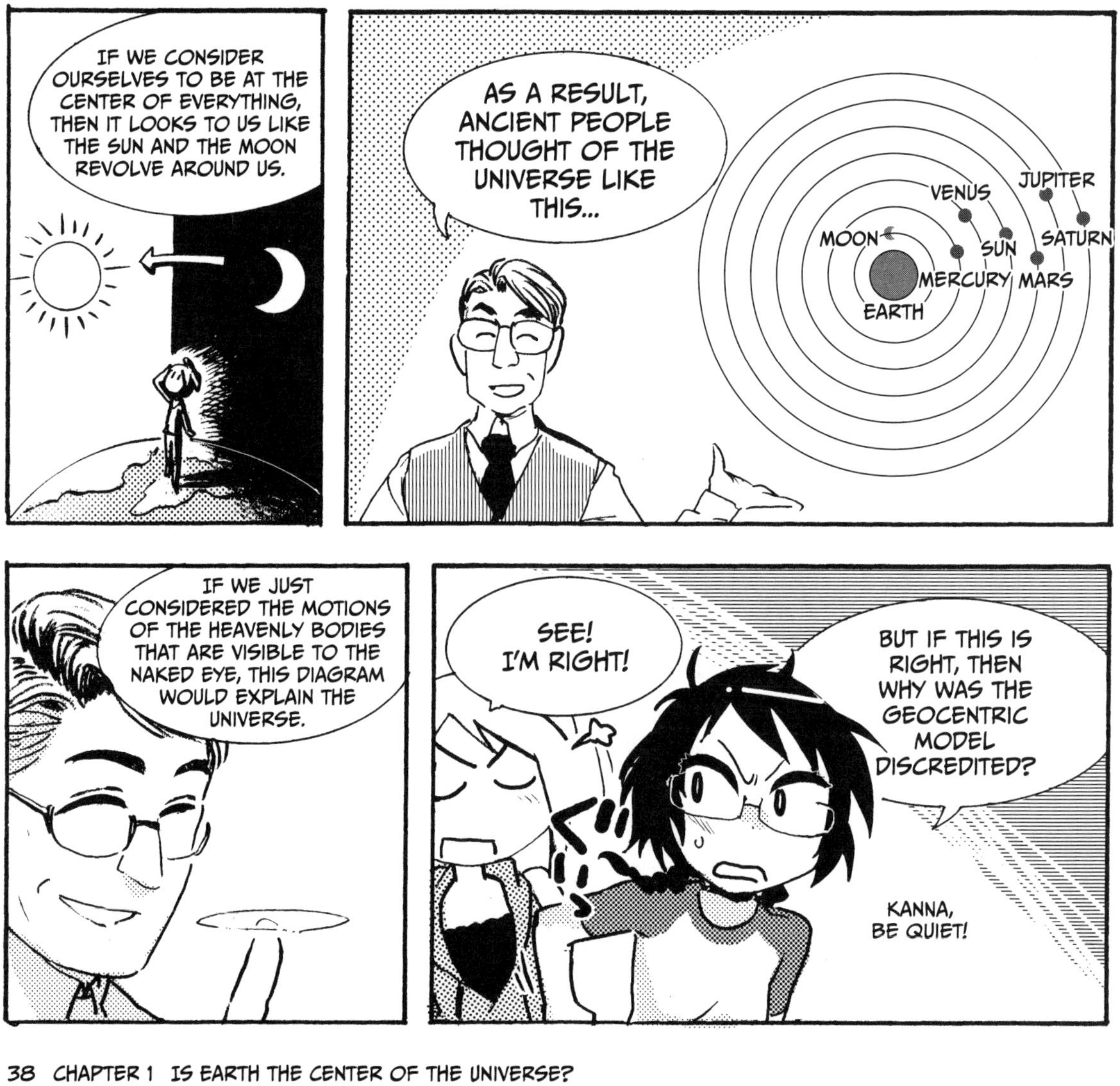
IF WE CONSIDER OURSELVES TO BE AT THE CENTER OF EVERYTHING, THEN IT LOOKS TO US LIKE THE SUN AND THE MOON REVOLVE AROUND US.
AS A RESULT, ANCIENT PEOPLE THOUGHT OF THE UNIVERSE LIKE THIS...
VENUS
JUPITER
MOON
SUN
SATURN
MERCURY
MARS
EARTH
IF WE JUST CONSIDERED THE MOTIONS OF THE HEAVENLY BODIES THAT ARE VISIBLE TO THE NAKED EYE, THIS DIAGRAM WOULD EXPLAIN THE UNIVERSE.
SEE! I'M RIGHT!
BUT IF THIS IS RIGHT, THEN WHY WAS THE GEOCENTRIC MODEL DISCREDITED?
KANNA, BE QUIET!

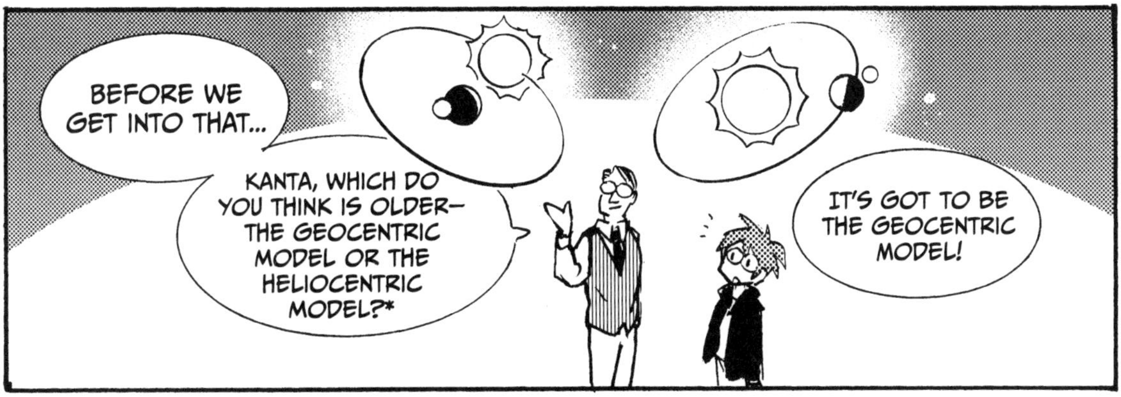
BEFORE WE GET INTO THAT...
KANTA, WHICH DO YOU THINK IS OLDER— THE GEOCENTRIC MODEL OR THE HELIOCENTRIC MODEL?*
IT'S GOT TO BE THE GEOCENTRIC MODEL!

* A *HELIOCENTRIC* MODEL OF OUR PLANETARY SYSTEM HAS THE SUN AT ITS CENTER.

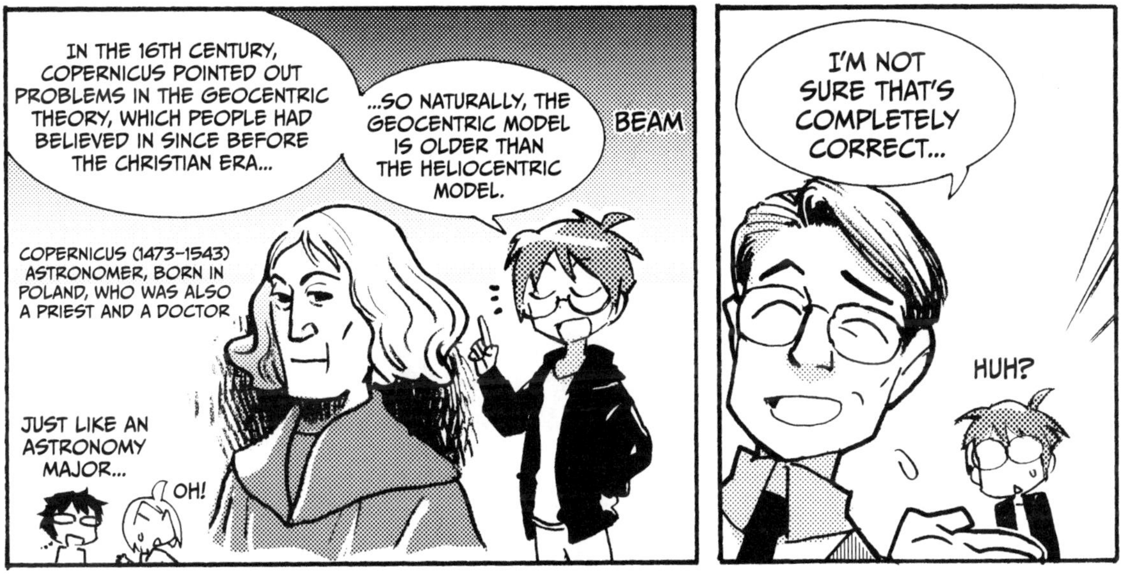
IN THE 16TH CENTURY, COPERNICUS POINTED OUT PROBLEMS IN THE GEOCENTRIC THEORY, WHICH PEOPLE HAD BELIEVED IN SINCE BEFORE THE CHRISTIAN ERA...
...SO NATURALLY, THE GEOCENTRIC MODEL IS OLDER THAN THE HELIOCENTRIC MODEL.
BEAM
COPERNICUS (1473–1543) ASTRONOMER, BORN IN POLAND, WHO WAS ALSO A PRIEST AND A DOCTOR
JUST LIKE AN ASTRONOMY MAJOR...
OH!
I'M NOT SURE THAT'S COMPLETELY CORRECT...
HUH?

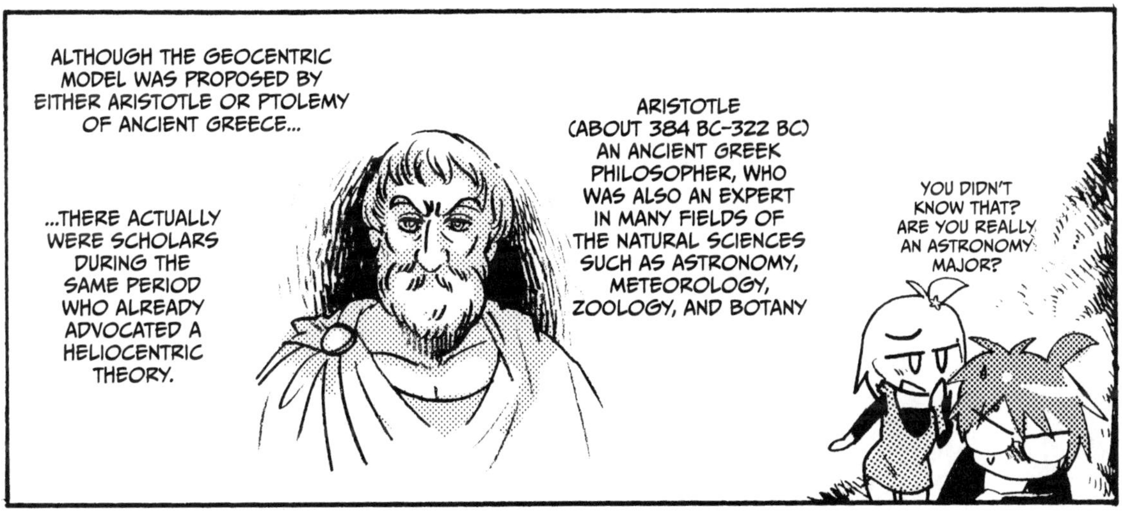
ALTHOUGH THE GEOCENTRIC MODEL WAS PROPOSED BY EITHER ARISTOTLE OR PTOLEMY OF ANCIENT GREECE...
...THERE ACTUALLY WERE SCHOLARS DURING THE SAME PERIOD WHO ALREADY ADVOCATED A HELIOCENTRIC THEORY.
ARISTOTLE (ABOUT 384 BC–322 BC) AN ANCIENT GREEK PHILOSOPHER, WHO WAS ALSO AN EXPERT IN MANY FIELDS OF THE NATURAL SCIENCES SUCH AS ASTRONOMY, METEOROLOGY, ZOOLOGY, AND BOTANY
YOU DIDN'T KNOW THAT? ARE YOU REALLY AN ASTRONOMY MAJOR?

A HELIOCENTRIC MODEL WAS PROPOSED 2,300 YEARS AGO

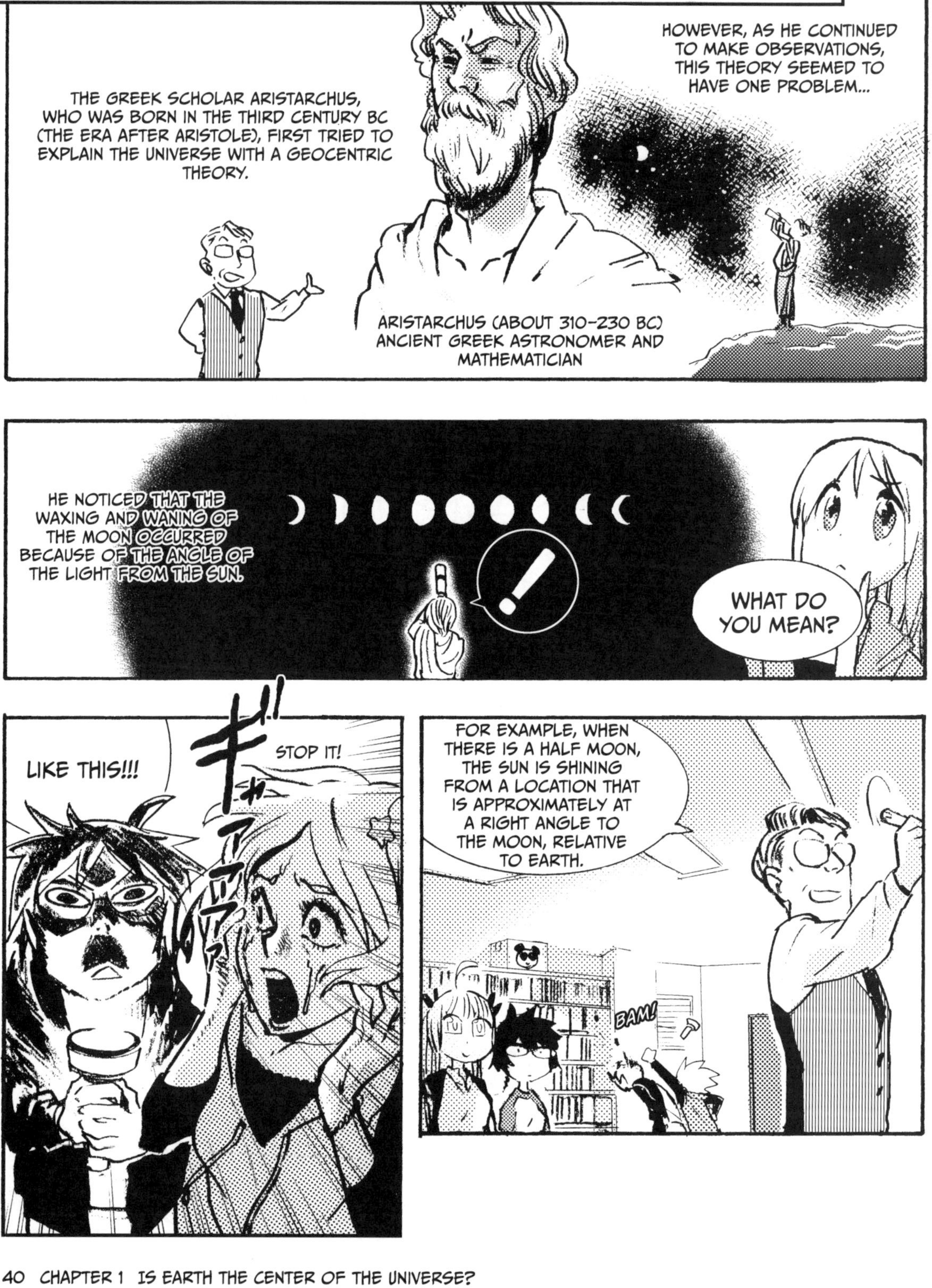

THE GREEK SCHOLAR ARISTARCHUS, WHO WAS BORN IN THE THIRD CENTURY BC (THE ERA AFTER ARISTOLE), FIRST TRIED TO EXPLAIN THE UNIVERSE WITH A GEOCENTRIC THEORY.
ARISTARCHUS (ABOUT 310–230 BC) ANCIENT GREEK ASTRONOMER AND MATHEMATICIAN
HOWEVER, AS HE CONTINUED TO MAKE OBSERVATIONS, THIS THEORY SEEMED TO HAVE ONE PROBLEM...
HE NOTICED THAT THE WAXING AND WANING OF THE MOON OCCURRED BECAUSE OF THE ANGLE OF THE LIGHT FROM THE SUN.
!
WHAT DO YOU MEAN?
LIKE THIS!!!
STOP IT!
FOR EXAMPLE, WHEN THERE IS A HALF MOON, THE SUN IS SHINING FROM A LOCATION THAT IS APPROXIMATELY AT A RIGHT ANGLE TO THE MOON, RELATIVE TO EARTH.
BAM!

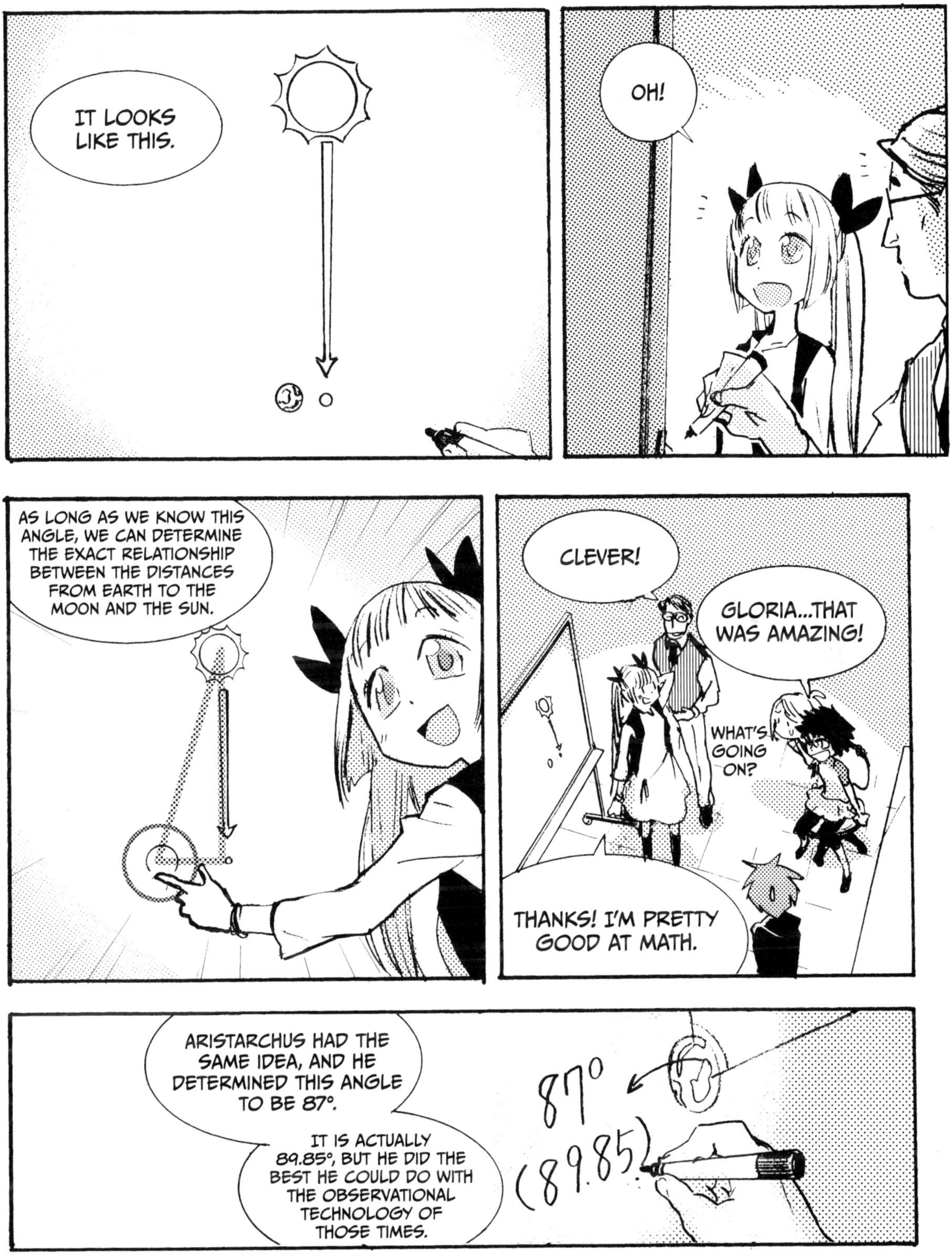
IT LOOKS LIKE THIS.
OH!
AS LONG AS WE KNOW THIS ANGLE, WE CAN DETERMINE THE EXACT RELATIONSHIP BETWEEN THE DISTANCES FROM EARTH TO THE MOON AND THE SUN.
CLEVER!
GLORIA...THAT WAS AMAZING!
WHAT'S GOING ON?
THANKS! I'M PRETTY GOOD AT MATH.
ARISTARCHUS HAD THE SAME IDEA, AND HE DETERMINED THIS ANGLE TO BE 87°.
IT IS ACTUALLY 89.85°, BUT HE DID THE BEST HE COULD DO WITH THE OBSERVATIONAL TECHNOLOGY OF THOSE TIMES.
87°
(89.85)

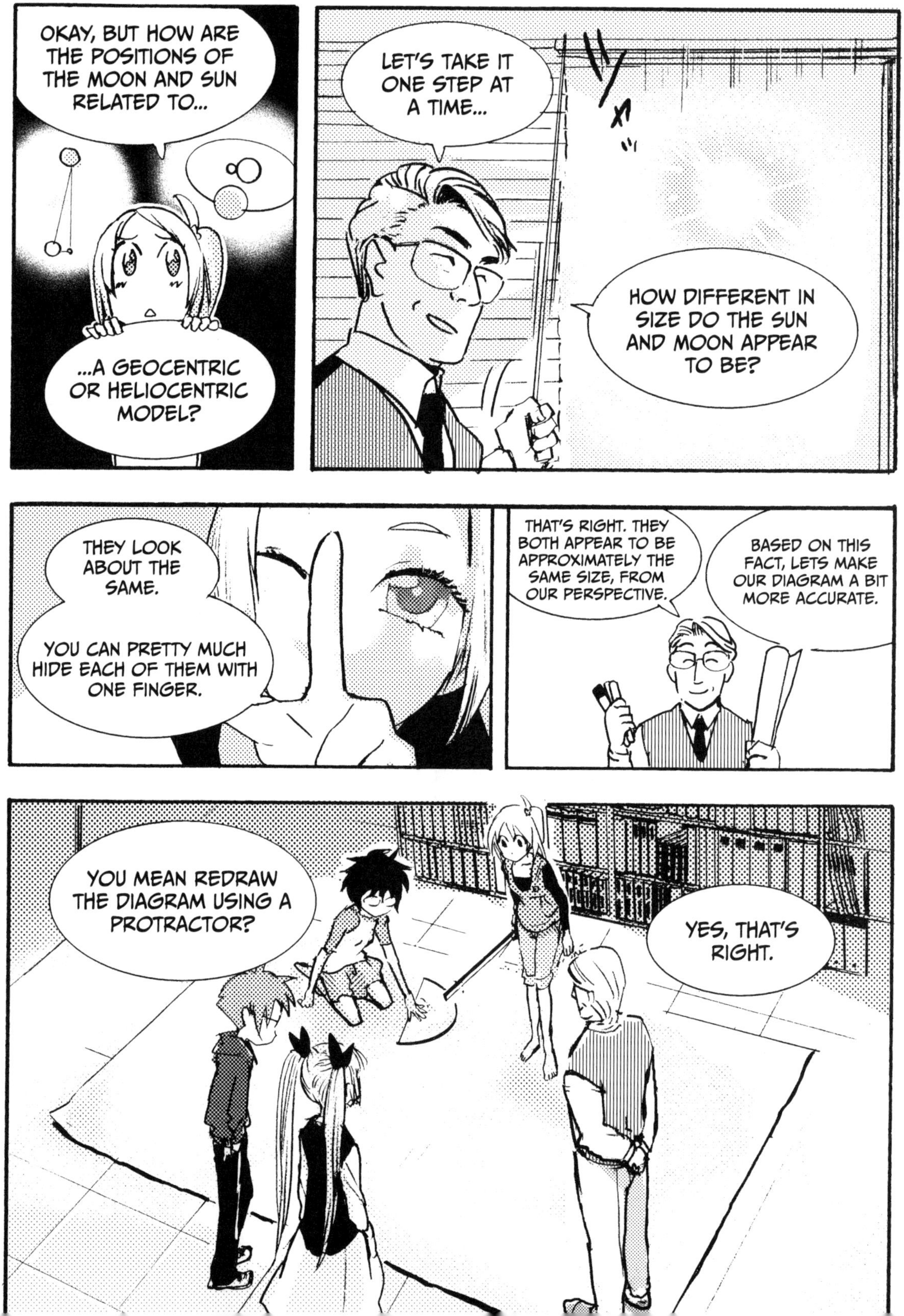
OKAY, BUT HOW ARE THE POSITIONS OF THE MOON AND SUN RELATED TO...
...A GEOCENTRIC OR HELIOCENTRIC MODEL?
LET'S TAKE IT ONE STEP AT A TIME...
HOW DIFFERENT IN SIZE DO THE SUN AND MOON APPEAR TO BE?
THEY LOOK ABOUT THE SAME.
YOU CAN PRETTY MUCH HIDE EACH OF THEM WITH ONE FINGER.
THAT'S RIGHT. THEY BOTH APPEAR TO BE APPROXIMATELY THE SAME SIZE, FROM OUR PERSPECTIVE.
BASED ON THIS FACT, LETS MAKE OUR DIAGRAM A BIT MORE ACCURATE.
YOU MEAN REDRAW THE DIAGRAM USING A PROTRACTOR?
YES, THAT'S RIGHT.

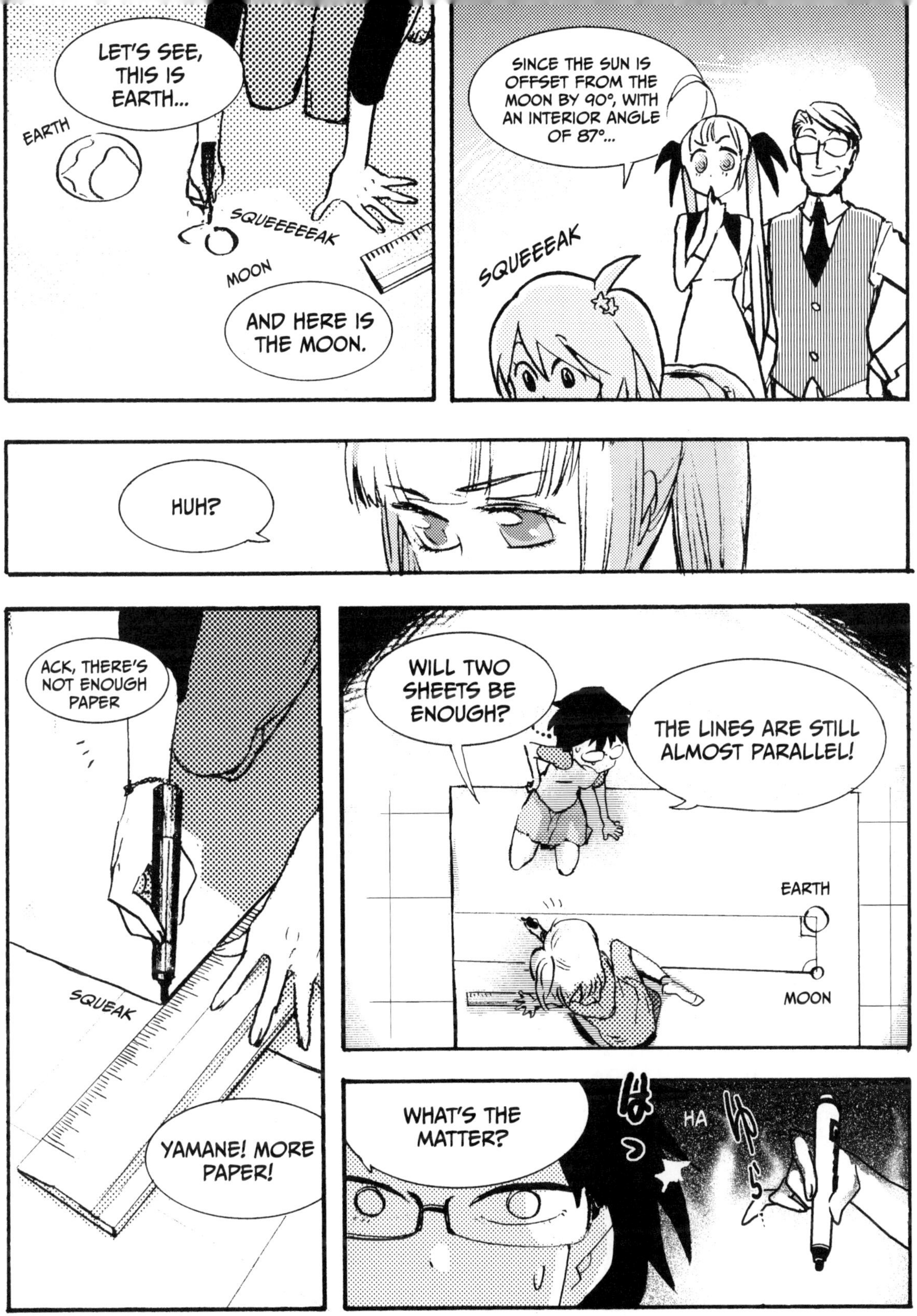
LET'S SEE, THIS IS EARTH...
EARTH
SQUEEEEEAK
MOON
AND HERE IS THE MOON.
SINCE THE SUN IS OFFSET FROM THE MOON BY 90°, WITH AN INTERIOR ANGLE OF 87°...
SQUEEEAK
HUH?
ACK, THERE'S NOT ENOUGH PAPER
SQUEAK
YAMANE! MORE PAPER!
WILL TWO SHEETS BE ENOUGH?
THE LINES ARE STILL ALMOST PARALLEL!
EARTH
MOON
WHAT'S THE MATTER?
HA

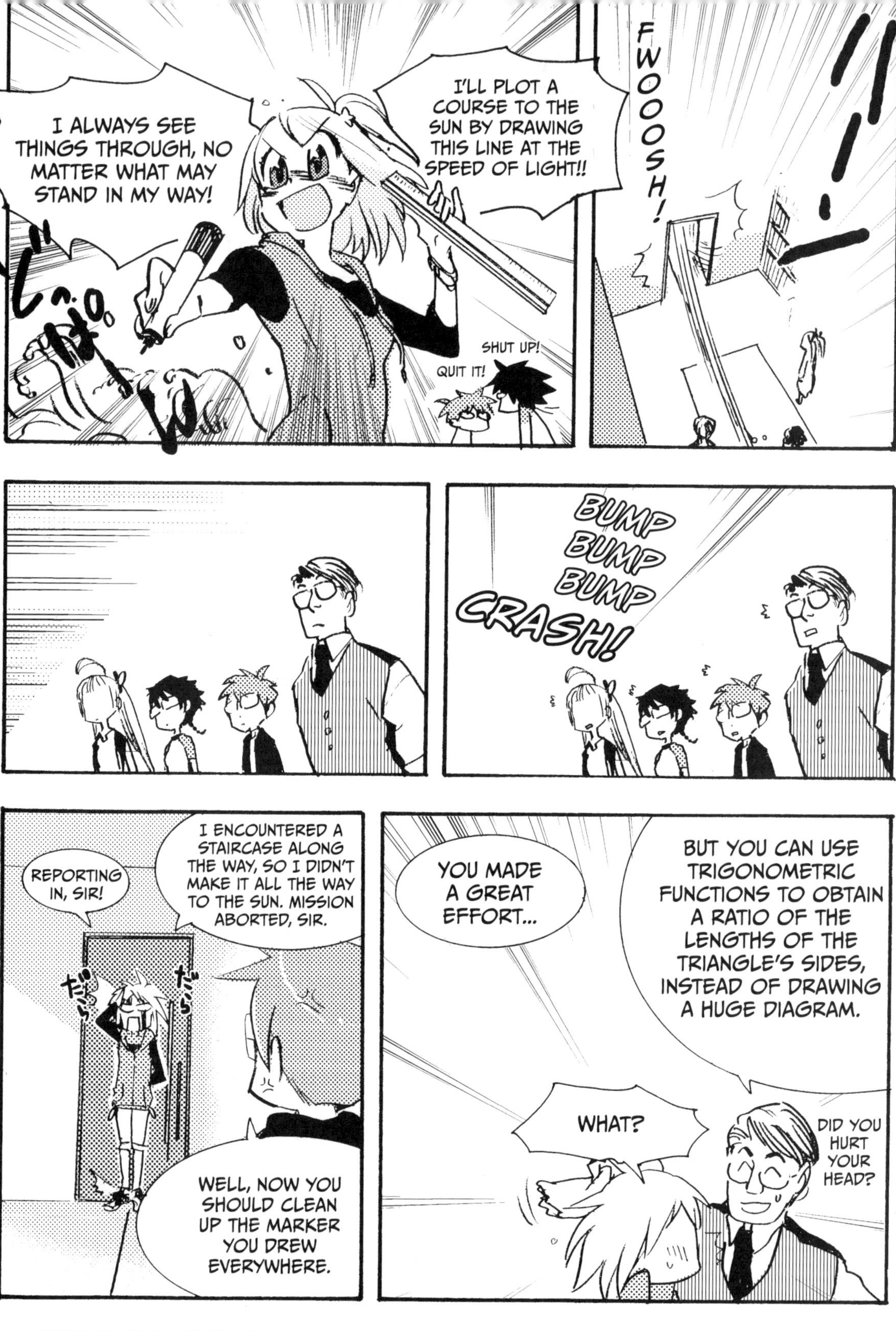

I ALWAYS SEE THINGS THROUGH, NO MATTER WHAT MAY STAND IN MY WAY!
I'LL PLOT A COURSE TO THE SUN BY DRAWING THIS LINE AT THE SPEED OF LIGHT!!
QUIT IT!
SHUT UP!
FWOOOOSH!
BUMP
BUMP
BUMP
CRASH!
REPORTING IN, SIR!
I ENCOUNTERED A STAIRCASE ALONG THE WAY, SO I DIDN'T MAKE IT ALL THE WAY TO THE SUN. MISSION ABORTED, SIR.
WELL, NOW YOU SHOULD CLEAN UP THE MARKER YOU DREW EVERYWHERE.
YOU MADE A GREAT EFFORT...
BUT YOU CAN USE TRIGONOMETRIC FUNCTIONS TO OBTAIN A RATIO OF THE LENGTHS OF THE TRIANGLE'S SIDES, INSTEAD OF DRAWING A HUGE DIAGRAM.
WHAT?
DID YOU HURT YOUR HEAD?

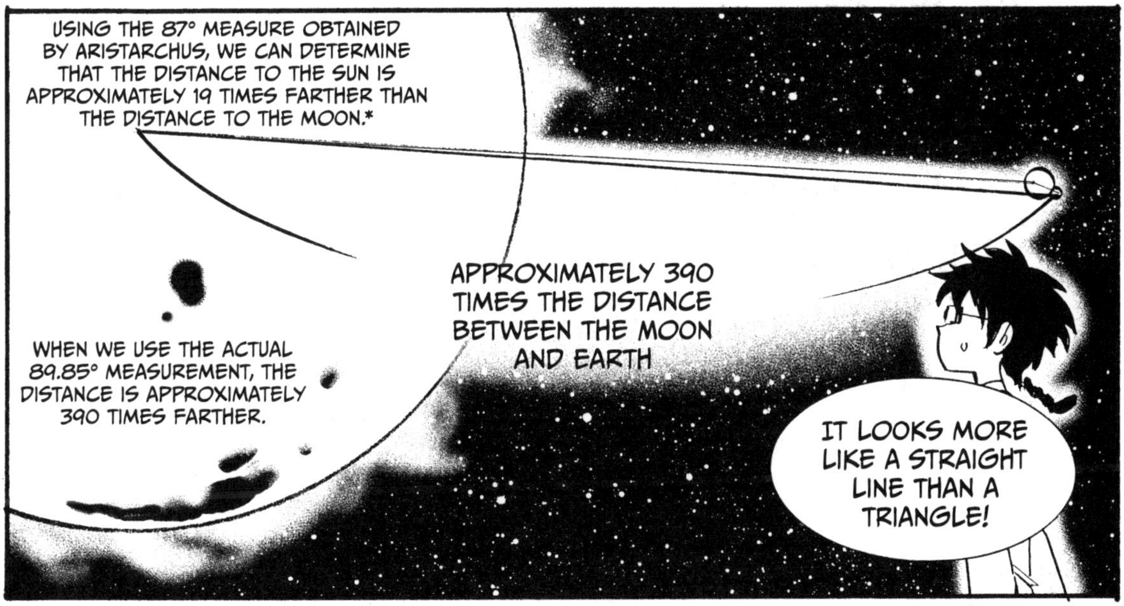
USING THE 87° MEASURE OBTAINED BY ARISTARCHUS, WE CAN DETERMINE THAT THE DISTANCE TO THE SUN IS APPROXIMATELY 19 TIMES FARTHER THAN THE DISTANCE TO THE MOON.*
WHEN WE USE THE ACTUAL 89.85° MEASUREMENT, THE DISTANCE IS APPROXIMATELY 390 TIMES FARTHER.
APPROXIMATELY 390 TIMES THE DISTANCE BETWEEN THE MOON AND EARTH
IT LOOKS MORE LIKE A STRAIGHT LINE THAN A TRIANGLE!

* THE TANGENT FUNCTION DESCRIBES THE RATIO OF OPPOSITE-TO-ADJACENT SIDES OF A RIGHT TRIANGLE (tan87° = 19).

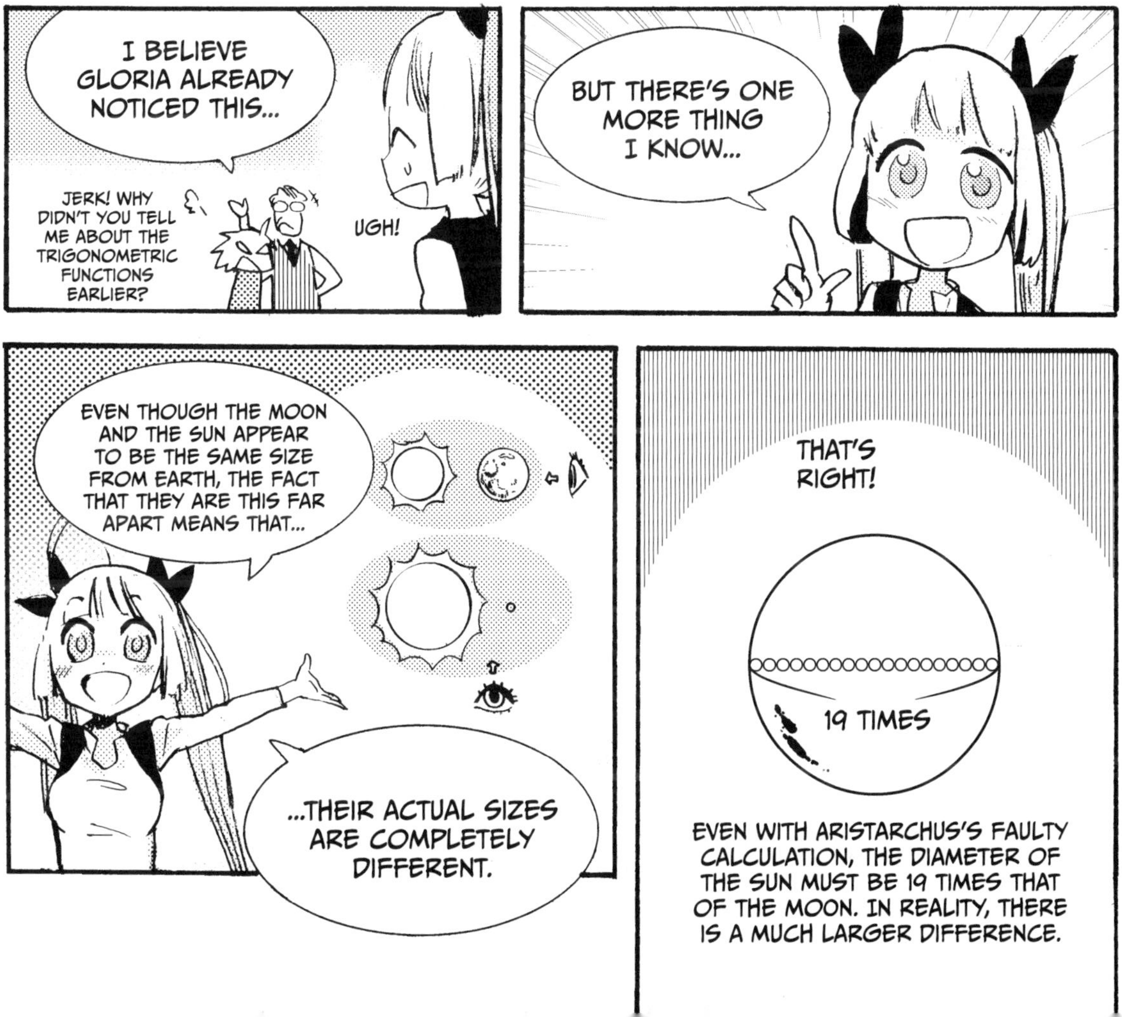
I BELIEVE GLORIA ALREADY NOTICED THIS...
JERK! WHY DIDN'T YOU TELL ME ABOUT THE TRIGONOMETRIC FUNCTIONS EARLIER?
UGH!
BUT THERE'S ONE MORE THING I KNOW...
EVEN THOUGH THE MOON AND THE SUN APPEAR TO BE THE SAME SIZE FROM EARTH, THE FACT THAT THEY ARE THIS FAR APART MEANS THAT...
...THEIR ACTUAL SIZES ARE COMPLETELY DIFFERENT.
THAT'S RIGHT!
19 TIMES
EVEN WITH ARISTARCHUS'S FAULTY CALCULATION, THE DIAMETER OF THE SUN MUST BE 19 TIMES THAT OF THE MOON. IN REALITY, THERE IS A MUCH LARGER DIFFERENCE.

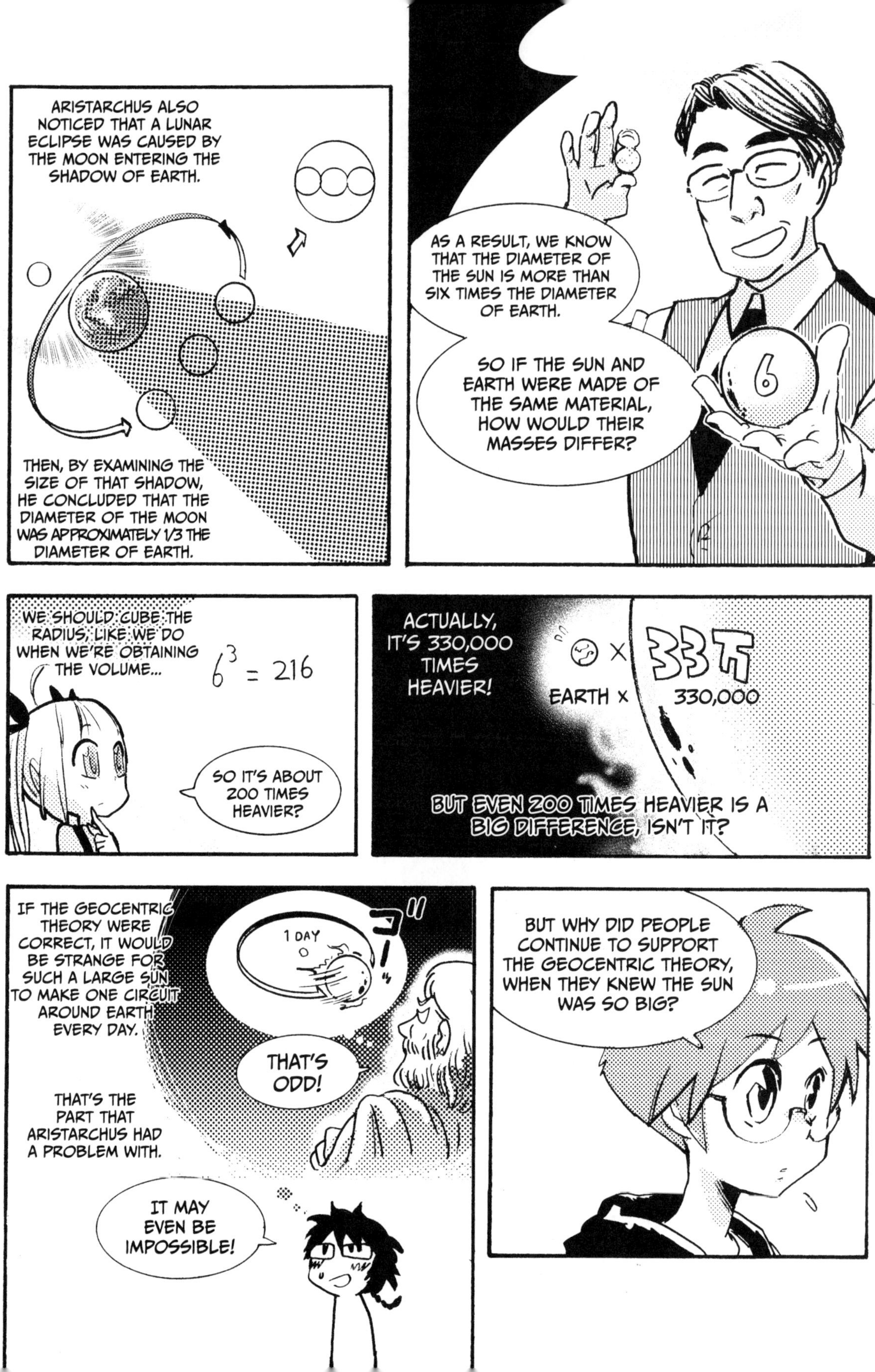
ARISTARCHUS ALSO NOTICED THAT A LUNAR ECLIPSE WAS CAUSED BY THE MOON ENTERING THE SHADOW OF EARTH.
THEN, BY EXAMINING THE SIZE OF THAT SHADOW, HE CONCLUDED THAT THE DIAMETER OF THE MOON WAS APPROXIMATELY 1/3 THE DIAMETER OF EARTH.
AS A RESULT, WE KNOW THAT THE DIAMETER OF THE SUN IS MORE THAN SIX TIMES THE DIAMETER OF EARTH.
SO IF THE SUN AND EARTH WERE MADE OF THE SAME MATERIAL, HOW WOULD THEIR MASSES DIFFER?
6
WE SHOULD CUBE THE RADIUS, LIKE WE DO WHEN WE'RE OBTAINING THE VOLUME...
6^3 = 216
SO IT'S ABOUT 200 TIMES HEAVIER?
ACTUALLY, IT'S 330,000 TIMES HEAVIER!
× 33万
EARTH × 330,000
BUT EVEN 200 TIMES HEAVIER IS A BIG DIFFERENCE, ISN'T IT?
IF THE GEOCENTRIC THEORY WERE CORRECT, IT WOULD BE STRANGE FOR SUCH A LARGE SUN TO MAKE ONE CIRCUIT AROUND EARTH EVERY DAY.
1 DAY
THAT'S ODD!
THAT'S THE PART THAT ARISTARCHUS HAD A PROBLEM WITH.
IT MAY EVEN BE IMPOSSIBLE!
BUT WHY DID PEOPLE CONTINUE TO SUPPORT THE GEOCENTRIC THEORY, WHEN THEY KNEW THE SUN WAS SO BIG?

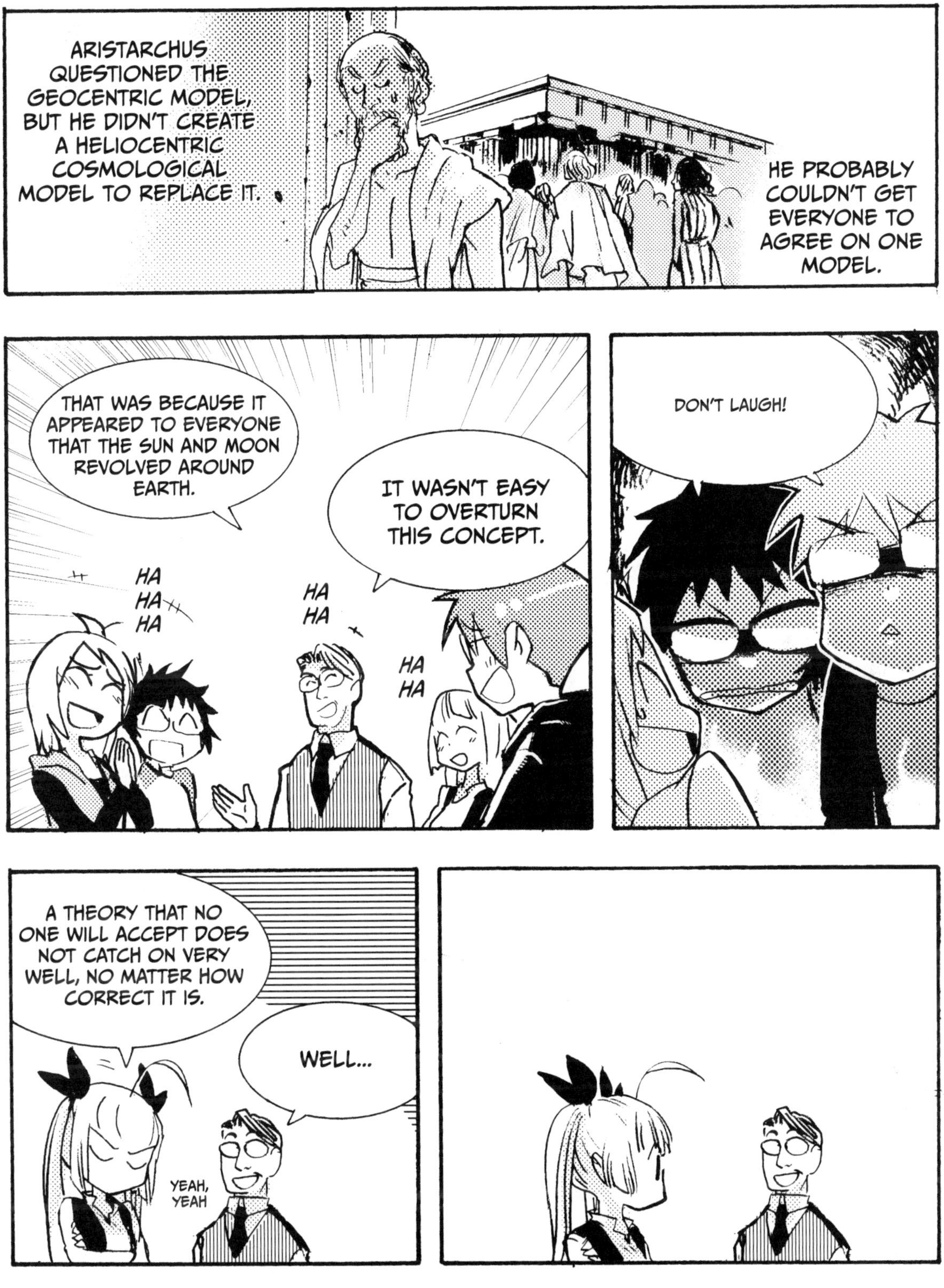
ARISTARCHUS QUESTIONED THE GEOCENTRIC MODEL, BUT HE DIDN'T CREATE A HELIOCENTRIC COSMOLOGICAL MODEL TO REPLACE IT.
HE PROBABLY COULDN'T GET EVERYONE TO AGREE ON ONE MODEL.
THAT WAS BECAUSE IT APPEARED TO EVERYONE THAT THE SUN AND MOON REVOLVED AROUND EARTH.
HA HA HA
HA HA
IT WASN'T EASY TO OVERTURN THIS CONCEPT.
HA HA
DON'T LAUGH!
A THEORY THAT NO ONE WILL ACCEPT DOES NOT CATCH ON VERY WELL, NO MATTER HOW CORRECT IT IS.
WELL...
YEAH, YEAH

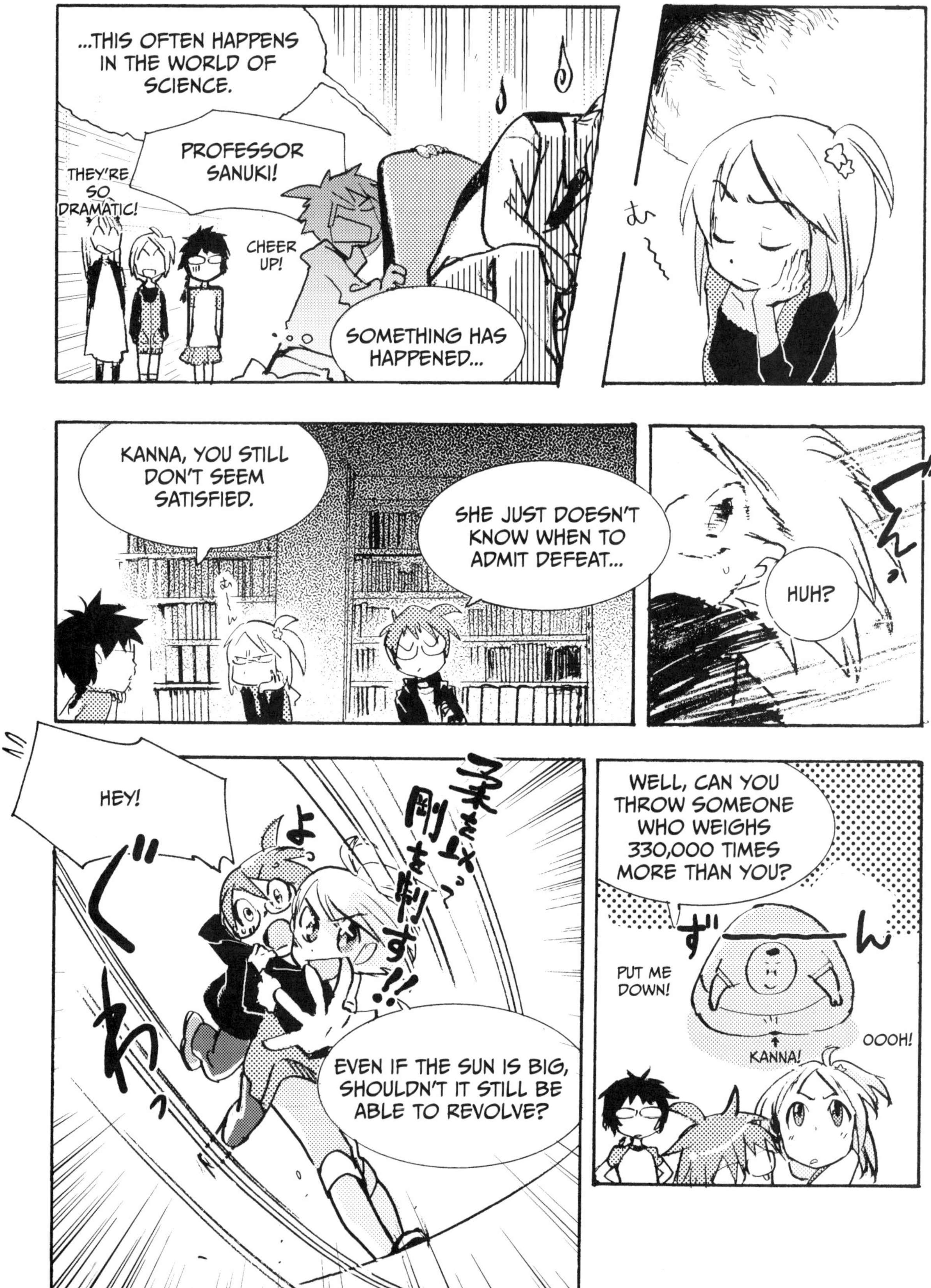

...THIS OFTEN HAPPENS IN THE WORLD OF SCIENCE.
PROFESSOR SANUKI!
THEY'RE SO DRAMATIC!
CHEER UP!
SOMETHING HAS HAPPENED...
KANNA, YOU STILL DON'T SEEM SATISFIED.
SHE JUST DOESN'T KNOW WHEN TO ADMIT DEFEAT...
HUH?
HEY!
EVEN IF THE SUN IS BIG, SHOULDN'T IT STILL BE ABLE TO REVOLVE?
WELL, CAN YOU THROW SOMEONE WHO WEIGHS 330,000 TIMES MORE THAN YOU?
PUT ME DOWN!
OOOH!
KANNA!

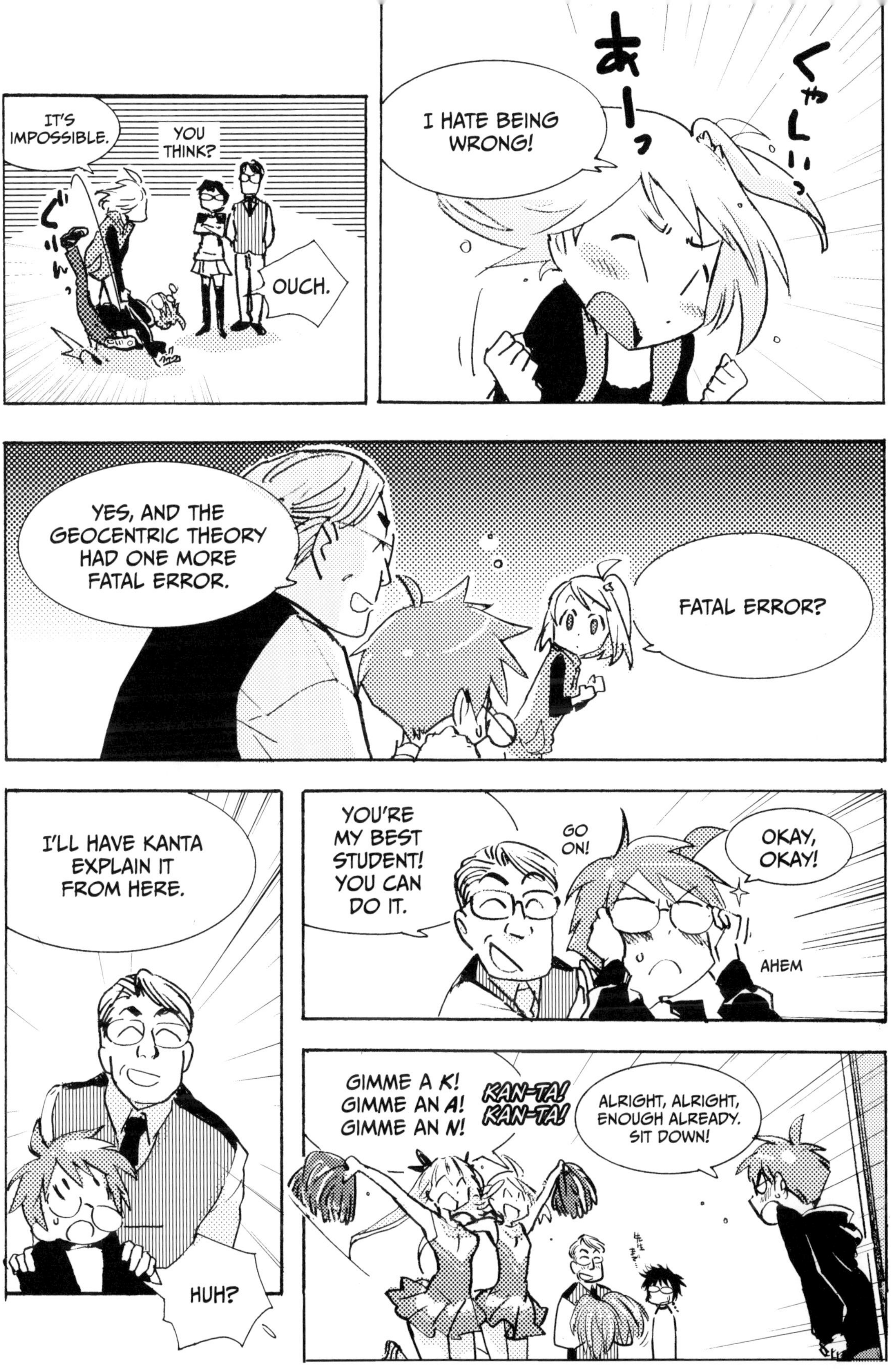
IT'S IMPOSSIBLE.
YOU THINK?
OUCH.
I HATE BEING WRONG!
あーっ
くやしいっ
YES, AND THE GEOCENTRIC THEORY HAD ONE MORE FATAL ERROR.
FATAL ERROR?
I'LL HAVE KANTA EXPLAIN IT FROM HERE.
HUH?
YOU'RE MY BEST STUDENT! YOU CAN DO IT.
GO ON!
OKAY, OKAY!
AHEM
GIMME A K! GIMME AN A! GIMME AN N!
KAN-TA! KAN-TA!
ALRIGHT, ALRIGHT, ENOUGH ALREADY. SIT DOWN!

FROM THE GEOCENTRIC THEORY TO THE HELIOCENTRIC THEORY

The mysterious behavior of planets puzzled people for a long time, as evidenced by the fact that etymologically, the word *planet* comes from the Greek word for *wandering star*. People thought that the planets wandered around and changed their locations, while the other stars—that is, the so-called fixed stars*—revolved while maintaining the same positional relationships on the celestial sphere.

Oh yeah, I heard that explanation at the planetarium once.

Yeah, I bet the name *planetarium* comes from the word *planet*.

You sure know a lot of things, don't you?

That's because I went to school in America.

Well, aren't you lucky?

Enough! I don't want to hear your bickering!

In the first geocentric model that Professor Sanuki drew, this kind of planetary behavior could not be explained. It failed to account for the fact that the Sun and the Moon don't move in the same way.

So although the geocentric theory should have been abandoned at this stage...

It didn't go away. Ptolemy arrived on the scene at that point.

* The stars do "move"—just not in relation to other stars in the sky.

Who? When did he live?

 Um...

 We're not exactly sure when he lived, but we know he was an astronomer and geographer in Roman Egypt during the second century. The world maps he left behind were used until the Middle Ages.
Only a very talented individual like Ptolemy could have come up with a method to explain the motion of the planets using the geocentric theory.

Claudius Ptolemy (about 90–168 AD) A Greek astronomer and geographer; the world maps he drew in his book called Geography used latitude and longitude for the first time and established the convention that "North is up" that we still use today.

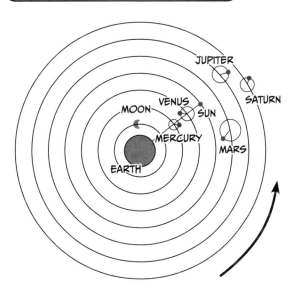

Model of the geocentric theory, proposed by Ptolemy

 That's right, that's right. The model shown here is his.

FROM THE GEOCENTRIC THEORY TO THE HELIOCENTRIC THEORY 51

The ancient Greeks believed that the Moon, the Sun, and the other planets revolved around Earth, and Ptolemy's geocentric theory didn't challenge this. However, the earlier geocentric theories from ancient Greece were unable to explain the retrograde motion of the planets.

What do we mean by retrograde? Planets appear to move in an eastward direction most of the time. However, planets sometimes appear to reverse directions in the night sky, which we call retrograde motion. All planets periodically exhibit this behavior. Ptolomy's geocentric model explained this motion as a revolution around a point in a planet's orbit. (In reality, this apparent backward motion is caused by Earth "lapping" the planet in a revolution around the sun.)

That's a clever idea, isn't it?

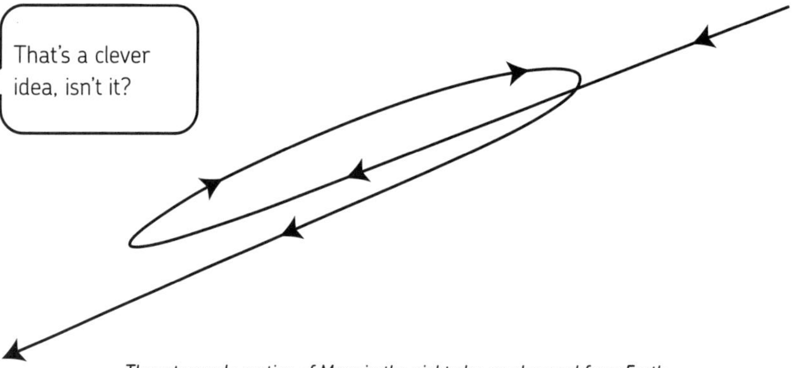

The retrograde motion of Mars in the night sky, as observed from Earth

Why are Mercury, Venus, Earth, and the Sun all lined up in a straight line in Ptolemy's model??

This is how Ptolemy explained why Mercury and Venus always appear closer to the Sun.

Of course. Since they revolve around Earth together with the Sun, they will always appear closer to the Sun.

Um...

What's the matter?

Doesn't it seem like he went way overboard in creating this diagram?

He certainly didn't make it believable, did he? If his model were true, the planets would all revolve like swirling eddies.

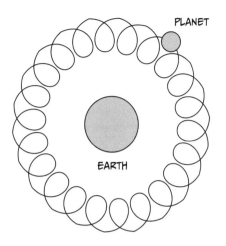

Planetary motion, according to Ptolemy's geocentric theory

It feels like he had to keep fudging the geocentric model—to try to keep it in line with new scientific discoveries.

It looks totally fake! It wouldn't fool anybody.

I wonder...I think this model is well constructed, but...

Why don't we try to compare our impressions of this model with our impressions of Copernicus's model of the heliocentric theory?

Good idea! Ptolemy's model was accepted as true for nearly 1,400 years. But then Copernicus challenged it in 1543 with his heliocentric theory, which he presented in his book *On the Revolutions of the Celestial Spheres*. For details, read "Just How Progressive Was Copernicus?" on page 71.

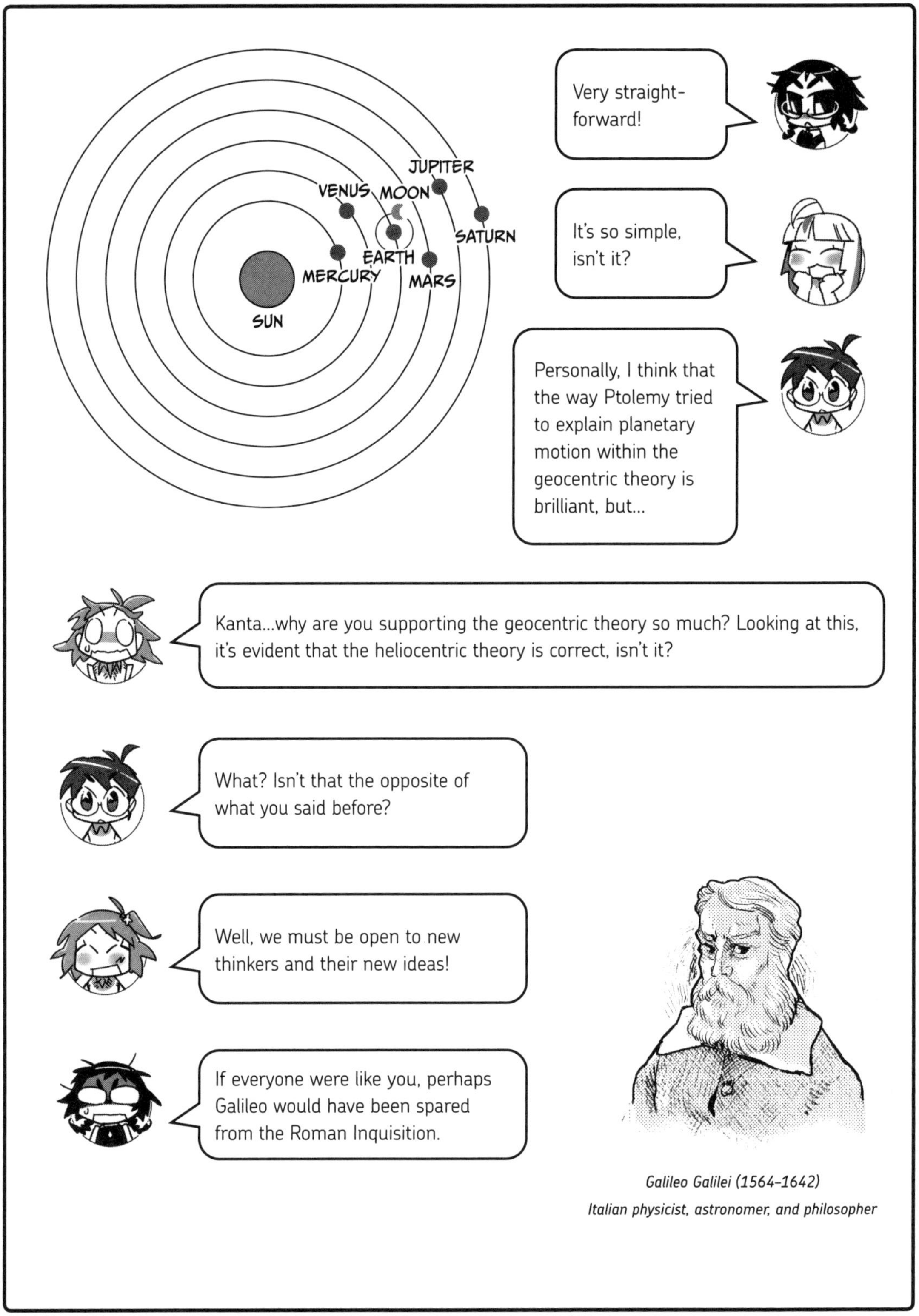
JUPITER
VENUS
MOON
SATURN
EARTH
MERCURY
MARS
SUN
Very straight-forward!
It's so simple, isn't it?
Personally, I think that the way Ptolemy tried to explain planetary motion within the geocentric theory is brilliant, but...
Kanta...why are you supporting the geocentric theory so much? Looking at this, it's evident that the heliocentric theory is correct, isn't it?
What? Isn't that the opposite of what you said before?
Well, we must be open to new thinkers and their new ideas!
If everyone were like you, perhaps Galileo would have been spared from the Roman Inquisition.

Galileo Galilei (1564–1642)
Italian physicist, astronomer, and philosopher

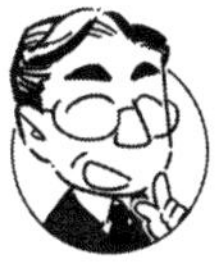

Well, just because the heliocentric theory is correct doesn't mean that it's the absolute truth. It's just more successful, compared with the geocentric theory. We could say that the geocentric theory is correct in that it can explain the observed planetary motions from the perspective of Earth.

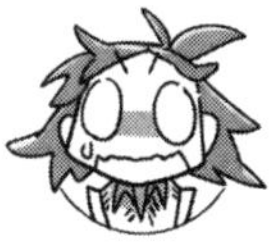

What? So the geocentric theory is also correct? But which one is better?

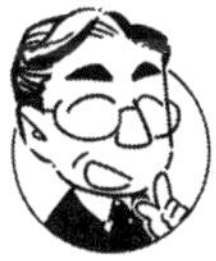

Does everyone know the expression *Occam's razor*?

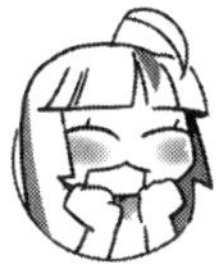

Yeah! It's the concept that says, "If two or more theories can explain the same phenomenon, then the simplest one is more likely to be correct." Is that right?

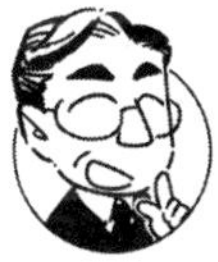

Exactly! If we consider applying this concept to the geocentric and heliocentric theories, the heliocentric theory is more likely to be correct, since it is the simpler one. We're only saying that the heliocentric theory is correct based on that fact.

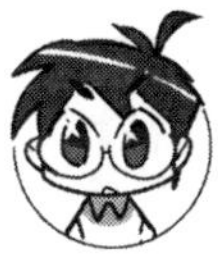

Well, Copernicus's diagram certainly is more straightforward.

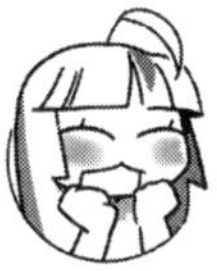

Simple is best!

GALILEO'S DISCOVERIES—AND TRIAL

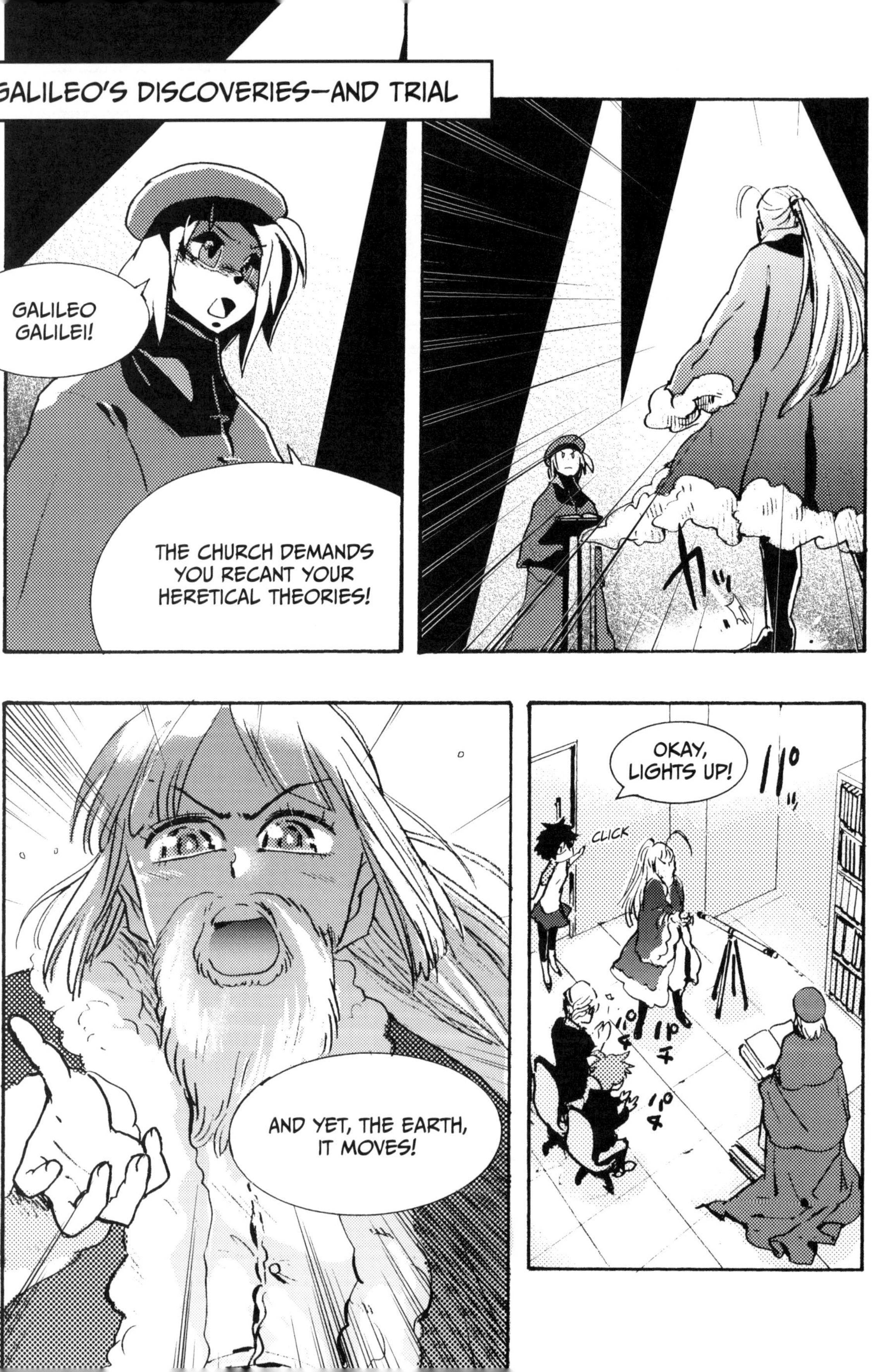
GALILEO GALILEI!
THE CHURCH DEMANDS YOU RECANT YOUR HERETICAL THEORIES!
AND YET, THE EARTH, IT MOVES!
OKAY, LIGHTS UP!
CLICK

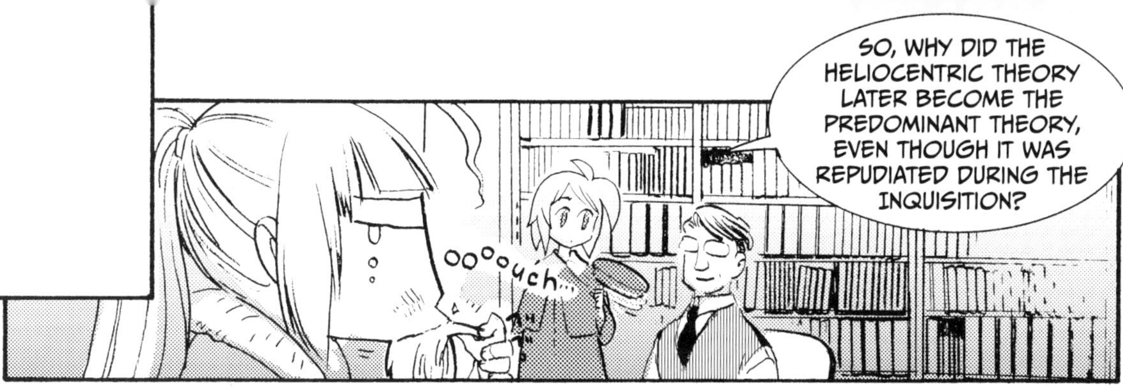
oooouch...
SO, WHY DID THE HELIOCENTRIC THEORY LATER BECOME THE PREDOMINANT THEORY, EVEN THOUGH IT WAS REPUDIATED DURING THE INQUISITION?

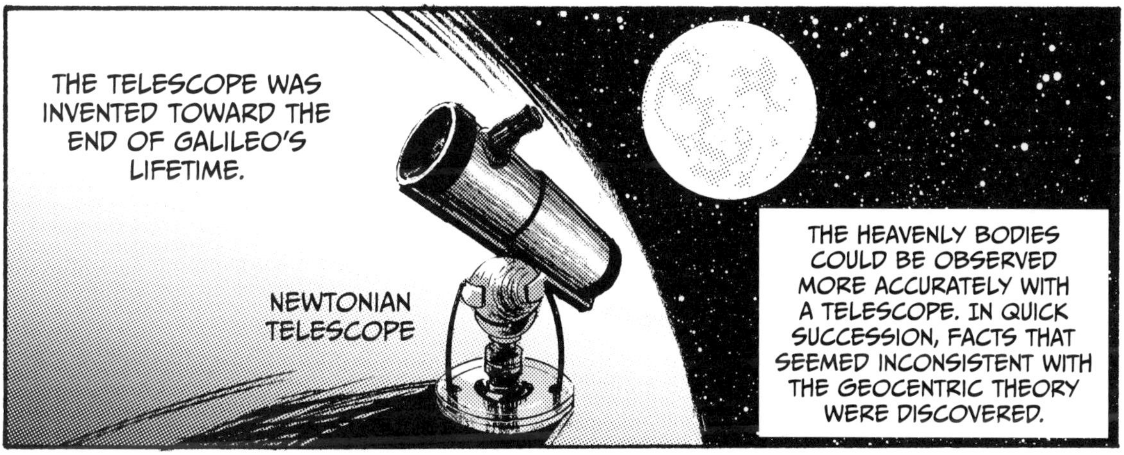
THE TELESCOPE WAS INVENTED TOWARD THE END OF GALILEO'S LIFETIME.
NEWTONIAN TELESCOPE
THE HEAVENLY BODIES COULD BE OBSERVED MORE ACCURATELY WITH A TELESCOPE. IN QUICK SUCCESSION, FACTS THAT SEEMED INCONSISTENT WITH THE GEOCENTRIC THEORY WERE DISCOVERED.

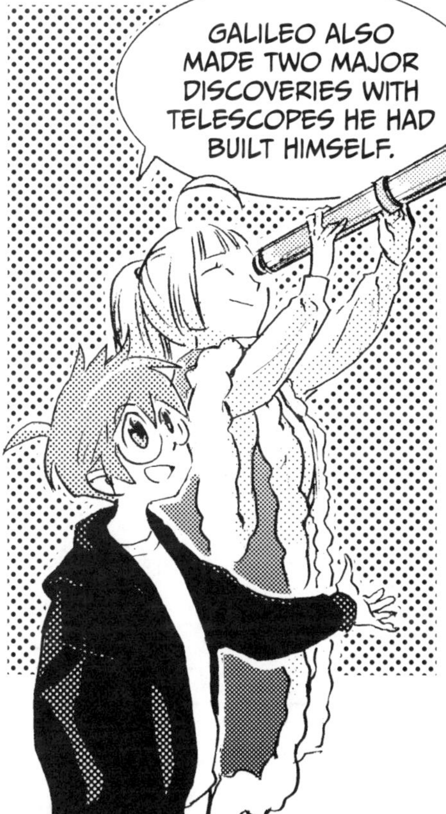
GALILEO ALSO MADE TWO MAJOR DISCOVERIES WITH TELESCOPES HE HAD BUILT HIMSELF.

GALILEO'S FIRST DISCOVERY: THE FOUR SATELLITES OF JUPITER

BY DISCOVERING THAT A PLANET OTHER THAN EARTH HAD SATELLITES, GALILEO SHOOK THE FOUNDATIONS OF THE GEOCENTRIC THEORY, WHICH WAS BASED ON THE IDEA THAT EARTH WAS UNIQUE.

GALILEO'S SECOND DISCOVERY: THE PHASES OF VENUS

GALILEO DISCOVERED THAT VENUS CHANGED IN ITS APPARENT SIZE AND, PERHAPS MORE IMPORTANTLY, THAT VENUS HAS PHASES SIMILAR TO THE MOON'S PHASES. THAT WOULD BE POSSIBLE ONLY IF BOTH EARTH AND VENUS WERE ORBITING AROUND THE SUN. HIS OBSERVATION EFFECTIVELY DISPROVED PTOLEMY'S GEOCENTRIC MODEL. THIS DISCOVERY WOULD HAVE BEEN IMPOSSIBLE USING ONLY THE NAKED EYE.

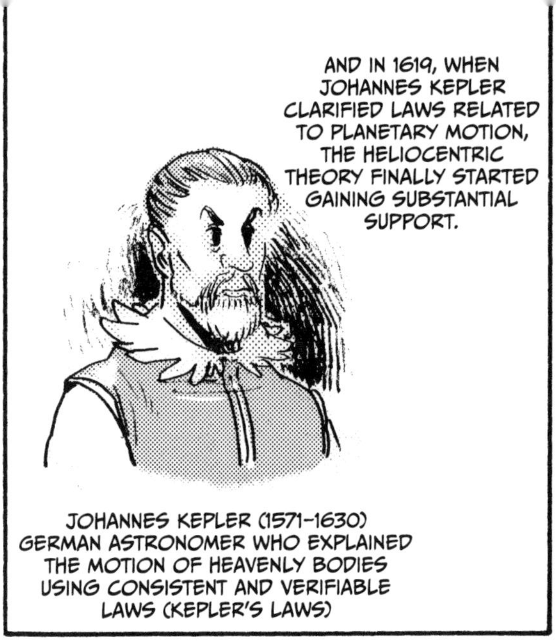
AND IN 1619, WHEN JOHANNES KEPLER CLARIFIED LAWS RELATED TO PLANETARY MOTION, THE HELIOCENTRIC THEORY FINALLY STARTED GAINING SUBSTANTIAL SUPPORT.
JOHANNES KEPLER (1571–1630) GERMAN ASTRONOMER WHO EXPLAINED THE MOTION OF HEAVENLY BODIES USING CONSISTENT AND VERIFIABLE LAWS (KEPLER'S LAWS)

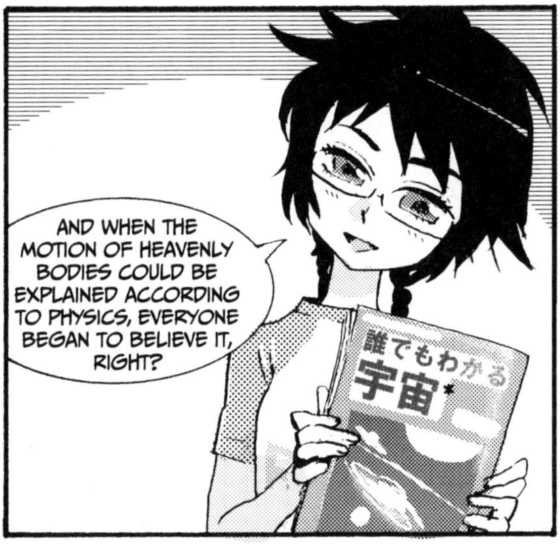
AND WHEN THE MOTION OF HEAVENLY BODIES COULD BE EXPLAINED ACCORDING TO PHYSICS, EVERYONE BEGAN TO BELIEVE IT, RIGHT?
誰でもわかる
宇宙*

* THE DULLARD'S GUIDE TO THE UNIVERSE

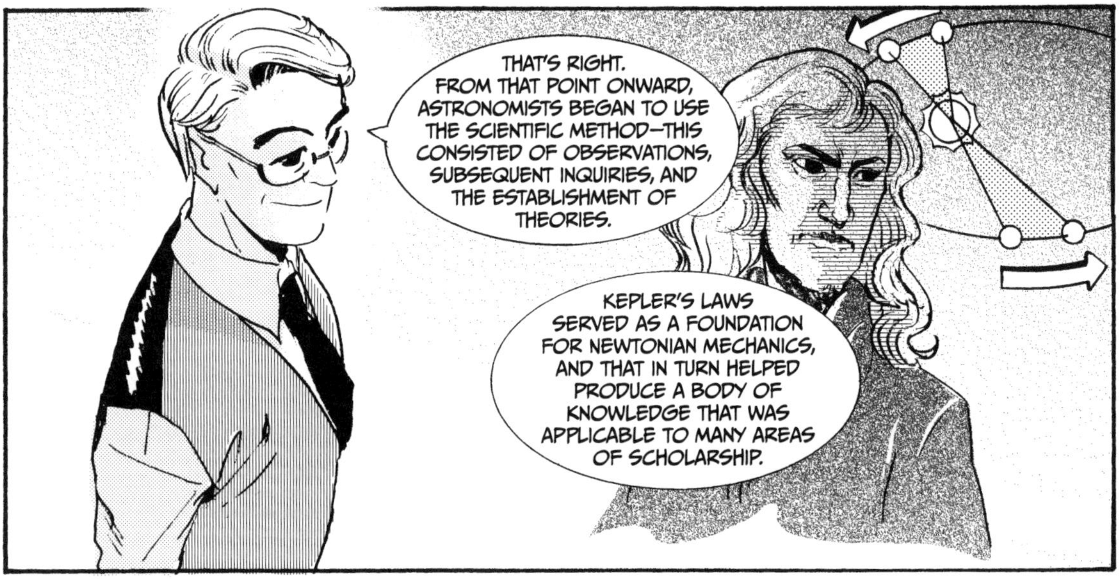
THAT'S RIGHT. FROM THAT POINT ONWARD, ASTRONOMISTS BEGAN TO USE THE SCIENTIFIC METHOD—THIS CONSISTED OF OBSERVATIONS, SUBSEQUENT INQUIRIES, AND THE ESTABLISHMENT OF THEORIES.
KEPLER'S LAWS SERVED AS A FOUNDATION FOR NEWTONIAN MECHANICS, AND THAT IN TURN HELPED PRODUCE A BODY OF KNOWLEDGE THAT WAS APPLICABLE TO MANY AREAS OF SCHOLARSHIP.

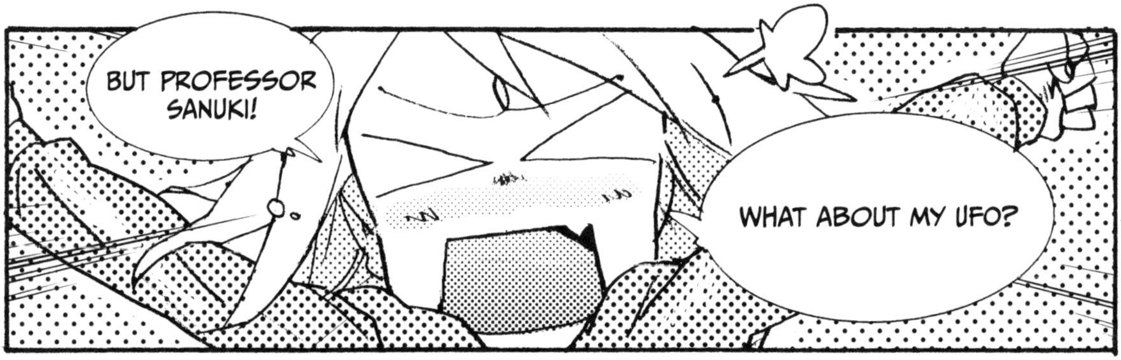
BUT PROFESSOR SANUKI!
WHAT ABOUT MY UFO?

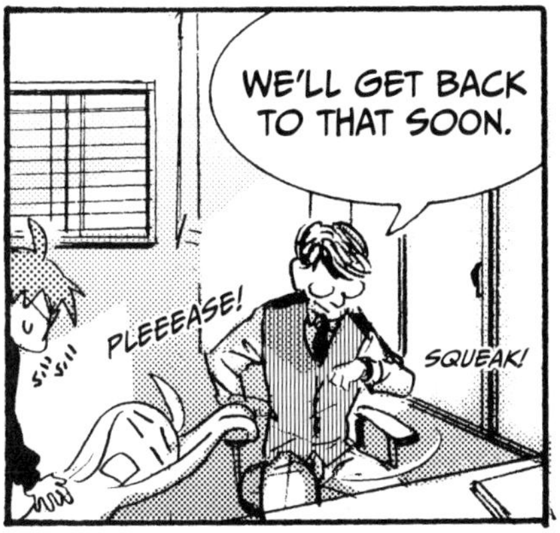
WE'LL GET BACK TO THAT SOON.
PLEEEASE!
SQUEAK!
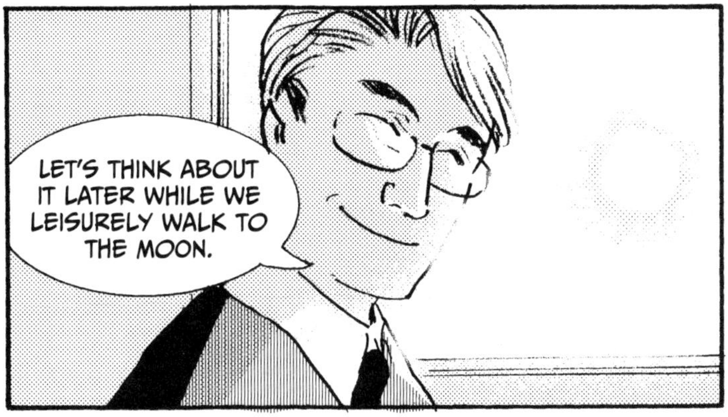
LET'S THINK ABOUT IT LATER WHILE WE LEISURELY WALK TO THE MOON.
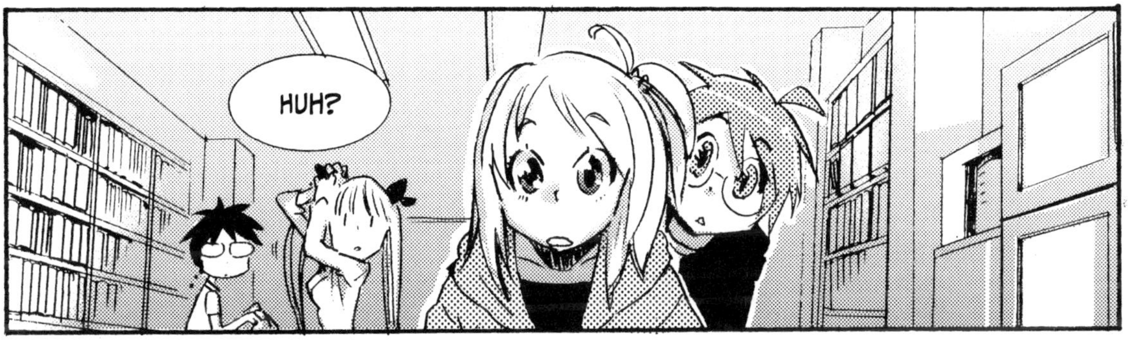
HUH?

PUTTING THINGS IN PERSPECTIVE

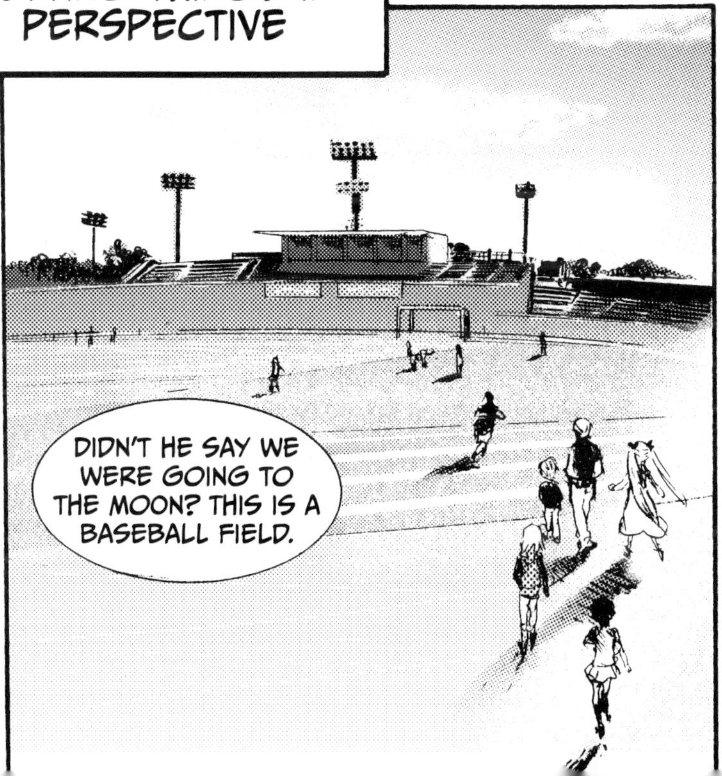
DIDN'T HE SAY WE WERE GOING TO THE MOON? THIS IS A BASEBALL FIELD.
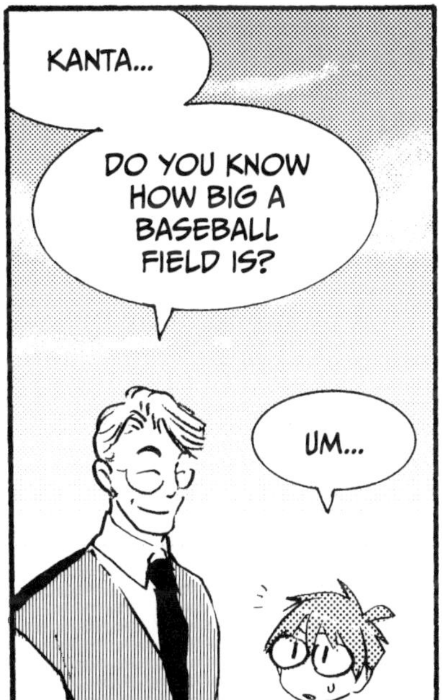
KANTA...
DO YOU KNOW HOW BIG A BASEBALL FIELD IS?
UM...

I'M NOT VERY GOOD AT SPORTS...
I KNOW!
IT'S ABOUT 120 METERS (131 YARDS) FROM HOME PLATE TO CENTER FIELD.
120m
120 M
(131 YARDS)
KANNA, YOU WERE THE ACE LITTLE LEAGUE PITCHER IN ELEMENTARY SCHOOL, RIGHT?
I BATTED CLEANUP, TOO.
FOR SIBLINGS, YOU TWO COULDN'T BE ANY MORE DIFFERENT.
LEAVE ME ALONE.
WELL, MISS CLEANUP-HITTING, NO-HITTER-PITCHING KANNA...
DO YOU KNOW THE DIAMETER OF EARTH?
I KNOW THAT!
AT THE EQUATOR, IT'S 12,756 KM (7,926 MILES)!
12,756 KM
(7,926 MILES)
THAT'S RIGHT!
SO A BASEBALL FIELD IS APPROXIMATELY ONE ONE-HUNDRED-THOUSANDTH THE SIZE OF EARTH.
1 / 100,000 →
12,756 KM
(7,926 MILES)
120 M
(131 YARDS)

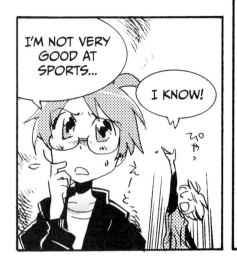

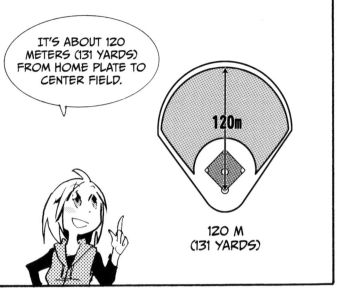

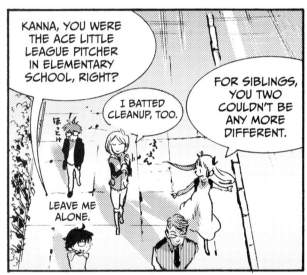

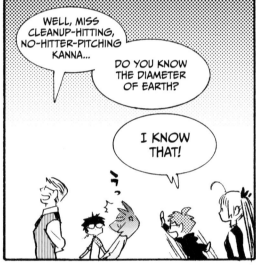

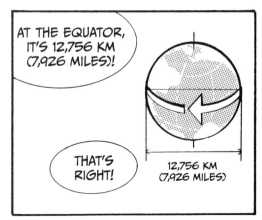

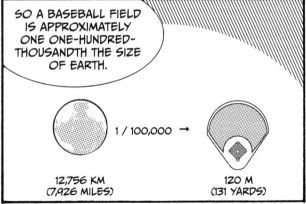

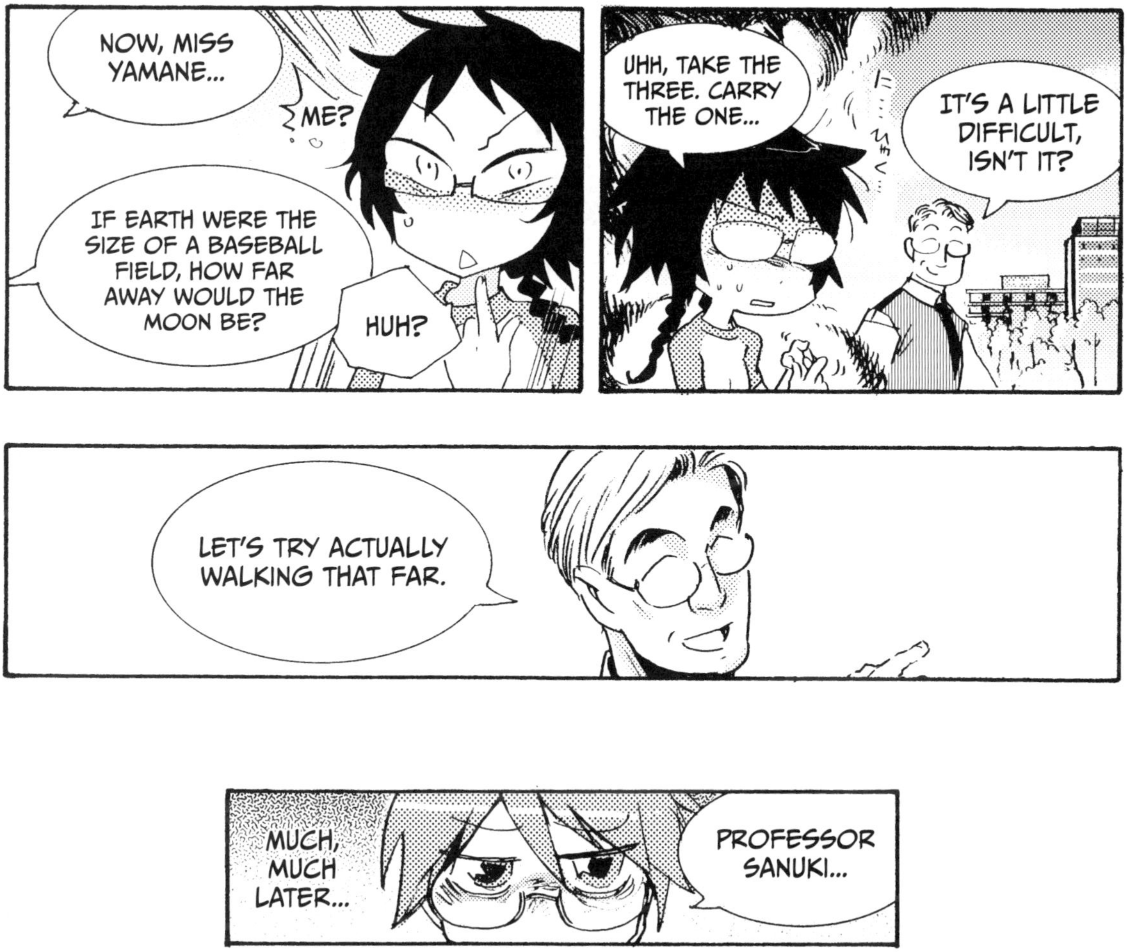
NOW, MISS YAMANE...
ME?
IF EARTH WERE THE SIZE OF A BASEBALL FIELD, HOW FAR AWAY WOULD THE MOON BE?
HUH?
UHH, TAKE THE THREE. CARRY THE ONE...
IT'S A LITTLE DIFFICULT, ISN'T IT?
LET'S TRY ACTUALLY WALKING THAT FAR.
MUCH, MUCH LATER...
PROFESSOR SANUKI...

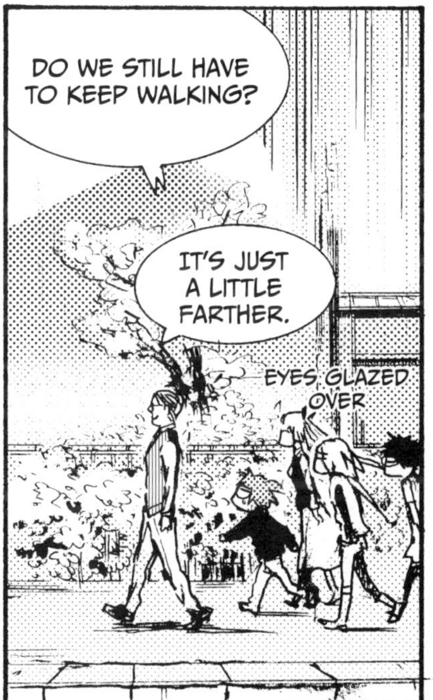
DO WE STILL HAVE TO KEEP WALKING?
IT'S JUST A LITTLE FARTHER.
EYES GLAZED OVER

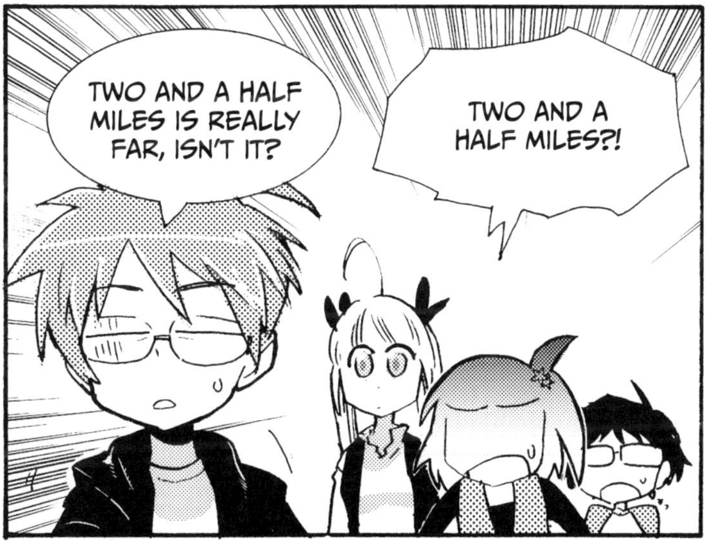
TWO AND A HALF MILES IS REALLY FAR, ISN'T IT?
TWO AND A HALF MILES?!

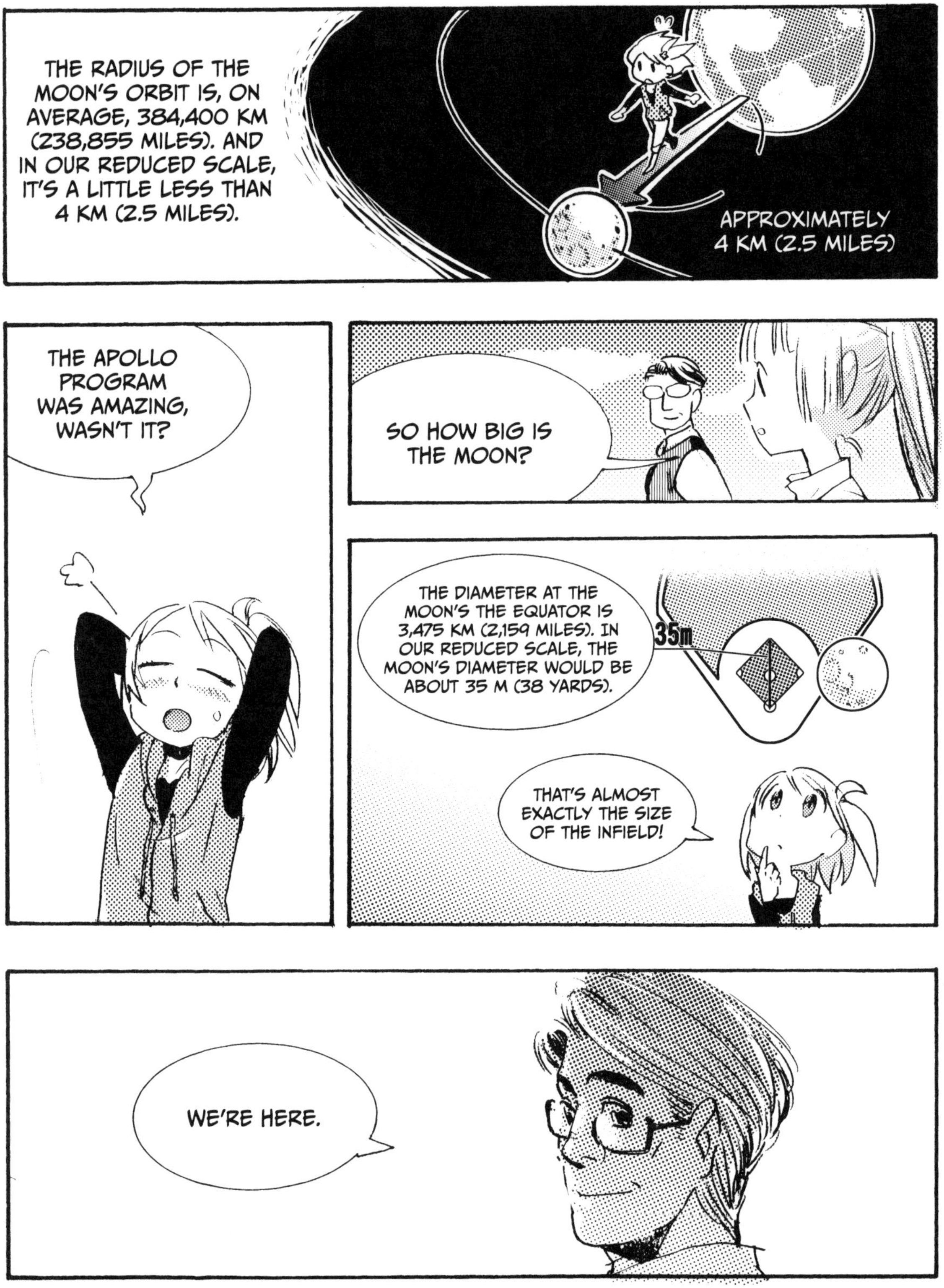
THE RADIUS OF THE MOON'S ORBIT IS, ON AVERAGE, 384,400 KM (238,855 MILES). AND IN OUR REDUCED SCALE, IT'S A LITTLE LESS THAN 4 KM (2.5 MILES).
APPROXIMATELY 4 KM (2.5 MILES)
THE APOLLO PROGRAM WAS AMAZING, WASN'T IT?
SO HOW BIG IS THE MOON?
THE DIAMETER AT THE MOON'S THE EQUATOR IS 3,475 KM (2,159 MILES). IN OUR REDUCED SCALE, THE MOON'S DIAMETER WOULD BE ABOUT 35 M (38 YARDS).
35m
THAT'S ALMOST EXACTLY THE SIZE OF THE INFIELD!
WE'RE HERE.

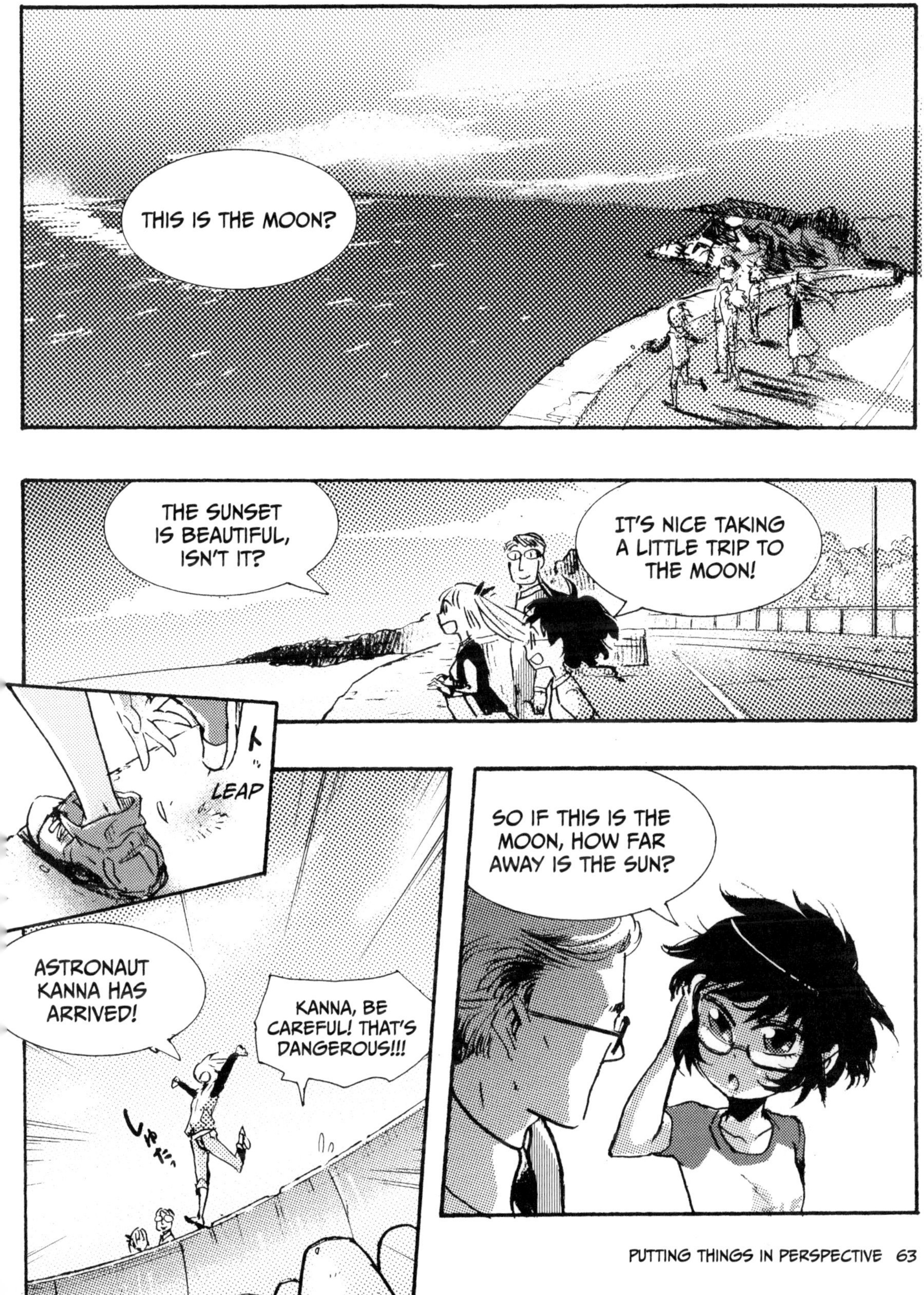
THIS IS THE MOON?
THE SUNSET IS BEAUTIFUL, ISN'T IT?
IT'S NICE TAKING A LITTLE TRIP TO THE MOON!
LEAP
ASTRONAUT KANNA HAS ARRIVED!
KANNA, BE CAREFUL! THAT'S DANGEROUS!!!
SO IF THIS IS THE MOON, HOW FAR AWAY IS THE SUN?

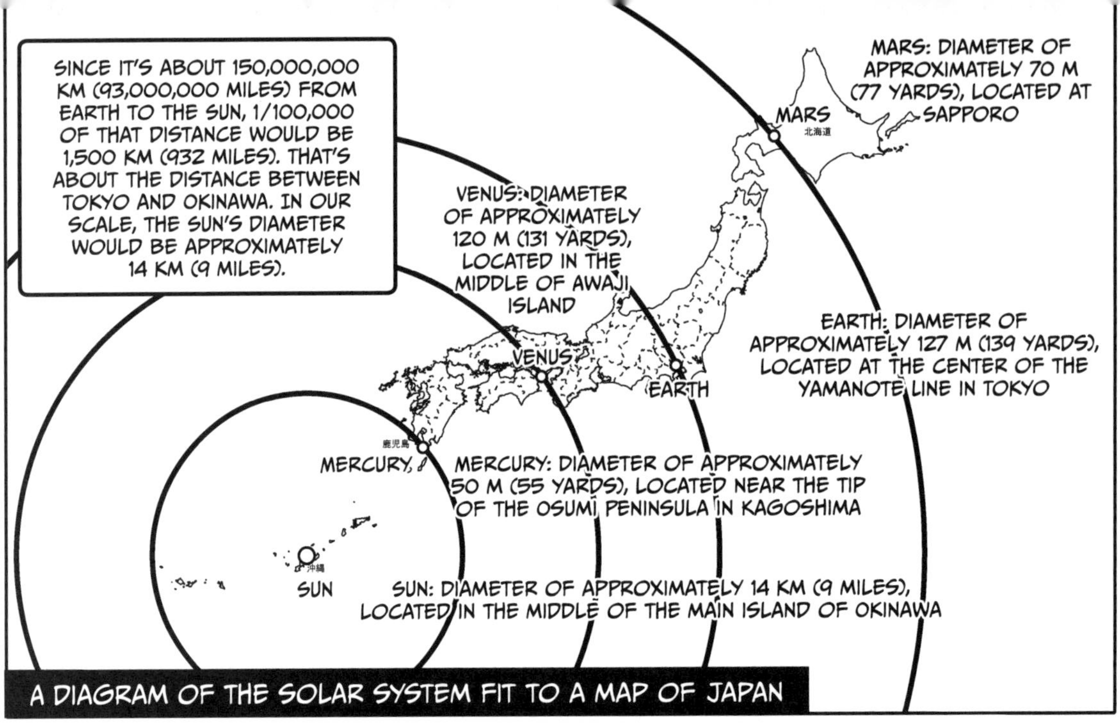
SINCE IT'S ABOUT 150,000,000 KM (93,000,000 MILES) FROM EARTH TO THE SUN, 1/100,000 OF THAT DISTANCE WOULD BE 1,500 KM (932 MILES). THAT'S ABOUT THE DISTANCE BETWEEN TOKYO AND OKINAWA. IN OUR SCALE, THE SUN'S DIAMETER WOULD BE APPROXIMATELY 14 KM (9 MILES).
VENUS: DIAMETER OF APPROXIMATELY 120 M (131 YARDS), LOCATED IN THE MIDDLE OF AWAJI ISLAND
MARS: DIAMETER OF APPROXIMATELY 70 M (77 YARDS), LOCATED AT SAPPORO
MARS
北海道
VENUS
EARTH
EARTH: DIAMETER OF APPROXIMATELY 127 M (139 YARDS), LOCATED AT THE CENTER OF THE YAMANOTE LINE IN TOKYO
鹿児島
MERCURY
MERCURY: DIAMETER OF APPROXIMATELY 50 M (55 YARDS), LOCATED NEAR THE TIP OF THE OSUMI PENINSULA IN KAGOSHIMA
沖縄
SUN
SUN: DIAMETER OF APPROXIMATELY 14 KM (9 MILES), LOCATED IN THE MIDDLE OF THE MAIN ISLAND OF OKINAWA
A DIAGRAM OF THE SOLAR SYSTEM FIT TO A MAP OF JAPAN

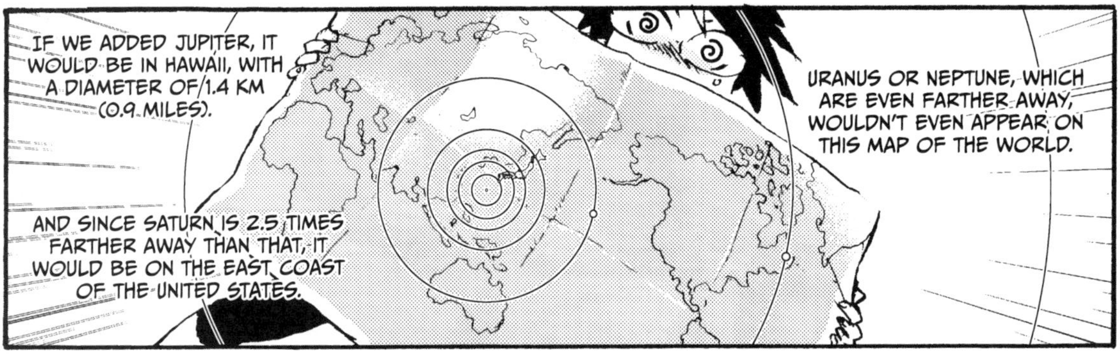
IF WE ADDED JUPITER, IT WOULD BE IN HAWAII, WITH A DIAMETER OF 1.4 KM (0.9 MILES).
AND SINCE SATURN IS 2.5 TIMES FARTHER AWAY THAN THAT, IT WOULD BE ON THE EAST COAST OF THE UNITED STATES.
URANUS OR NEPTUNE, WHICH ARE EVEN FARTHER AWAY, WOULDN'T EVEN APPEAR ON THIS MAP OF THE WORLD.

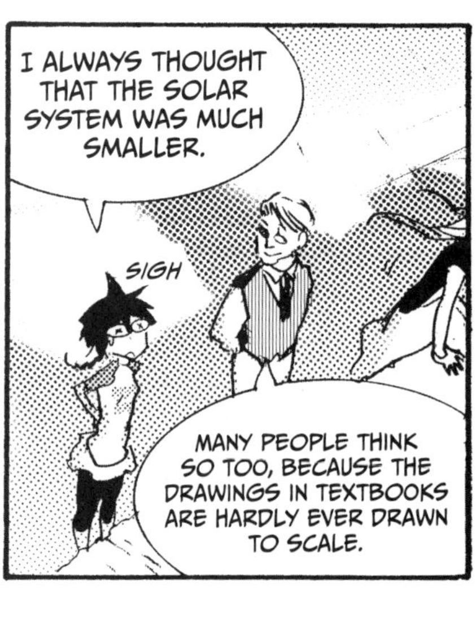
I ALWAYS THOUGHT THAT THE SOLAR SYSTEM WAS MUCH SMALLER.
SIGH
MANY PEOPLE THINK SO TOO, BECAUSE THE DRAWINGS IN TEXTBOOKS ARE HARDLY EVER DRAWN TO SCALE.

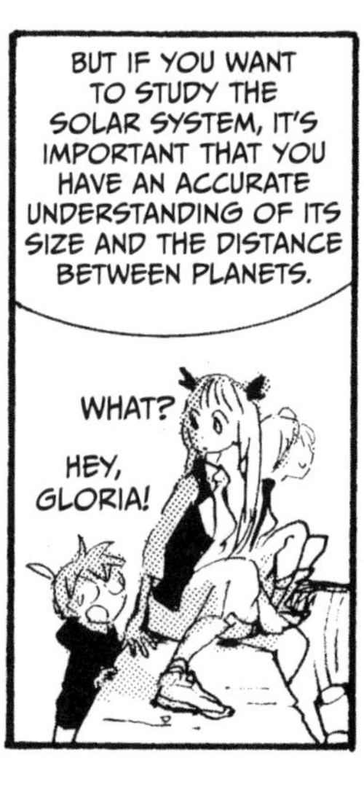
BUT IF YOU WANT TO STUDY THE SOLAR SYSTEM, IT'S IMPORTANT THAT YOU HAVE AN ACCURATE UNDERSTANDING OF ITS SIZE AND THE DISTANCE BETWEEN PLANETS.
WHAT?
HEY, GLORIA!

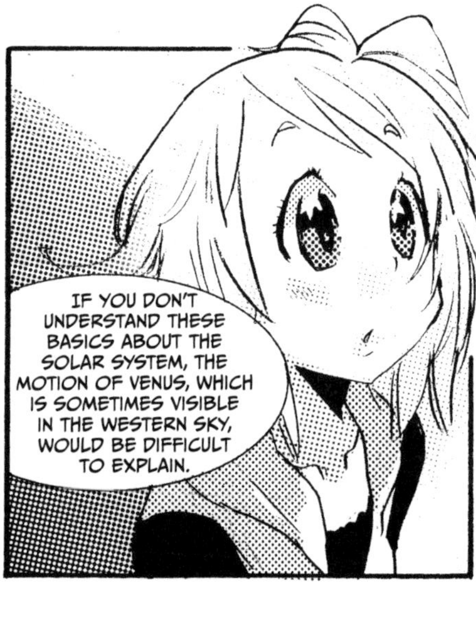
IF YOU DON'T UNDERSTAND THESE BASICS ABOUT THE SOLAR SYSTEM, THE MOTION OF VENUS, WHICH IS SOMETIMES VISIBLE IN THE WESTERN SKY, WOULD BE DIFFICULT TO EXPLAIN.

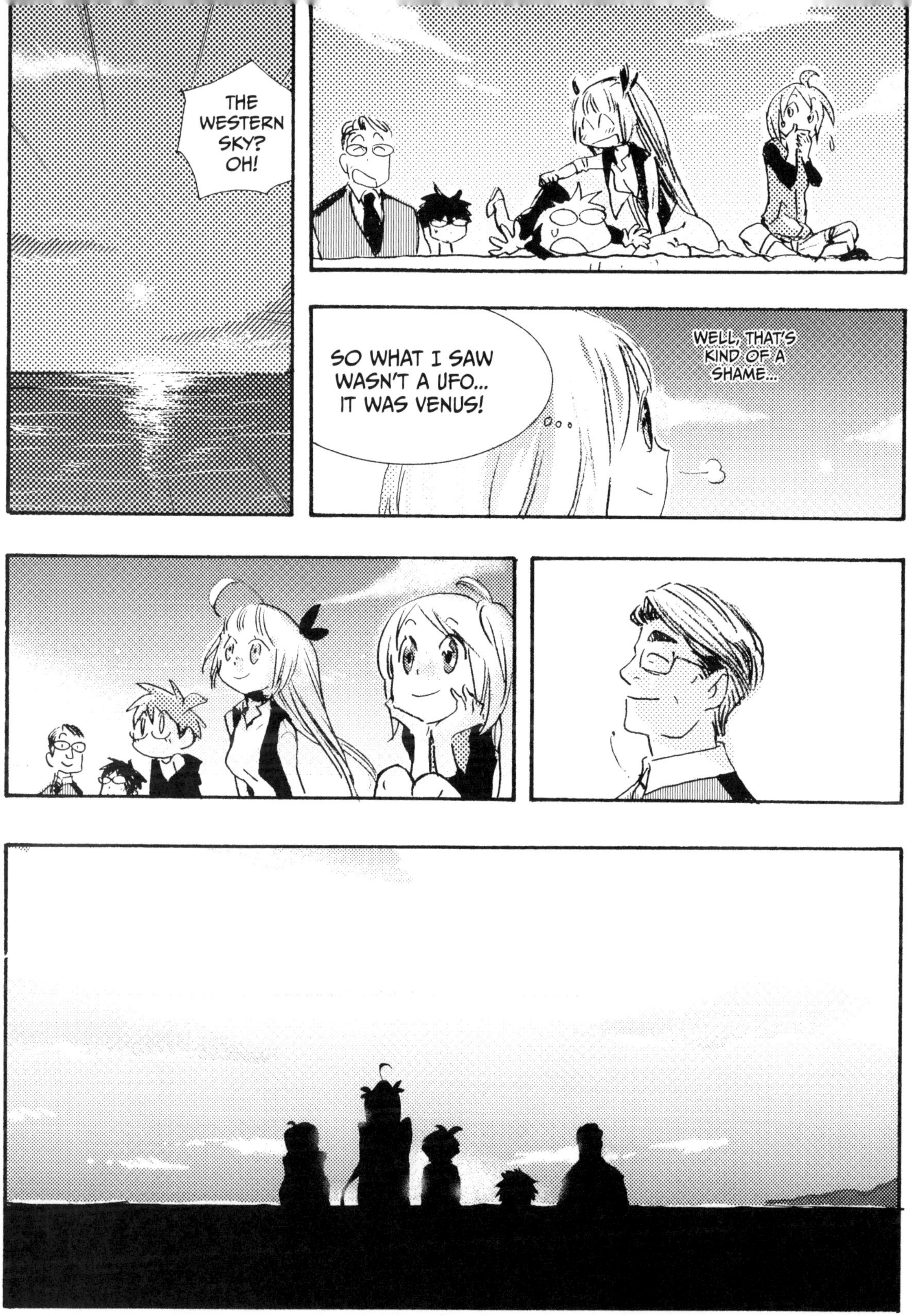
THE WESTERN SKY? OH!
SO WHAT I SAW WASN'T A UFO... IT WAS VENUS!
WELL, THAT'S KIND OF A SHAME...

WHAT IS THE APPROXIMATE DISTANCE TO THE HORIZON?

If Earth was flat and the air was perfectly clear, you could see for an unlimited distance. Therefore, the fact that there is a *horizon*—that is, an apparent line that separates Earth from the sky—is evidence that Earth is round.

So approximately how far is it to the horizon?

If we let r represent the radius of Earth and let h represent the height of the observer's viewpoint, then the relationship to the desired distance L will be as follows, according to the Pythagorean theorem:

$$(r+h)^2 = r^2 + L^2$$

Therefore,

$$\begin{aligned} L &= \sqrt{(r+h)^2 - r^2} \\ &= \sqrt{2rh + h^2} \end{aligned}$$

Since the radius of Earth (r) is approximately 6,378 km (which is 3,963 miles), if we assume that the height of a normal person's eye (h) is 0.0015 km (equal to 5 feet), then we can calculate L as approximately 4.4 km (2.75 miles). In other words, if we gaze at the ocean from the beach, we will only be able to see the ocean's surface out to a little more than 4.4 km (2.75 miles).

Incidentally, even when looking out of the window of a plane flying at an altitude of approximately 10 km (33,000 feet), the distance to the horizon is approximately 360 km (224 miles). Since this is no more than the distance from New York City to Washington, DC, we can't see very far even at that great height.

h
L
This is the horizon.
r
Earth

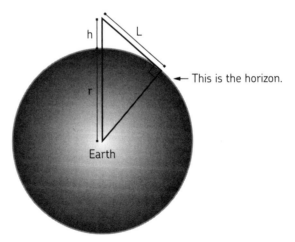

Distance to the horizon

MEASURING THE SIZE OF THE UNIVERSE: HOW FAR TO THE MOON?

The Moon is currently moving away from Earth at a rate of approximately 3.8 cm (1.5 inches) each year. The mean distance between Earth and the Moon is approximately 385,000 km (239,228 miles), so this distance will increase by 1 percent in approximately 100 million years, if movement continues at this rate. Isn't it remarkable that the distances between heavenly bodies can be measured in millimeters?

CORNER CUBE MIRRORS

A mirror for measuring distance, which was placed on the surface of the Moon

We can make such precise measurements thanks to the Apollo space missions that were launched beginning in 1969. Apollo 11, 14, and 15 placed special mirrors on the surface of the Moon to reflect laser light emitted from Earth. These mirrors, which are different from ordinary household ones, are called *corner cube mirrors*. The surface of a corner cube mirror is constructed so that any light that hits it will be accurately reflected back to the source in a parallel path, no matter what direction it comes from.

Scientists can use this kind of mirror to accurately measure distance by shining a light at the mirror and recording the time it takes the light's reflection to return. Since the speed of light is always constant at 299,792,458 meters per second, it is extremely well suited as a "ruler" for measuring distance.

HOW A CORNER CUBE MIRROR WORKS

If we use an easy-to-understand, two-dimensional model, a corner cube mirror would be made up of two mirrors joined at a right angle, as shown in the figure. Light that hits the mirror will reflect at the same angle as the angle of incidence. The working principle is the same in three dimensions.

The principle behind the corner cube mirror is not unusual—it is the same principle used in reflectors on cars or road signs.

The mirror that the Apollo spacecraft brought to the Moon uses a prism to reflect light, so it is more accurately called a *corner cube prism*.

Light ray
Light ray
Light ray

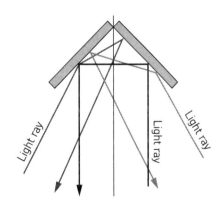

Corner cube mirror (two-dimensional)

Light ray
Exits at the same angle

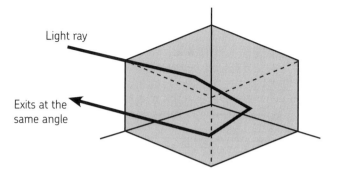

Corner cube prism (three-dimensional)

BEFORE THE CORNER CUBE PRISM

Before the arrival of the Apollo spacecraft, what did scientists have to do to find the distance to the Moon?

Triangulation is the most general method of measuring the distance to a place that you cannot actually go. This technique involves observing an object at a distance from two different places. In the figure to the right, the line at the bottom (known as the baseline) represents the distance between the two observation points *A* and *B*, and *C* is the object being observed. In order for triangulation to work, one must know the correct length of line *AB*. One then uses the angles drawn from points *A* and *B* to point *C* in trigonometric functions to determine the distances to point *C*. This practice is said to have already been established in ancient Egypt around 3,000 BC. It was also actively used in Greece during the era in which the sciences of geography and astronomy developed rapidly (from 2,500 to 2,000 years ago). One famous example of the successful use of triangulation was Eratosthenes' measurement of the size of Earth (see page 20).

C
A
B

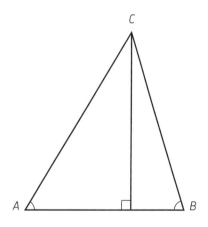

If the length of AB is known, the lengths of AC and BC can be determined by finding ∠BAC and ∠ABC.

Hipparchus (who lived around 190–120 BC) was a Greek astronomer who measured the distance to the Moon a generation after Eratosthenes did. Unfortunately, the method he used to do this is no longer known. Scientists speculate that he probably measured the angles at which the Moon was visible at the same time of day from two points a known distance apart, but since there were no clocks in his day, he must have used a solar or lunar eclipse to know to mark the exact same time at two different locations.

Hipparchus concluded that the distance to the Moon was about 59 to 72.3 times the radius of Earth. Since we now know that the distance to the Moon is actually about 60 times the radius of Earth, his calculation was pretty good, all things considered.

Until the Apollo spacecraft placed a corner cube prism on the surface of the Moon, modern scientists were still using the same method that ancient Greeks had used to calculate the distance to the Moon. This method is easy to understand—just find one location (*A*) from which the center of the Moon's surface is visible at the zenith and another location (*B*) from which the center of the Moon's surface is visible at the horizon, at the same time of day. If points *A* and *B* are at the same longitude, the difference in latitude between them will be $\angle BOC$. Then, you can use the radius of Earth (*BO*) to calculate the distance to the Moon by figuring out *BO* × tan(difference in latitude), where O is the center of Earth.

B
Difference in latitude
A
O
C
Moon
Earth

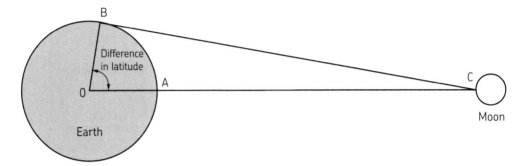

One method of measuring the distance from Earth to the Moon

GEOCENTRIC THEORY VS. HELIOCENTRIC THEORY—THE OUTCOME OF A BATTLE ROYALE

We know from many experiences that what we observe with our own eyes is not necessarily the truth. The best example of this is a mirror—when you gaze into it, you see a person who is the spitting image of yourself. However, no one would look into a mirror and get all excited, saying, "There's another me over there!"

You immediately understand when you turn over a mirror that there is not another world on the other side of it. You can comprehend that it is only your own image reflected, even if you don't think about the physical phenomenon of the reflection of light. However, when it's the universe that is being observed, people have a hard time buying any explanation that is different from what they can see.

The Sun, the Moon, and many stars that shine in the evening sky all certainly seem to revolve around our Earth. Therefore, it's natural that early models of the universe conformed to the geocentric theory.

It seems that if Earth were moving, we wouldn't be able to stand on the ground, would we? We'd be thrown off and fly somewhere into space! Before physics was developed, it certainly wasn't easy to answer these questions.

Since no one even considered the possibility of a heliocentric theory except some thinkers in ancient Greece, the geocentric theory had a monopoly until Nicolaus Copernicus (1473–1543) appeared on the scene.

WHAT KIND OF ORBIT DID A PLANET TRACE IN THE GEOCENTRIC THEORY?

The most likely reason that the geocentric theory was the mainstream model of the universe for so long was that every time someone made an observation of the heavenly bodies that might overturn it, someone else made an argument that explained why the new observation was still consistent with the geocentric theory. Ptolemy's map of space, which explained the motion of the planets in a way that accounted for their apparent position or brightness varying subtly, is certainly a representative example of this.

Planet
Earth

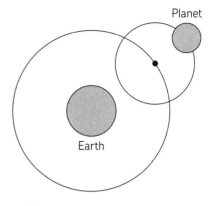

Planetary motion according to the geocentric theory (Ptolemy)

A part of this map is given here. If this is the only theory you look at, it can be persuasive. However, if we continue to develop a theory of planetary motion based on this diagram, it is apparent that the planet will move in an orbit shaped like a stretched-out spring, as shown in the next diagram. Why would the planets move in a spiral, while the Moon has a circular orbit? Upon careful consideration, this seems like an awfully contrived and complicated explanation.

Planet
Earth

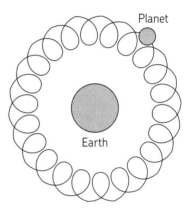

According to this theory, planets swirled around Earth.

Even so, people seemed to want to stick to the geocentric theory, and even though new observation results continued to emerge, people did not readily accept the heliocentric theory.

THE TYCHONIC SYSTEM THAT EMBELLISHED THE GEOCENTRIC THEORY

What might be called the last gasp of the geocentric theory was the astronomy diagram of Tycho Brahe (1549–1601).

Tycho Brahe was a Danish astronomer who lived a little before Galileo Galilei (who lived 1564–1642). Tycho proposed a model of the universe that was a compromise between the geocentric and heliocentric theories. The outer edges of the model looked almost exactly like a diagram drawn in accordance with the heliocentric theory, but the model was drawn centered on Earth, since Tycho firmly believed that Earth was at rest. The result was a rather attractive astronomy diagram.

Jupiter
Saturn
Mars
Venus
Mercury
Sun
Earth
Moon
Fixed stars

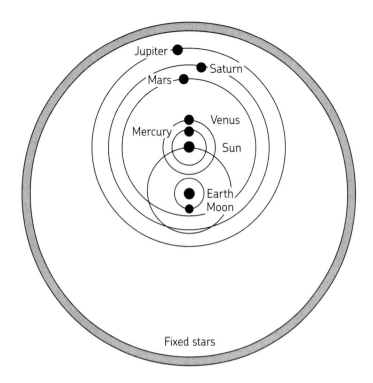

Tychonic system

JUST HOW PROGRESSIVE WAS COPERNICUS?

As you have learned, the heliocentric theory did not gain a wide following at first.

There is a saying that the Copernican Revolution turned our way of viewing everything upside down. However, there are still questions about how progressive a scholar Copernicus actually was.

He did advocate the heliocentric theory in his book *On the Revolutions of the Celestial Spheres*, but it was published the year he died, so Copernicus never endured criticism for his view. Furthermore, the heliocentric theory described in his book was still incomplete, so it did not generate a great deal of controversy at first.

In the final stages of the geocentric theory, the model of the universe contained corrections on top of corrections, and it could account for all the observed motion of the heavenly bodies, except comets. Copernicus seemed to believe that the orbits of the planets were all perfect circles—we now know that the orbits are actually ellipses, since they are influenced by other planets—and he, too, was completely unable to explain the motion of comets. Therefore, Copernicus's heliocentric theory didn't really explain anything that the geocentric theory couldn't already prove.

KEPLER COMPLETED THE HELIOCENTRIC THEORY

The person responsible for a mainstream cosmological model that brought about the Copernican Revolution in its true sense was the German astronomer Johannes Kepler (1571–1630). Kepler showed that the motion of the planets followed a distorted circle (or *ellipse*) and ultimately established three laws that came to be known as *Kepler's Laws*. These laws showed for the first time that the heliocentric theory was more logical, accurate, clear-cut, and easy to understand than any of the previous theories.

First Law: The orbit of every planet is an ellipse with the Sun at one of its foci.

Second Law: A line joining a planet and the Sun sweeps out equal areas during equal intervals of time.

Third Law: The square of the orbital period of a planet is directly proportional to the cube of the semimajor axis of its orbit.

Kepler's Laws are described in more detail starting on the next page.

Kepler was able to capture the motion of the planets, which was thought to be very complex, in these simple laws because he had been an assistant to Tycho Brahe and was able to use the massive amount of observational data left by his teacher. Tycho was an extremely diligent scientist, and his observational records are said to be the most accurate and precise of any collected before the telescope was invented. The refracting telescope known as a *Galilean telescope* was invented just after Tycho died. If it had been invented before his death, Tycho might have been one of the pioneers of the heliocentric theory.

WHAT DID GALILEO DO?

Generally, when it comes to the heliocentric theory, Galileo is more famous than Kepler.

But first, a quick note on Italian language and culture. Since Galileo's full name is Galileo Galilei, his surname is Galilei. Therefore, by all rights, we should call him *Galilei*. Why then, do we generally call him *Galileo*? Also, why are his last name and first name so similar?

In the Tuscany region of Italy, where Galileo has born, the first name of the eldest son was often a singularized form of his last name. So this naming convention carries a connotation like, "Mr. Sato, who represents the Satos." Therefore, for people in Galileo's time, *Galileo* was equivalent to *oldest son of the Galilei family*.

Now, although Galileo was treated by the Inquisition as a representative of the heliocentric theory faction, he made many blunders—such as believing that the planetary orbits were circles, as Copernicus did. In any case, Galileo was a rather stubborn man who continued to insist that the planets did not move in elliptical orbits, even after Kepler's Laws were published (both men lived at almost the same time).

Nevertheless, there is no doubt that Galileo, who did research in the various fields of medicine, mathematics, astronomy, and physics and who created the refracting telescope for observing heavenly bodies, was a genius. In particular, his establishment of the scientific methodology in use today—in which we mathematically analyze experimental results to construct theories—is surely a worthy achievement for someone referred to as the father of modern science.

WHAT HAS THE HELIOCENTRIC THEORY TAUGHT US?

It is generally accepted that the debate over the geocentric versus the heliocentric theory finally ended in 1619, when Kepler's Third Law was published (the first and second were published in 1609). However, even today, many people do not seem to understand the true significance of the heliocentric theory.

An example of this is illustrated by time machines that appear in science fiction novels and movies. Whether or not a time machine can actually be created is beside the point. In a typical story, a person who boards a time machine travels through time without changing location—it's a convention that the person will reappear at the same place, but in a different era.

However, in space, there is no true "same place" throughout time. Earth rotates while revolving around the Sun. In addition, as you will learn in a later chapter, the solar system itself is revolving within the Milky Way galaxy, and the Milky Way also does not stay in one fixed place. In other words, no matter what scale we use to view the universe, there is no place in it that is ever at rest.

Therefore, we cannot indicate a specific place in space. Since everything is moving and there is no reference point, it is impossible to identify even the position of Earth after several seconds.

So while the heliocentric theory is said to be the starting point for modern cosmological theory (in which the universe is always moving and changing), the theory advocated by Copernicus was actually just a Sun-centered theory and not a true heliocentric theory. Nevertheless, we've become conscious of philosophical concepts beyond science that posit that neither Earth nor the Sun is the center of the universe and everything is moving and changing in perpetual motion.

A SOMEWHAT COMPLICATED EXPLANATION OF KEPLER'S LAWS

Let's take a closer look at Kepler's Laws.

FIRST LAW: THE ORBIT OF EVERY PLANET IS AN ELLIPSE WITH THE SUN AT THE FOCUS

With this law, Kepler clearly indicated that the orbits of the planets are ellipses rather than circles. Moreover, the Sun is positioned at one of the two foci rather than at the center of the ellipse. This relationship is illustrated in the following figure.

Planet
Minor axis
Major axis
(focus 2)
Perihelion
Aphelion
Sun
(focus 1)
Orbit of planet

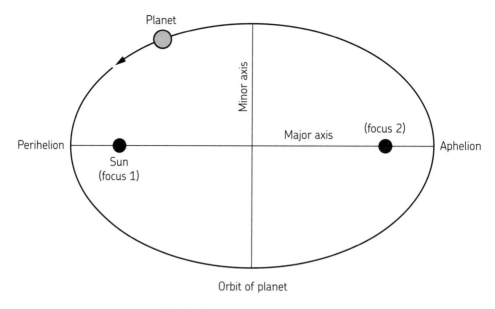

Orbit of a planet according to Kepler's First Law

Although Earth's orbit is nearly circular, the orbits of other plants, like Mars, are more elliptical. As a result, Copernicus's theory that every planet moves in a circular orbit could not accurately explain observational results for Mars. That's why Kepler introduced this law first.

The degree of ellipticality is indicated by the eccentricity. Using the diagram here, we can define eccentricity as follows:

$$\text{eccentricity} = \frac{\text{distance between foci}}{\text{length of major axis}}$$

The *major axis*, which is the longer axis of the ellipse, is the line connecting the perihelion and aphelion in the diagram. The *semimajor axis* is half that length, and it's also the mean (or average) distance of the planet from the Sun.

A true circle has only one focus (at the center of the circle), so its eccentricity is 0. The higher the eccentricity of a planet's orbit, the more elliptical its shape. The following table shows the orbital eccentricity of every planet in the solar system.

ECCENTRICITY OF EACH PLANET IN THE SOLAR SYSTEM

Planet	Mercury	Venus	Earth	Mars	Jupiter	Saturn	Uranus	Neptune
Eccentricity	0.2056	0.0068	0.0167	0.0934	0.0485	0.0555	0.0463	0.0090

As a quick but important side note, in mathematics, the eccentricity does not necessarily represent *only* the degree of ellipticality. We have a true circle or an ellipse only when the eccentricity (normally represented by e) is greater than or equal to 0 and less than 1. For other values, we will have a parabola or hyperbola.

Eccentricity (e) = 0	True circle
0 < Eccentricity (e) < 1	Ellipse
Eccentricity (e) = 1	Parabola
1 < Eccentricity (e)	Hyperbola

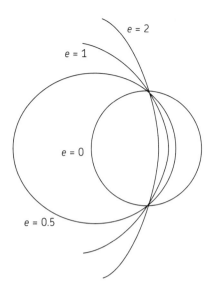

The relationship between eccentricity and geometric shape

SECOND LAW: A LINE JOINING A PLANET AND THE SUN SWEEPS OUT EQUAL AREAS DURING EQUAL INTERVALS OF TIME

Or in other words, a planet moving along an elliptical orbit will move faster when it is closer to the Sun and slower when it is farther away from the Sun. If we illustrate this, the areas of the shaded portions in the following figure are all the same.

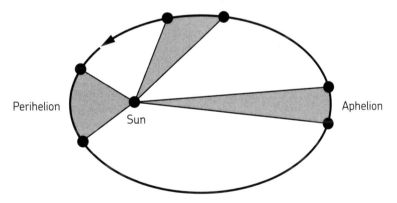

Orbit of a planet according to Kepler's Second Law

This is the same as the conservation of angular momentum in Newtonian mechanics. Although it is a little difficult to prove mathematically, it is intuitively similar to the rotation of a figure skater—a skater who begins to turn with her arms extended will turn faster as she brings her arms closer to her body.

A SOMEWHAT COMPLICATED EXPLANATION OF KEPLER'S LAWS 75

We can also consider a situation in which we attach a weight to a string and swing it around in a circle. When the string is longer, the weight will be harder to rotate, and its speed will be slower.

Conceptually, the following explanation may be easier to understand. When no external force acts on a body, uniform linear motion will continue, due to the law of conservation of momentum. In the diagram here, the body will move as follows: $P_1 \rightarrow P_2 \rightarrow P_3$, where the length from P_1 to P_2 is the same as the length from P_2 to P_3.

P_3'
f
P_3
f
P_2
S
P_1
Sun
Planet

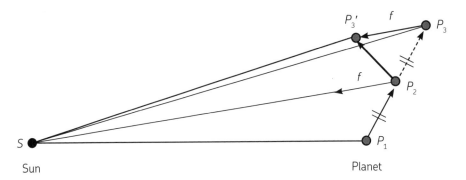

Orbit of a planet according to Kepler's Second Law

However, a planet does not move with linear motion, since it is affected by the gravity of the Sun at point S. Although a planet is continuously pulled toward the Sun and moves with circular motion, to make the explanation easier to understand here, we'll assume that gravity continuously pulls the planet toward the Sun when the planet moves from P_2 to P_3. The planet's motion will be pulled to the left by the force of gravity f to arrive at position P_3' (composition of the force of gravity f and the force that tries to continue uniform linear motion). Since momentum does not change, the lengths of P_2P_3 and P_2P_3' are the same.

Now, if we compare the triangles that were formed, since $P_1P_2 = P_2P_3$ due to uniform linear motion, $\triangle SP_1P_2$ and $\triangle SP_2P_3$ will be triangles with equal length sides and equal heights, so their areas will also be equal.

Next, since $\triangle SP_2P_3$ and $\triangle SP_2P_3'$ share base SP_2 and have the same height (since f is the force of gravity at P_2, the arrow used in the composition of forces will also be parallel to SP_2), and their areas will be equal. In other words, the following holds true:

$$\triangle SP_2P_3 = \triangle SP_2P_3'$$

Since this relationship holds regardless of the positions of the Sun and planet, a line joining a planet and the Sun sweeps out equal areas during equal intervals of time.

THIRD LAW: THE SQUARE OF THE ORBITAL PERIOD OF A PLANET IS DIRECTLY PROPORTIONAL TO THE CUBE OF THE SEMIMAJOR AXIS OF ITS ORBIT

This is the law that may be the most difficult to understand. In short, it means that the length of the orbital period depends only on the semimajor axis of the orbit. Since it does not depend on the eccentricity of the elliptical orbit, the period will be the same as long as the semimajor axis is the same, regardless of whether the motion is circular or elliptical.

Incidentally, the *semimajor axis* is half of the major axis, which appeared in the description of the first law. In other words, it is the mean distance between the planet and the Sun.

Even before Kepler, scientists observed that planets with larger orbits took longer to make one circuit (that is, their orbital period was larger). However, the mathematical relationship between that period and the semimajor axis of the orbit was not understood until Kepler.

If we assume P is the orbital period in years and a is the semimajor axis (the mean distance between a planet and the Sun) in astronomical units (or AUs, with 1 AU defined as the distance between Earth and the Sun), then the following equation holds for Earth (where $a = 1$ AU):

$$\frac{a^3}{P^2} = 1$$

If we calculate this ratio for any planet in the solar system, we will get the following table, which confirms Kepler's Third Law.

SEMIMAJOR AXIS OF A PLANET'S ORBIT AND ORBITAL PERIOD

Planet	Semimajor axis of orbit a (AUs)	a^3	Orbital period relative to the fixed star's P (solar years)	P^2	a^3/P^2
Mercury	0.3871	0.05800555	0.2409	0.05803281	0.9995
Venus	0.7233	0.37840372	0.6152	0.37847104	0.9998
Earth	1.0000	1	1.0000	1	1.0000
Mars	1.5237	3.53751592	1.8809	3.53778481	0.9999
Jupiter	5.2026	140.819017	11.8620	150.707044	1.0008
Saturn	9.5549	872.32524	29.4580	867.773764	1.0052
Uranus	19.2184	7098.25644	84.0220	7049.69648	1.0055
Neptune	30.1104	27299.1783	164.7740	27150.4711	1.0055

2
FROM THE SOLAR SYSTEM TO THE MILKY WAY

WHAT IF KAGUYA-HIME CAME FROM A PLANET IN OUR SOLAR SYSTEM?

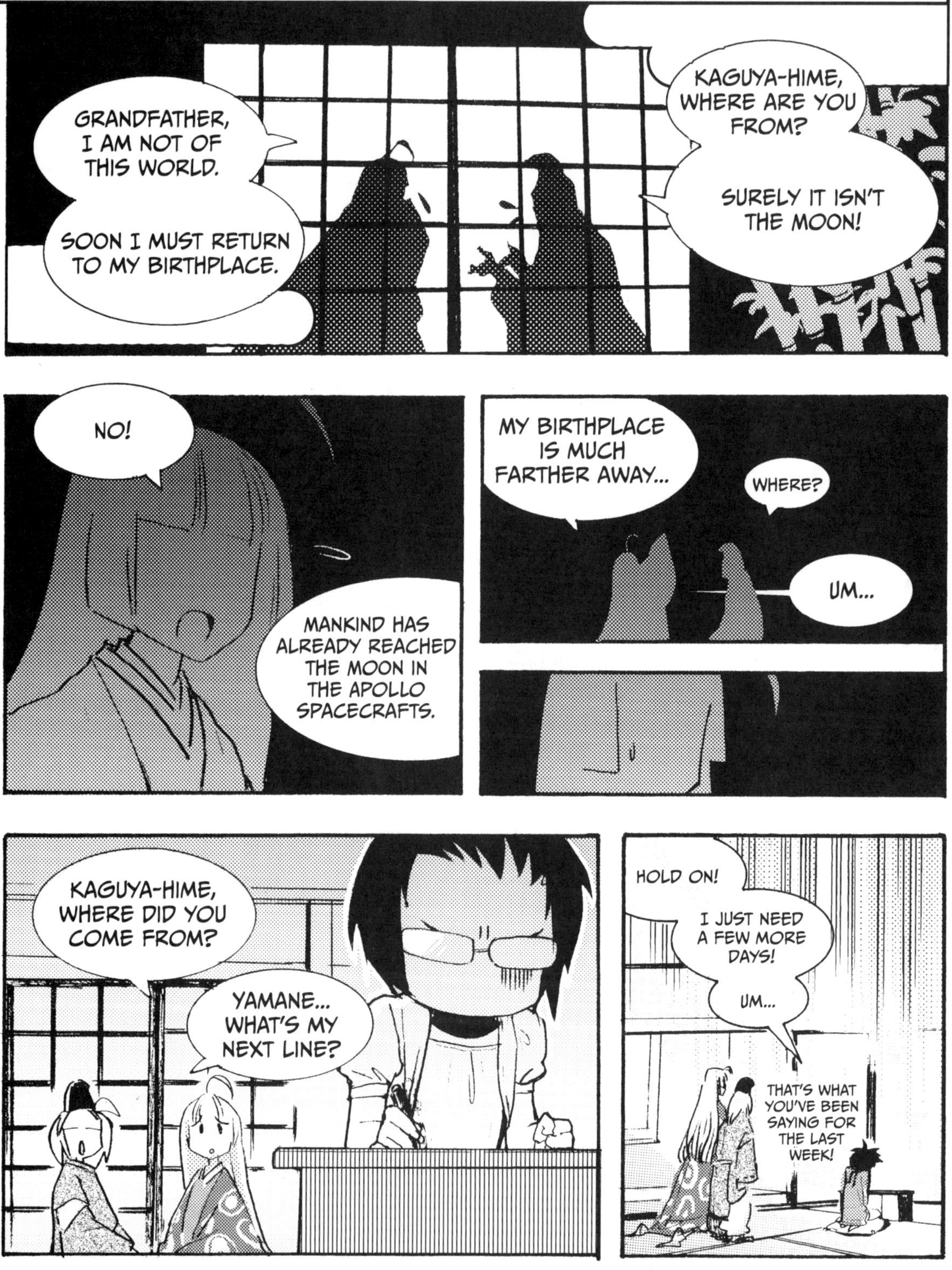
GRANDFATHER, I AM NOT OF THIS WORLD.
SOON I MUST RETURN TO MY BIRTHPLACE.
KAGUYA-HIME, WHERE ARE YOU FROM?
SURELY IT ISN'T THE MOON!
NO!
MANKIND HAS ALREADY REACHED THE MOON IN THE APOLLO SPACECRAFTS.
MY BIRTHPLACE IS MUCH FARTHER AWAY...
WHERE?
UM...
KAGUYA-HIME, WHERE DID YOU COME FROM?
YAMANE... WHAT'S MY NEXT LINE?
HOLD ON!
I JUST NEED A FEW MORE DAYS!
UM...
THAT'S WHAT YOU'VE BEEN SAYING FOR THE LAST WEEK!

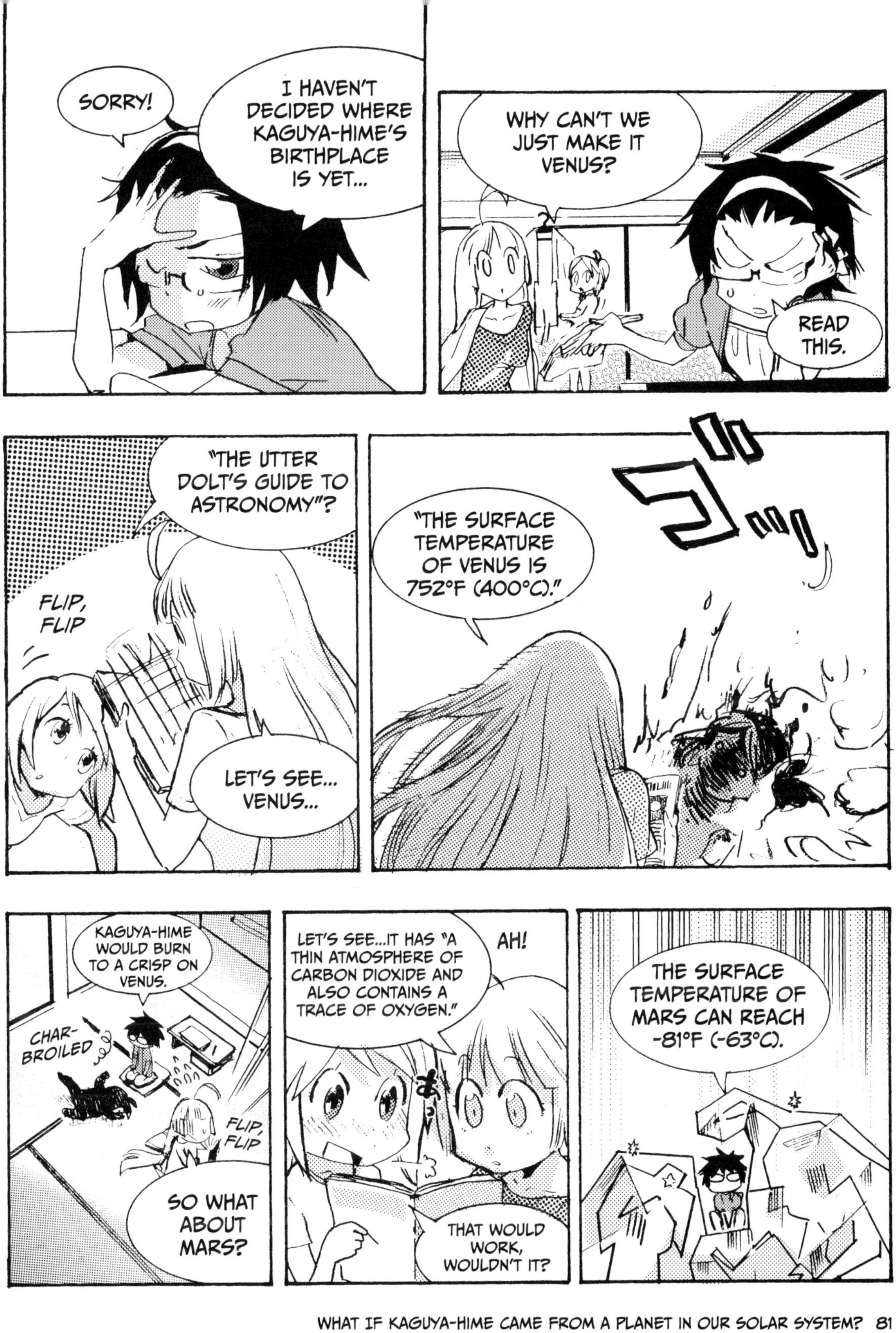
SORRY!
I HAVEN'T DECIDED WHERE KAGUYA-HIME'S BIRTHPLACE IS YET...
WHY CAN'T WE JUST MAKE IT VENUS?
READ THIS.
"THE UTTER DOLT'S GUIDE TO ASTRONOMY"?
FLIP, FLIP
LET'S SEE... VENUS...
"THE SURFACE TEMPERATURE OF VENUS IS 752°F (400°C)."
KAGUYA-HIME WOULD BURN TO A CRISP ON VENUS.
CHAR-BROILED
FLIP, FLIP
SO WHAT ABOUT MARS?
LET'S SEE...IT HAS "A THIN ATMOSPHERE OF CARBON DIOXIDE AND ALSO CONTAINS A TRACE OF OXYGEN."
AH!
あっ
THAT WOULD WORK, WOULDN'T IT?
THE SURFACE TEMPERATURE OF MARS CAN REACH -81°F (-63°C).

KAGUYA-HIME AND THE SOLAR SYSTEM

Let's try to determine which planet should be Kaguya-hime's birthplace. Can we find a suitable birthplace for Kaguya-hime in our solar system?

Mars
Mercury
Earth
Sun
Venus
Neptune
Uranus
Jupiter
Saturn
[Pluto]

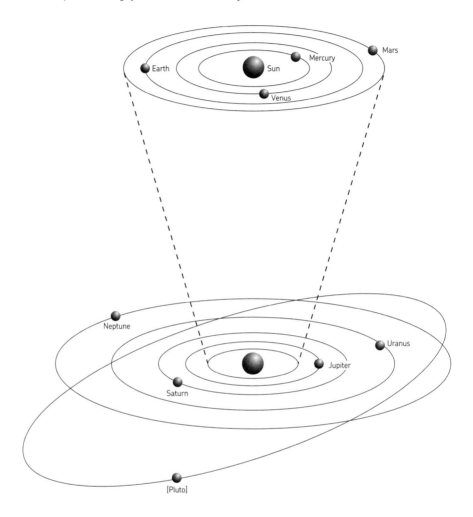

The orbits of planets in our solar system

(In August 2006, Pluto was reclassified and removed from the list of planets.)

MERCURY

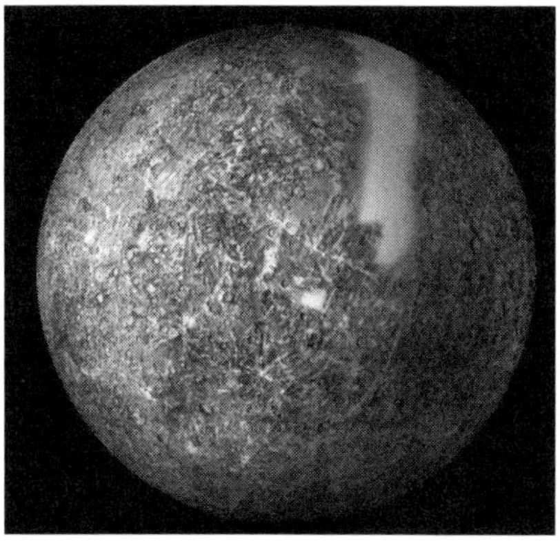

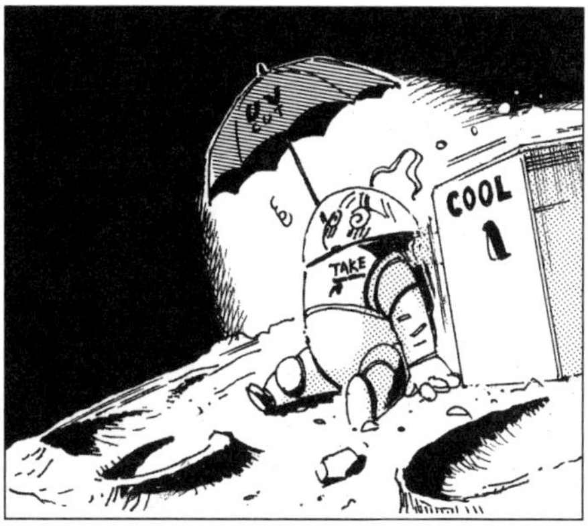
UV CUT
COOL
TAKE

Size Approximately 0.38 times the size of Earth (equatorial radius of 1,516 miles, or 2,440 km)

Mass Approximately 0.055 times the mass of Earth

Surface gravity Approximately 0.38 times the gravity of Earth

Satellites None

Mean distance from the Sun The mean distance between Earth and the Sun is called an astronomical unit, or 1 AU. Mercury's mean distance from the Sun is 0.39 AU.

Orbital period (how long it takes to complete one circuit around the Sun)
Approximately 88 days

Rotation period (how long the planet takes to rotate once on its own axis)
Approximately 59 days

WHAT'S IT LIKE THERE?

Since Mercury's orbit is the closest to the Sun of any planet in the solar system, the amount of energy it receives from the Sun is intense—approximately 6.7 times what Earth receives, per unit area. Kaguya-hime would definitely need to protect herself from ultraviolet rays on Mercury, but actually, she'd probably just burn up first—the maximum surface temperature is 800°F (427°C).

Mercury has hardly any atmosphere because of its low gravity, so the environment is nearly a vacuum. Unless Kaguya-hime wore a completely airtight spacesuit with a high-power cooling unit attached, she would be sunburned in an instant. The northern polar region is said to contain ice, and it is very likely that the temperature there drops to approximately -292° F (-180°C).

If it weren't for Mercury's extremely hot and inhospitable environment, Kaguya-hime might feel quite at home there, as the surface scenery of Mercury consists of deserts and craters much like the Moon's.

VENUS

Size Approximately 0.95 times the size of Earth (equatorial radius of 3,761 miles, or 6,052 km)

Mass Approximately 0.82 times the mass of Earth

Surface gravity Approximately 0.91 times the gravity of Earth

Satellites None

Mean distance from the Sun 0.72 AU

Orbital period Approximately 225 days

Rotation period Approximately 243 days

WHAT'S IT LIKE THERE?

Venus is the closest planet to Earth, and the two planets have many features in common. The gravity on Venus is practically the same as gravity on Earth. Venus has an atmosphere, and it also has active volcanic activity that spews out hydrogen sulfide, nitrogen, and sulfurous acid gas.

Since Venus is covered with thick clouds consisting of concentrated sulfuric acid, and the main component of its atmosphere (approximately 96 percent) is carbon dioxide, its surface temperature is between 752°F and 932°F (400°C–500°C) because of the intense greenhouse effect. Also, the atmospheric pressure at the planet's surface is approximately 92 times the pressure at Earth's surface.

Although this is a minor point, the direction of Venus's rotation is opposite to that of Earth, so there the Sun rises in the west and sets in the east.

MARS

Size Approximately 0.53 times the size of Earth (equatorial radius of 2,110 miles, or 3,396 km)

Mass Approximately 0.11 times the mass of Earth

Surface gravity Approximately 0.38 times the gravity of Earth

Satellites 2: Phobos and Deimos

Mean distance from the Sun 1.52 AU

Orbital period Approximately 687 days

Rotation period Approximately 1 day

WHAT'S IT LIKE THERE?

The surface of Mars is entirely desert, except for two polar ice caps. The planet has a thin, carbon dioxide atmosphere. The existence of water there has been confirmed, and among the planets of the solar system, its atmosphere is the most similar to Earth's. Its highest atmospheric temperature is approximately 68°F (20°C), and Kaguya-hime herself could easily live there as long as she had an *Apollo*-type spacesuit.

The geological composition of Mars is also similar to that of Earth, and since the length of one Mars day is also about the same as one Earth day, a person's daily living patterns would not be disrupted. However, since one of Mars's two satellites, Phobos, revolves around Mars faster than Mars rotates, it rises and sets twice each night.

Mars is the home of Olympus Mons, which is said to be the largest mountain in the solar system. Its height is approximately 82,021 feet (25,000 m), which is approximately three times the height of Mount Everest! Mars also has the largest canyon in the solar system, and its terrain includes many plains.

Tornado-like whirlwinds sometimes blow across the surface of Mars. One of these was photographed by the *Mars Pathfinder* spacecraft, which was launched by the United States in 1996. In 2004, two robotic vehicles called *Opportunity* and *Spirit* landed on Mars. These NASA rovers have made many important discoveries about Mars and have sent back more than 133,000 images of the planet's surface.*

---

* Want to see those photographs and hear more about the Mars rovers? Visit *http://marsrovers.jpl.nasa.gov/*.

JUPITER

Size Approximately 11.2 times the size of Earth (equatorial radius of 44,423 miles, or 71,492 km)

Mass Approximately 318 times the mass of Earth

Surface gravity Approximately 2.37 times the gravity of Earth

Satellites At least 63

Mean distance from the Sun 5.2 AU

Orbital period Approximately 12 years

Rotation period Approximately 9 hours, 50 minutes

WHAT'S IT LIKE THERE?

While Mercury, Venus, Earth, and Mars, which are composed of rocks or metals, are referred to as the *terrestrial* planets, Jupiter and Saturn, whose primary component is gas, are referred to as the *Jovian* planets (or the *gas giants*). The density of Jupiter is only 1/4 that of Earth, and since it is composed primarily of gaseous hydrogen and helium, it does not have a "surface" like Earth does. Since the center (approximately 1/11 of the radius) is a rocky core and the rest is just a collection of gaseous or liquid materials, Jupiter differs from our ordinary image of a planet. Being on Jupiter would be like being in the middle of a cloud.

Jupiter has a mass of more than 300 times that of Earth, and its gravity is more than twice that of Earth. If she lived on Jupiter, Kaguya-hime would feel as if she were always carrying around twice her own weight.

Jupiter's most noticeable feature is the Great Red Spot—a massive, persistent storm on the planet's surface. It was discovered by the astronomer Robert Hooke in 1664 with one of the first telescopes. The Great Red Spot has a larger diameter than Earth! Scientists still don't know exactly why the spot is red.

Some scientists believe that Europa, one of Jupiter's moons, is one of the most likely places to find extraterrestrial life in the solar system due to the heat-generating tidal forces in the liquid water oceans believed to lie under Europa's icy mantel. The conditions in Europa's subsurface oceans may be similar to those that were the catalyst for life's beginning on Earth. Scientists hope to observe Europa and Jupiter's other satellites more closely via space missions set to launch in 2020.

SATURN

Size   Approximately 9.45 times the size of Earth (equatorial radius of 37,449 miles, or 60,268 km)

Mass   Approximately 95.2 times the mass of Earth

Surface gravity   Approximately 0.94 times the gravity of Earth

Satellites   Approximately 200 observed and at least 62 with secure orbits

Mean distance from the Sun   9.55 AU

Orbital period   Approximately 29.5 years

Rotation period   Approximately 10 hours, 34 minutes

WHAT'S IT LIKE THERE?

Saturn is a Jovian planet that is composed primarily of gaseous hydrogen and helium. Like Jupiter, it has no true surface. Saturn has a *specific gravity* of 0.69. Specific gravity is the ratio of a substance's density as compared to another substance's density—usually water. Saturn's low specific gravity means it is light enough to float in water! But since it's volume is more than 800 times that of Earth, its gravity is nearly the same as Earth's.

Saturn is far from the Sun—from Saturn, the Sun would only appear to be 1/10 the size that it appears from Earth. Naturally, the temperature on Saturn is low. The mean temperature is approximately -266°F (-130°C). For some reason, the polar regions have the highest temperature, and if Kaguya-hime lived there, she would see the characteristic rings near the horizon, rather than farther up in the sky. Since Saturn has the greatest number of satellites among the planets of the solar system, Kaguya-hime would be able to have lots of Otsukimi moon-viewing parties. However, since each day is only 10.5 hours long, morning would come quickly!

URANUS

 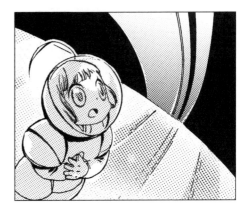

Size Approximately 4.01 times the size of Earth (equatorial radius of 15,882 miles, or 25,559 km)

Mass Approximately 14.5 times the mass of Earth

Surface gravity Approximately 0.89 times the gravity of Earth

Satellites At least 27

Mean distance from the Sun 19.2 AU

Orbital period Approximately 84 years

Rotation period Approximately 17 hours, 14 minutes

WHAT'S IT LIKE THERE?

Although Uranus was previously classified as a Jovian planet because of its size and location, it is mostly made up of ices of frozen water, methane, and ammonia. Thus, it is more accurately categorized as an *ice giant*. Uranus has a blue color that seems translucent, and its surface scenery may be similar to that of the South Pole on Earth. However, it is an extremely cold planet with temperatures below –328°F (–200°C).

Uranus has rings that are thinner than Saturn's. It also has 27 satellites, so the night sky must be quite lively.

The most unique thing about Uranus is that its axis of rotation is nearly parallel to the *ecliptic* (that is, the plane of the solar system), whereas the other planets have an axis of rotation perpendicular to the ecliptic. In fact, Uranus's axis actually dips below the ecliptic. Therefore, depending on the season, the north or the south pole almost directly faces the Sun and is in daylight all day, while the other pole is in darkness.

NEPTUNE

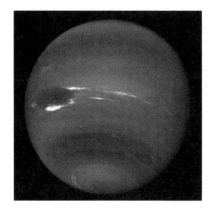 

Size Approximately 3.88 times the size of Earth (equatorial radius of 15,388 miles, or 24,764 km)

Mass Approximately 17.2 times the mass of Earth

Surface gravity Approximately 1.11 times the gravity of Earth

Satellites At least 13

Mean distance from the Sun 30.1 AU

Orbital period Approximately 165 years

Rotation period Approximately 16 hours

WHAT'S IT LIKE THERE?

The semimajor axis of Neptune's orbit is approximately 30 times that of Earth's semimajor axis of orbit. Because the amount of energy that reaches an object is inversely proportional to the square of the distance between the object and the energy source (in this case, the Sun), the energy that Neptune receives from the Sun is only 1/900 as strong as the energy that Earth receives.

Neptune has a thick atmosphere with hydrogen as its main constituent, and typhoon-like storms sometimes occur. Neptune's Great Dark Spot was discovered when the *Voyager 2* spacecraft went past the planet in 1989. When Neptune was observed years later by the Hubble telescope, the storm was gone.

PLUTO

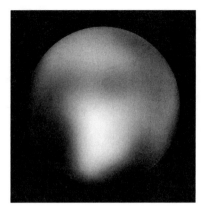

Credit: NASA, ESA, and M. Buie (Southwest Research Institute)

Size   Approximately 0.18 times the size of Earth (equatorial radius of 744 miles, or 1,197 km)

Mass   Approximately 0.0023 times the mass of Earth

Surface gravity   Approximately 0.07 times the gravity of Earth

Satellites   At least 3

Mean distance from the Sun   Approximately 30 to 50 AUs (the distance from the Sun varies so widely because Pluto has a very eccentric elliptical orbit)

Orbital period   Approximately 248 years

Rotation period   Approximately 6.4 days

WHAT'S IT LIKE THERE?

Although Pluto was formally considered the ninth planet of the solar system after its discovery by Clyde Tombaugh in 1930, it was found not to meet the definition of a planet during the 2006 meeting of the International Astronomical Union (IAU). We now call Pluto a *dwarf planet*. Since it has an elliptical orbit that is quite different from those of the other planets, and it differs from Uranus and Neptune in size and structure, many astronomers have said that this decision was long overdue.

Photographs of Pluto's atmosphere from the Hubble telescope suggest that the atmosphere of the planet itself changes quickly over time due to *sublimation*, the process by which a solid turns directly into gas vapor without passing through the liquid phase first. Pluto does not have much atmosphere because of its intensely cold temperatures (-382°F, or -280°C), and it is no doubt a harsh environment for life.

EARTH

Size Equatorial radius of 3,963 miles, or 6,378 km

Mass 5.976×10^{24} kg

Surface gravity 9.82 m/s^2

Satellites 1

Mean distance from the Sun 1 AU (approximately 149.6 million km)

Orbital period 1 year (approximately 365.25 Earth days)

Rotation period 1 Earth day (approximately 24 hours)

WHAT'S IT LIKE THERE?

Our home planet is often called *Mother Earth*, but 4.6 billion years ago when it was formed, it was not the kind of environment in which living creatures thrived. Its surface was covered by molten rock known as *magma*, and water only existed in gaseous form (water vapor). Also, the atmospheric pressure was approximately 300 times greater than it is now. Mars today is an environment much more conducive to life than Earth was then.

Earth's environment became moderate enough to support life as we know it approximately 550 million years ago. This was around the beginning of the Paleozoic Era, when the number of species began to increase rapidly. From the perspective of the history of Earth, that's the very recent past.

THE MOON

Size Approximately 0.27 times the size of Earth (equatorial radius of 1,080 miles, or 1,738 km)

Mass Approximately 0.012 times the mass of Earth

Surface gravity Approximately 0.17 times the gravity of Earth

Orbital period Approximately 27.3 days

Rotation period Approximately 27.3 days

HOW WAS THE MOON FORMED?

Earth and its satellite, the Moon, seem to have a parent-child relationship—the Moon is composed of materials that are nearly identical to those of Earth's mantle. As a result, one hypothesis presented by astronomers in the past was that the centrifugal force of Earth's rotation caused a chunk of Earth to break off while it was being formed. This chunk formed the Moon, and the hole that remained is the Pacific Ocean.

This hypothesis has a certain degree of persuasive power, but it also has several problems. The rotation speed of Earth today is not fast enough for such a strong centrifugal force to have been produced—it is impossible for the amount of material that formed the Moon to have been ejected suddenly while Earth rotated slowly enough to maintain its atmosphere (air). There is also no evidence that the rotation of Earth slowed down at some point after the chunk would have broken off.

If we delve into this a little deeper, we find that its greatest flaw is that the Moon is too large. The diameter of the Moon (we will always mean the equatorial diameter throughout the following discussion) is approximately 3,474 km (2,159 miles), which is approximately 1/4 the diameter of Earth. This is disproportionately large for a satellite, and as a result, some astronomers claim that the Moon should be considered a double planet with Earth, rather than a satellite of it.

Incidentally, although Ganymede (a satellite of Jupiter) is the largest satellite in the solar system, its diameter is only approximately 1/27 the diameter of its host planet. Titan, the largest satellite of Saturn, has a diameter that is approximately 1/25 the diameter of its host planet. Taking these numbers into consideration, it seems that the Moon is abnormally large. The theory that such a large object broke off from Earth due to centrifugal force created by the planet's rotation is implausible.

MAYBE THEY'RE BROTHER AND SISTER? OR NO RELATION AT ALL?

The *sister hypothesis* posits that Earth became its current size through the collision and coalescence of tiny dust particles and gas that existed when the solar system was formed. It says that the Moon was created at the same time and became a satellite by chance. Unfortunately, Earth and the Moon differ enough in density and composition to throw this hypothesis into doubt. If they were created at the same time, in the same area, from the same material, it is very unlikely that they would have ended up with differing densities and compositions.

The *capture hypothesis* could easily resolve the problems of the sister hypothesis. It claims that the Moon came near to Earth by accident and was captured by Earth's gravity. There are two main counterarguments to this idea. The first is that the Moon is too large relative to Earth to have been captured by Earth's gravity. The second is that although the Moon and Earth are significantly different in composition and density, they are still far too similar to have formed entirely independent of each other. These material similarities of Earth and the Moon have been proven through evidence collected by the *Apollo* missions and many other space exploration projects.

THE GIANT IMPACT

Finally, there is the *giant impact hypothesis*, which is still the most popular one today. This hypothesis proposes that a large heavenly body crashed into Earth shortly after our planet was formed, approximately 4.6 billion years ago. Reasoning from the amount of energy involved, this heavenly body might have been about the size of Mars (with a diameter approximately half that of Earth). Scientists call this theoretical planet *Theia*. This collision was truly a major event!

If a body that size had collided with Earth, a large amount of material from Earth would naturally have flown off into space, and the body that had collided with our planet would have disintegrated. The reasoning behind the giant impact hypothesis is that once this material broke off from Earth, it would have coalesced over time to form the Moon.

This hypothesis is able to explain why the materials composing the Moon and Earth are similar, why the Moon is a satellite that is "too big" relative to the size of Earth, and why the Moon began to revolve around Earth.

Also, volatile elements like water, carbon dioxide, carbon monoxide, and others are depleted on the Moon, and one explanation is that they were lost when the collison occurred.

For these reasons, the giant impact hypothesis is currently the dominant view, but conclusive proof for it has still not been found.

IT'S NOT SUCH A BAD THING THAT THE MOON IS SO LARGE

Let's put off the question of whether or not a giant impact occurred until we have more data from future lunar exploration. Instead, let's touch upon the effects that the Moon's great size has on Earth.

First, our enjoyment of the graceful appearance of the Moon at an Ostukimi moon-viewing party is undoubtedly because of its size. All of the other satellites in the solar system, including Ganymede, appear no larger than half the size of our Moon when viewed from their host planets.

The dramatic natural phenomenon of the ebb and flow of the tides also occurs because Earth has such an enormous satellite. The distance between Earth and the Moon is approximately 380,000 km (236,121 miles), which is approximately 30 times the diameter of Earth. (This is illustrated to scale below.)

Approximately 30 times the diameter of Earth
Earth
Moon

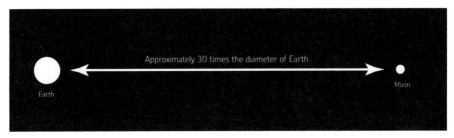

Model of the distance between Earth and the Moon

The tides are mainly caused by the effect of gravity from the Moon. Many cyclical living patterns evolved in sea creatures because of these tides, and mankind has used them to develop fishing techniques. Gathering shellfish at low tide is one example.

The ebb and flow of tides cause changes in the natural scenery that we have come to see as beautiful. If this is the result of a giant impact, we should be thankful for it, shouldn't we?

THE SUN

Size Approximately 109 times the size of Earth (equatorial radius of 432,474 miles, or 696,000 km)

Mass Approximately 332,946 times the mass of Earth

Surface gravity Approximately 28.01 times the gravity of Earth

Volume Approximately 1,300,000 times that of Earth

WHAT'S IT LIKE THERE?

The Sun that shines brightly in the sky is the heavenly body whose presence we feel the most. But how large is it, really?

First, its mass accounts for 99.9 percent of the entire solar system. In other words, the planets such as Earth and Jupiter, the satellites revolving around them, and all the various small solar system bodies exist almost as an afterthought, constituting a combined mass of only 1/1000 that of the Sun.

The Sun is approximately 150,000,000 km (93,205,679 miles) from Earth. Since the maximum velocity of a space shuttle is approximately 28,000 km per hour (17,398 miles per hour), it would take nearly 7.5 months to reach the Sun at full speed. An ordinary passenger jet would take around 20 years to get to the Sun (but, of course, a jet plane cannot fly in space).

However, it would be terrible for us if the Sun were not this far away. The daytime temperature on Mercury, which is less than 1/4 as far from the Sun as is Earth, is greater than 752°F (400°C). The energy that reaches Earth from the Sun is equivalent to the amount of energy generated by 200 million nuclear power plants (assuming their output is 100 megawatts on average).

THE SUN CONTINUES TO BURN HYDROGEN AS FUEL

Earlier, we mentioned the diameter of the Sun, but the Sun differs from an Earthlike planet in that it actually has no ground surface. It is a star with a rotation period of approximately 25 days at the equator and approximately 36 days near the poles. The Sun, just like most other stars in the universe, is composed of material in a plasma state. That means the gas that makes up the Sun and that is held together by the Sun's massive gravity is ionized (missing an electron or having an extra electron) and highly energetic.

The Sun's core, which is at the center of its structure, has a high pressure—200 to 250 billion times Earth's atmospheric pressure—because of the gravity that is produced by the Sun's mass, which is 330,000 times Earth's mass. This high pressure is the catalyst for the nuclear fusion reactions, which convert smaller atoms into bigger atoms (for example, turning hydrogen into helium).

In these fusion reactions, six protons interact to eventually create one helium atom, one neutrino, two protons, and a large amount of energy in the form of gamma rays. These fusion reactions are the Sun's power source and the origin of the 200 million nuclear power plants' worth of solar energy that reaches Earth every day. Also, the Sun's energy creates a radiative zone around its core, and gas above that radiative zone produces a convective zone. When we look at the Sun, what we really see is a thin surface layer that covers the convective zone. This opaque layer is called the photosphere. Its temperature is approximately 5,577 Kelvin (9,579°F or 5,304°C). Sunspots occur in the photosphere. (Sunspots look like dark blemishes on the Sun. But really, they are just areas of the photosphere that are cooler than the surrounding area and therefore appear darker.)

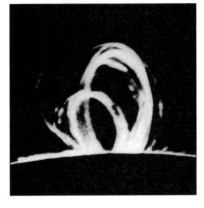

A solar prominence, a large loop of plasma that can be as long as 28 times the diameter of Earth

Surrounding the photosphere is the Sun's atmosphere, which is made up of the temperature-minimum layer, the chromosphere, and the corona. The corona is the Sun's outermost layer, and for reasons still unknown to science, it is millions of degrees hotter than the surface of the Sun.

The Sun's corona

The hydrogen in the core of the Sun is used as fuel and will eventually be entirely consumed. However, according to theoretical calculations, the lifetime of the Sun is approximately 10 billion years. Since the Sun was formed approximately 4.6 billion years ago (at almost the same time Earth was formed), it should continue to shine like it does today for at least another 5 billion years.

IS THE SUN MADE OF RECYCLED STARS?

The formation of a star like the Sun begins when a cloud of gas, plasma, and dust starts to gather in the void of space. This kind of *molecular cloud* is held together in a state called *hydrostatic equilibrium*, in which the gravitational force of a volume of gas is balanced by the "outward" force of pressure. When the gravitational force of the cloud becomes larger, due to a collision with another molecular cloud or an increase in pressure from the explosion of a nearby star, the cloud undergoes a gravitational collapse. As the molecular cloud collapses, it begins to form into a spinning sphere of gas. The huge increase in the gas's

Corona
Chromosphere
Convective Zone
Radiative Zone
Photosphereic Surface
Core, in Which Nuclear Fusion Reactions Occur

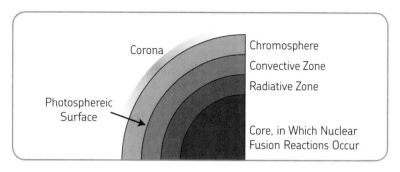

Internal structure of the Sun

temperature and pressure due to the gravitational collapse results in a nuclear fusion reaction—a highly energetic reaction in which two or more nuclei join together to form a heavier nucleus. The fusion of hydrogen within the core of the star produces heavier forms of hydrogen, which in turn can fuse and produce isotopes of helium. These fusion reactions create energy and the light that you see every day from the Sun.

Since hydrogen and helium are the most basic and abundant elements in the universe, it can be surmised that stars should also be made up of these two substances. However, spectral analysis of the Sun has verified the existence of heavy elements such as iron, gold, and uranium. How is that possible?

The entirety of a star's lifetime is a battle against gravity. Gravitational forces created by the mass of the star are trying to collapse the star, thus creating pressure that causes the fusion reactions described above. This in turn creates a *radiation pressure* that balances out the gravitational force, keeps the star at hydrostatic equilibrium, and prevents the star from further collapsing. As the hydrogen fuel at the star's core begins to run out, the star begins to burn helium to maintain the internal pressure that keeps it from collapsing. For the rest of their lives, stars that have similar masses to our own Sun will only burn hydrogen, periodically shedding their atmospheres until just the core remains. When the fuel runs out, a star slowly cools and fades. However, stars more massive than the Sun can continue fusing heavier and heavier elements to maintain equilibrium. Eventually, the massive star is similar to an onion, with the outermost layer being hydrogen and the innermost layer being iron.

Fusing iron requires more energy than it creates, and soon the equilibrium is disrupted and gravity wins, collapsing the core and creating an extremely energetic stellar explosion called a *supernova*. A supernova can radiate as much energy, if not more, than the Sun ever will during its entire lifetime and can often outshine the galaxy it is in. The explosion that creates a supernova also creates elements heavier than iron and expels those elements, plus the star's atmosphere, into the surrounding space, thus seeding other molecular clouds with those heavy elements.

The Sun is believed to have been created from interstellar matter from just such an explosion—the remains of a former star. That explains why those heavy elements (gold, uranium, and so on) are found in our own star. In fact, all elements heavier than hydrogen and helium anywhere in the universe, including within ourselves, were created within a star or by a star's explosive death. So, not only is the Sun a "recycled" star, but we ourselves are made of recycled stardust.

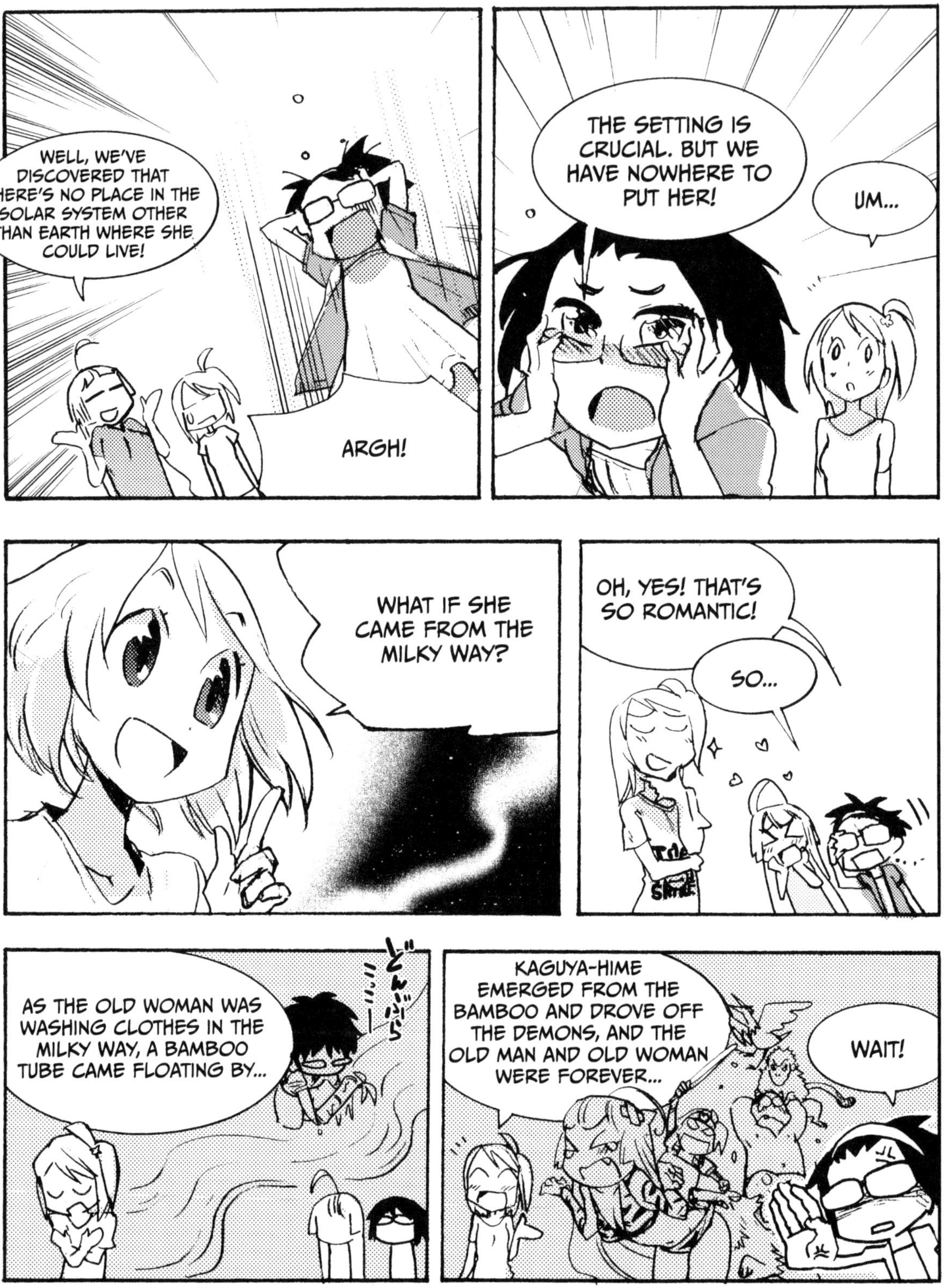
WELL, WE'VE DISCOVERED THAT THERE'S NO PLACE IN THE SOLAR SYSTEM OTHER THAN EARTH WHERE SHE COULD LIVE!
ARGH!
THE SETTING IS CRUCIAL. BUT WE HAVE NOWHERE TO PUT HER!
UM...
WHAT IF SHE CAME FROM THE MILKY WAY?
OH, YES! THAT'S SO ROMANTIC!
SO...
AS THE OLD WOMAN WAS WASHING CLOTHES IN THE MILKY WAY, A BAMBOO TUBE CAME FLOATING BY...
KAGUYA-HIME EMERGED FROM THE BAMBOO AND DROVE OFF THE DEMONS, AND THE OLD MAN AND OLD WOMAN WERE FOREVER...
WAIT!

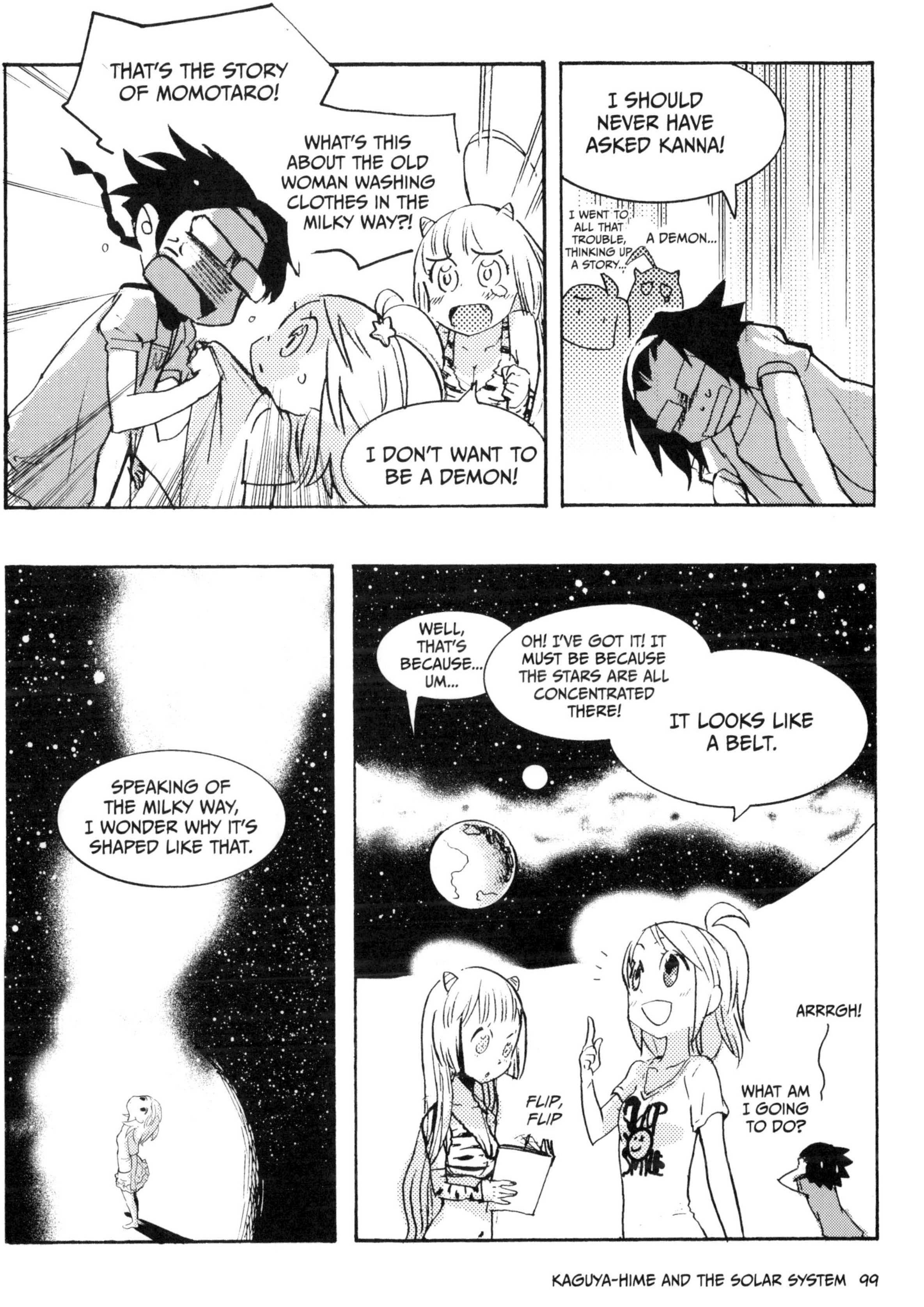
THAT'S THE STORY OF MOMOTARO!
WHAT'S THIS ABOUT THE OLD WOMAN WASHING CLOTHES IN THE MILKY WAY?!
I DON'T WANT TO BE A DEMON!
I SHOULD NEVER HAVE ASKED KANNA!
I WENT TO ALL THAT TROUBLE, THINKING UP A STORY...
A DEMON...
SPEAKING OF THE MILKY WAY, I WONDER WHY IT'S SHAPED LIKE THAT.
WELL, THAT'S BECAUSE... UM...
OH! I'VE GOT IT! IT MUST BE BECAUSE THE STARS ARE ALL CONCENTRATED THERE!
IT LOOKS LIKE A BELT.
FLIP, FLIP
ARRRGH!
WHAT AM I GOING TO DO?

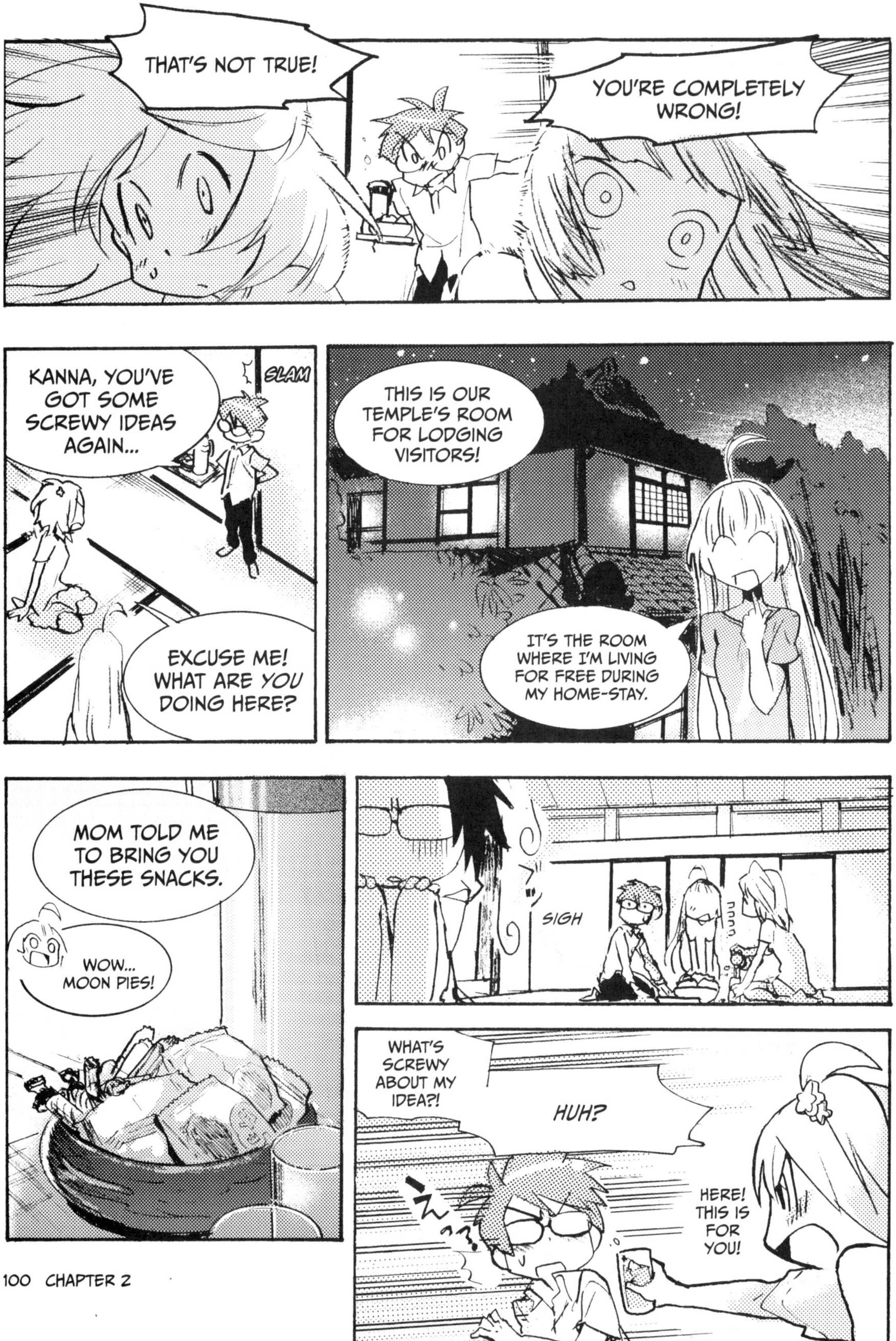
THAT'S NOT TRUE!
YOU'RE COMPLETELY WRONG!
KANNA, YOU'VE GOT SOME SCREWY IDEAS AGAIN...
SLAM
EXCUSE ME! WHAT ARE *YOU* DOING HERE?
THIS IS OUR TEMPLE'S ROOM FOR LODGING VISITORS!
IT'S THE ROOM WHERE I'M LIVING FOR FREE DURING MY HOME-STAY.
MOM TOLD ME TO BRING YOU THESE SNACKS.
WOW... MOON PIES!
SIGH
WHAT'S SCREWY ABOUT MY IDEA?!
HUH?
HERE! THIS IS FOR YOU!

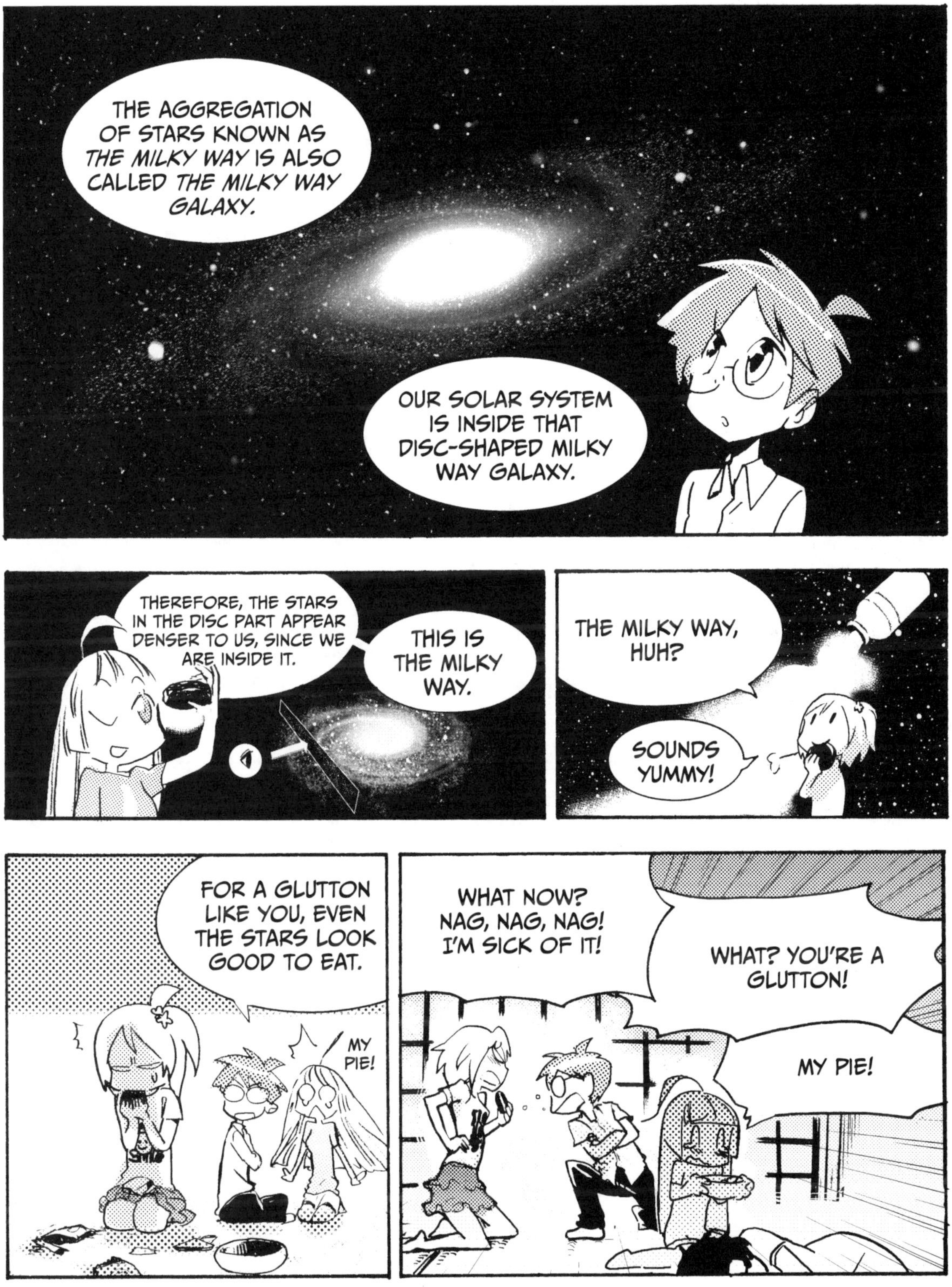
THE AGGREGATION OF STARS KNOWN AS *THE MILKY WAY* IS ALSO CALLED *THE MILKY WAY GALAXY*.
OUR SOLAR SYSTEM IS INSIDE THAT DISC-SHAPED MILKY WAY GALAXY.
THEREFORE, THE STARS IN THE DISC PART APPEAR DENSER TO US, SINCE WE ARE INSIDE IT.
THIS IS THE MILKY WAY.
THE MILKY WAY, HUH?
SOUNDS YUMMY!
FOR A GLUTTON LIKE YOU, EVEN THE STARS LOOK GOOD TO EAT.
MY PIE!
WHAT NOW? NAG, NAG, NAG! I'M SICK OF IT!
WHAT? YOU'RE A GLUTTON!
MY PIE!

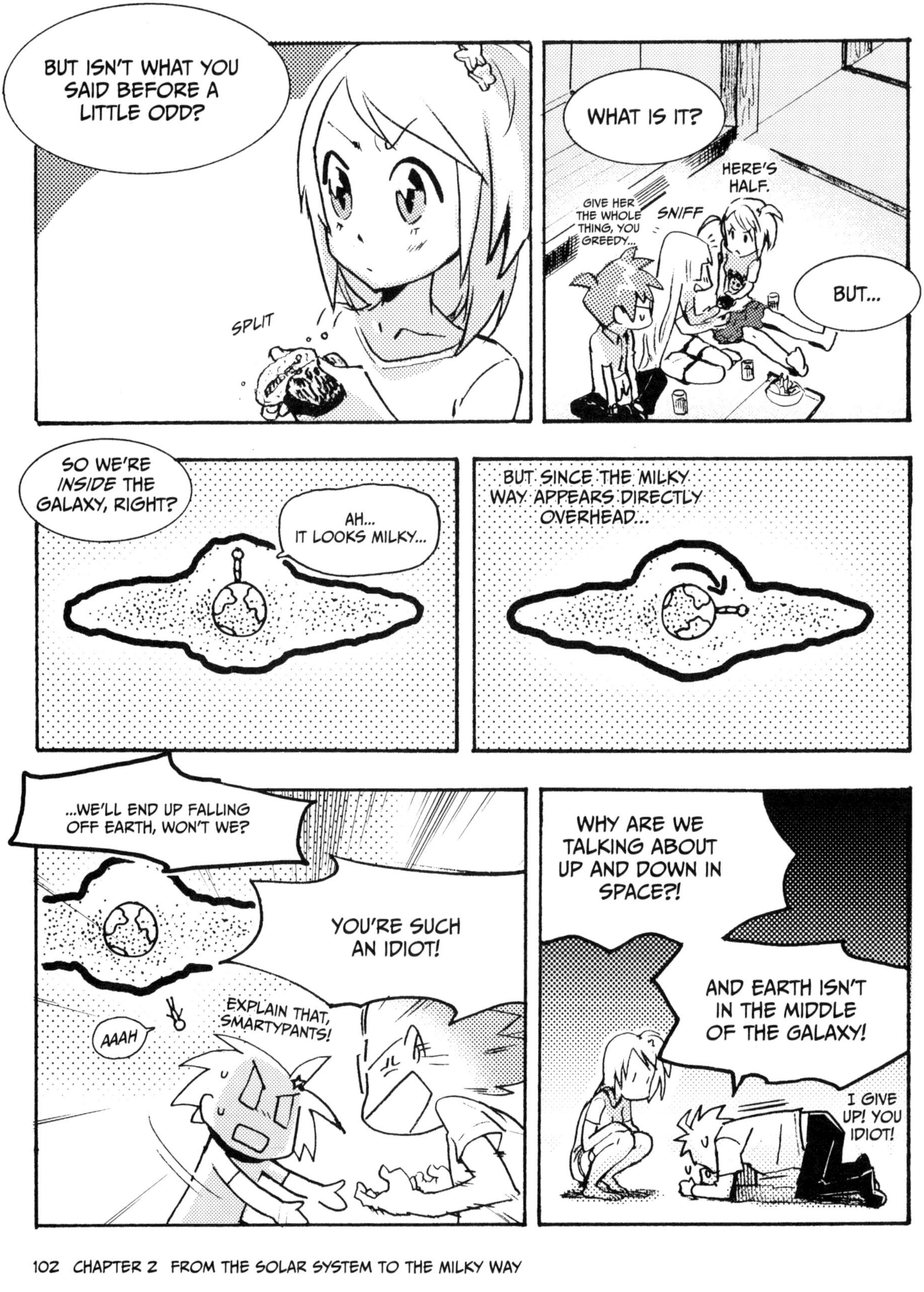
BUT ISN'T WHAT YOU SAID BEFORE A LITTLE ODD?
SPLIT
WHAT IS IT?
HERE'S HALF.
GIVE HER THE WHOLE THING, YOU GREEDY...
SNIFF
BUT...
SO WE'RE *INSIDE* THE GALAXY, RIGHT?
AH... IT LOOKS MILKY...
BUT SINCE THE MILKY WAY APPEARS DIRECTLY OVERHEAD...
...WE'LL END UP FALLING OFF EARTH, WON'T WE?
YOU'RE SUCH AN IDIOT!
EXPLAIN THAT, SMARTYPANTS!
AAAH
WHY ARE WE TALKING ABOUT UP AND DOWN IN SPACE?!
AND EARTH ISN'T IN THE MIDDLE OF THE GALAXY!
I GIVE UP! YOU IDIOT!

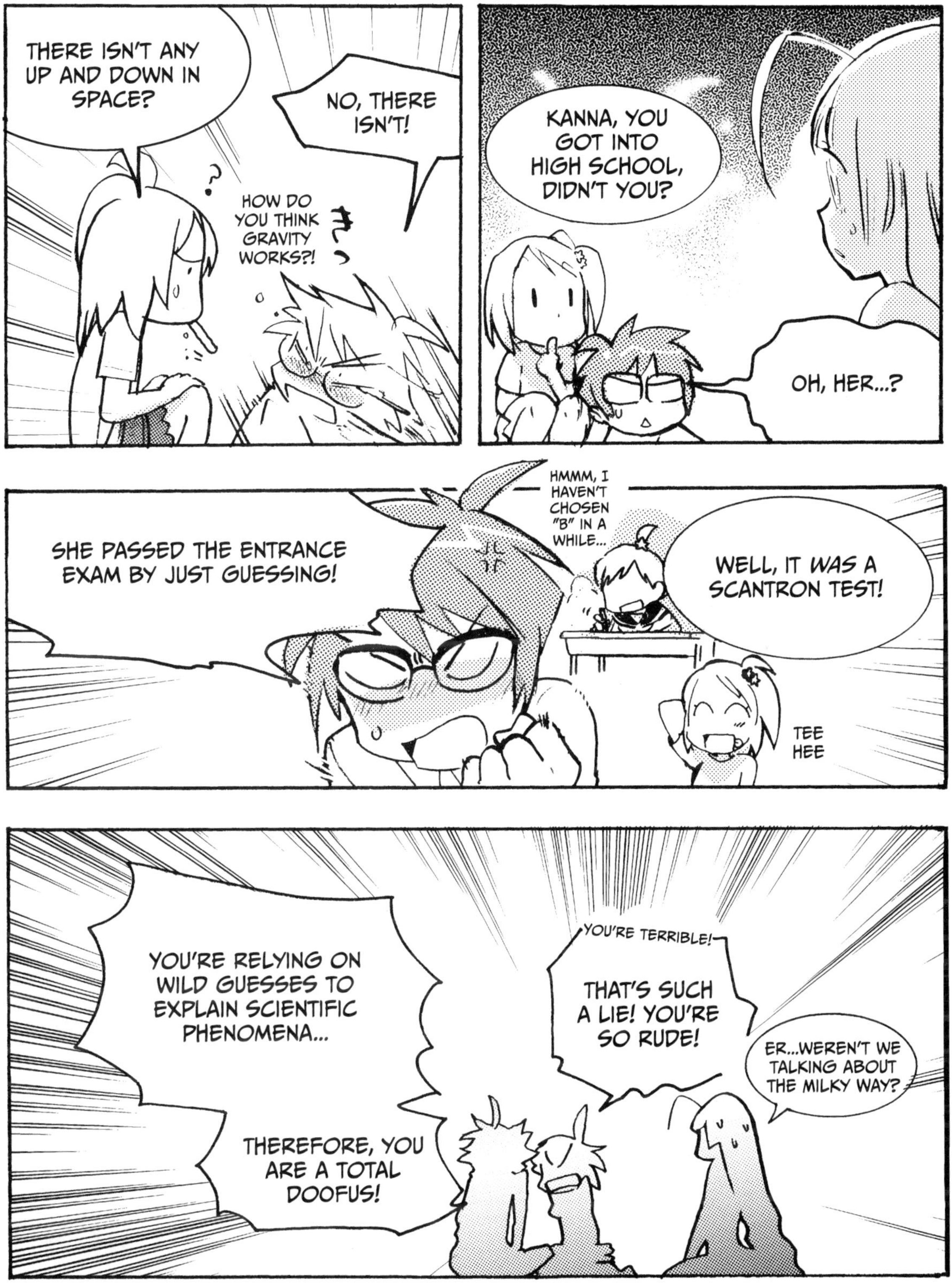
THERE ISN'T ANY UP AND DOWN IN SPACE?
NO, THERE ISN'T!
HOW DO YOU THINK GRAVITY WORKS?!
きっ
KANNA, YOU GOT INTO HIGH SCHOOL, DIDN'T YOU?
OH, HER...?
SHE PASSED THE ENTRANCE EXAM BY JUST GUESSING!
HMMM, I HAVEN'T CHOSEN "B" IN A WHILE...
WELL, IT *WAS* A SCANTRON TEST!
TEE HEE
YOU'RE RELYING ON WILD GUESSES TO EXPLAIN SCIENTIFIC PHENOMENA...
THEREFORE, YOU ARE A TOTAL DOOFUS!
YOU'RE TERRIBLE!
THAT'S SUCH A LIE! YOU'RE SO RUDE!
ER...WEREN'T WE TALKING ABOUT THE MILKY WAY?

THE SIZE OF THE MILKY WAY GALAXY

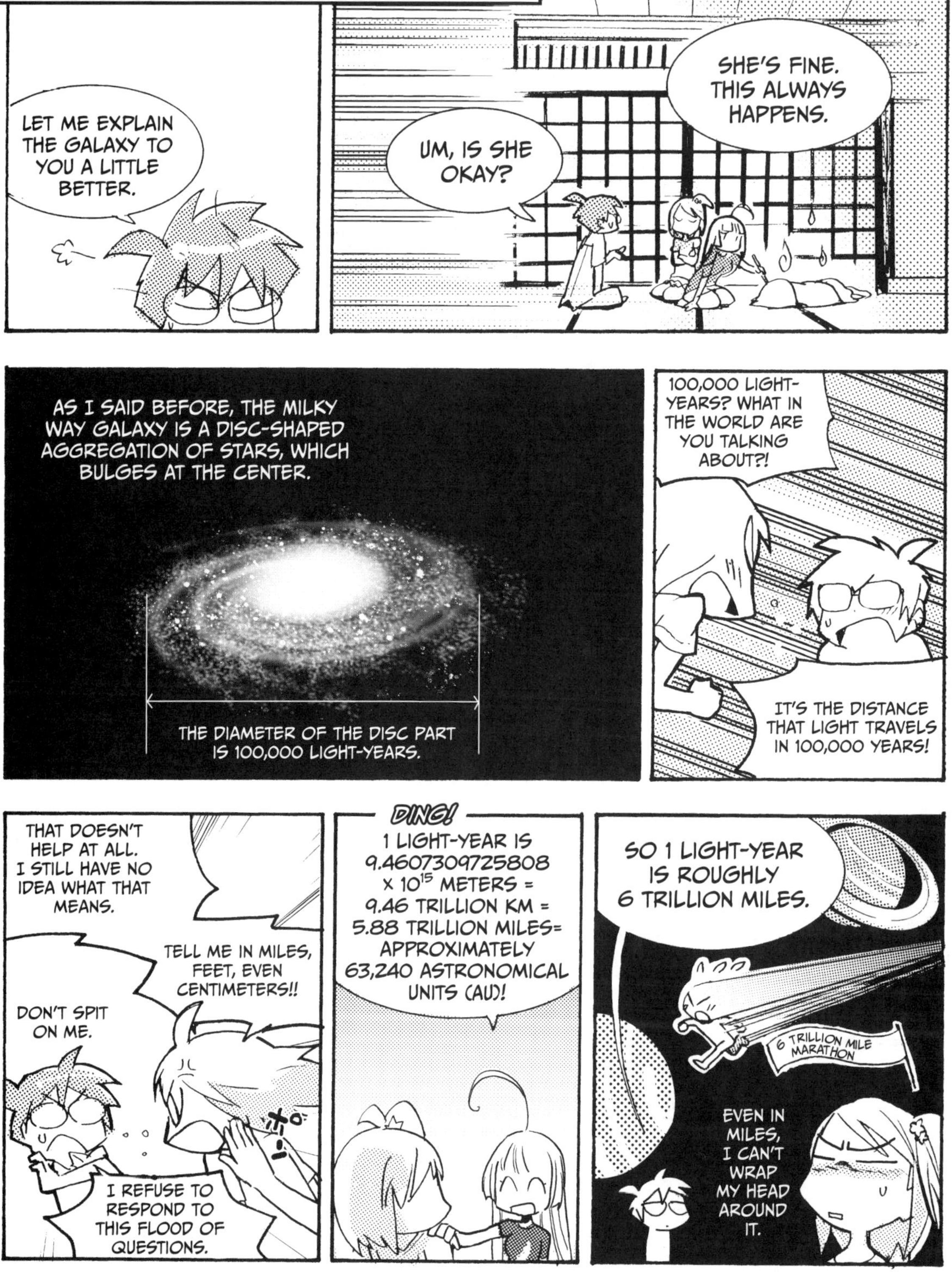

LET ME EXPLAIN THE GALAXY TO YOU A LITTLE BETTER.
UM, IS SHE OKAY?
SHE'S FINE. THIS ALWAYS HAPPENS.
AS I SAID BEFORE, THE MILKY WAY GALAXY IS A DISC-SHAPED AGGREGATION OF STARS, WHICH BULGES AT THE CENTER.
THE DIAMETER OF THE DISC PART IS 100,000 LIGHT-YEARS.
100,000 LIGHT-YEARS? WHAT IN THE WORLD ARE YOU TALKING ABOUT?!
IT'S THE DISTANCE THAT LIGHT TRAVELS IN 100,000 YEARS!
THAT DOESN'T HELP AT ALL. I STILL HAVE NO IDEA WHAT THAT MEANS.
TELL ME IN MILES, FEET, EVEN CENTIMETERS!!
DON'T SPIT ON ME.
I REFUSE TO RESPOND TO THIS FLOOD OF QUESTIONS.
DING!
1 LIGHT-YEAR IS 9.4607309725808 x 10^{15} METERS = 9.46 TRILLION KM = 5.88 TRILLION MILES= APPROXIMATELY 63,240 ASTRONOMICAL UNITS (AU)!
SO 1 LIGHT-YEAR IS ROUGHLY 6 TRILLION MILES.
6 TRILLION MILE MARATHON
EVEN IN MILES, I CAN'T WRAP MY HEAD AROUND IT.

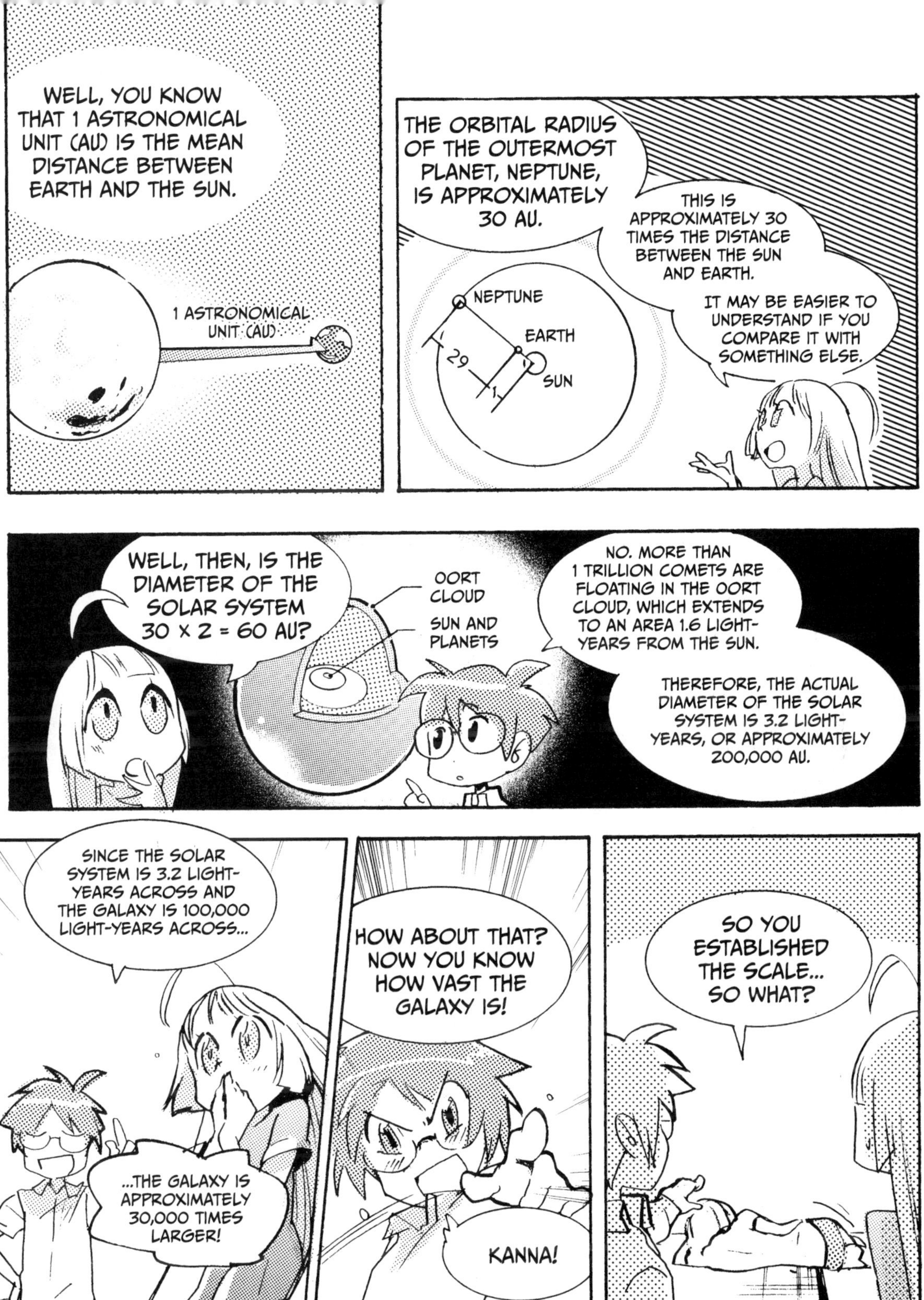
WELL, YOU KNOW THAT 1 ASTRONOMICAL UNIT (AU) IS THE MEAN DISTANCE BETWEEN EARTH AND THE SUN.
1 ASTRONOMICAL UNIT (AU)
THE ORBITAL RADIUS OF THE OUTERMOST PLANET, NEPTUNE, IS APPROXIMATELY 30 AU.
THIS IS APPROXIMATELY 30 TIMES THE DISTANCE BETWEEN THE SUN AND EARTH.
NEPTUNE
EARTH
SUN
29
1
IT MAY BE EASIER TO UNDERSTAND IF YOU COMPARE IT WITH SOMETHING ELSE.
WELL, THEN, IS THE DIAMETER OF THE SOLAR SYSTEM 30 × 2 = 60 AU?
OORT CLOUD
SUN AND PLANETS
NO. MORE THAN 1 TRILLION COMETS ARE FLOATING IN THE OORT CLOUD, WHICH EXTENDS TO AN AREA 1.6 LIGHT-YEARS FROM THE SUN.
THEREFORE, THE ACTUAL DIAMETER OF THE SOLAR SYSTEM IS 3.2 LIGHT-YEARS, OR APPROXIMATELY 200,000 AU.
SINCE THE SOLAR SYSTEM IS 3.2 LIGHT-YEARS ACROSS AND THE GALAXY IS 100,000 LIGHT-YEARS ACROSS...
...THE GALAXY IS APPROXIMATELY 30,000 TIMES LARGER!
HOW ABOUT THAT? NOW YOU KNOW HOW VAST THE GALAXY IS!
KANNA!
SO YOU ESTABLISHED THE SCALE... SO WHAT?

WHAT'S IN THE MIDDLE OF THE GALAXY?

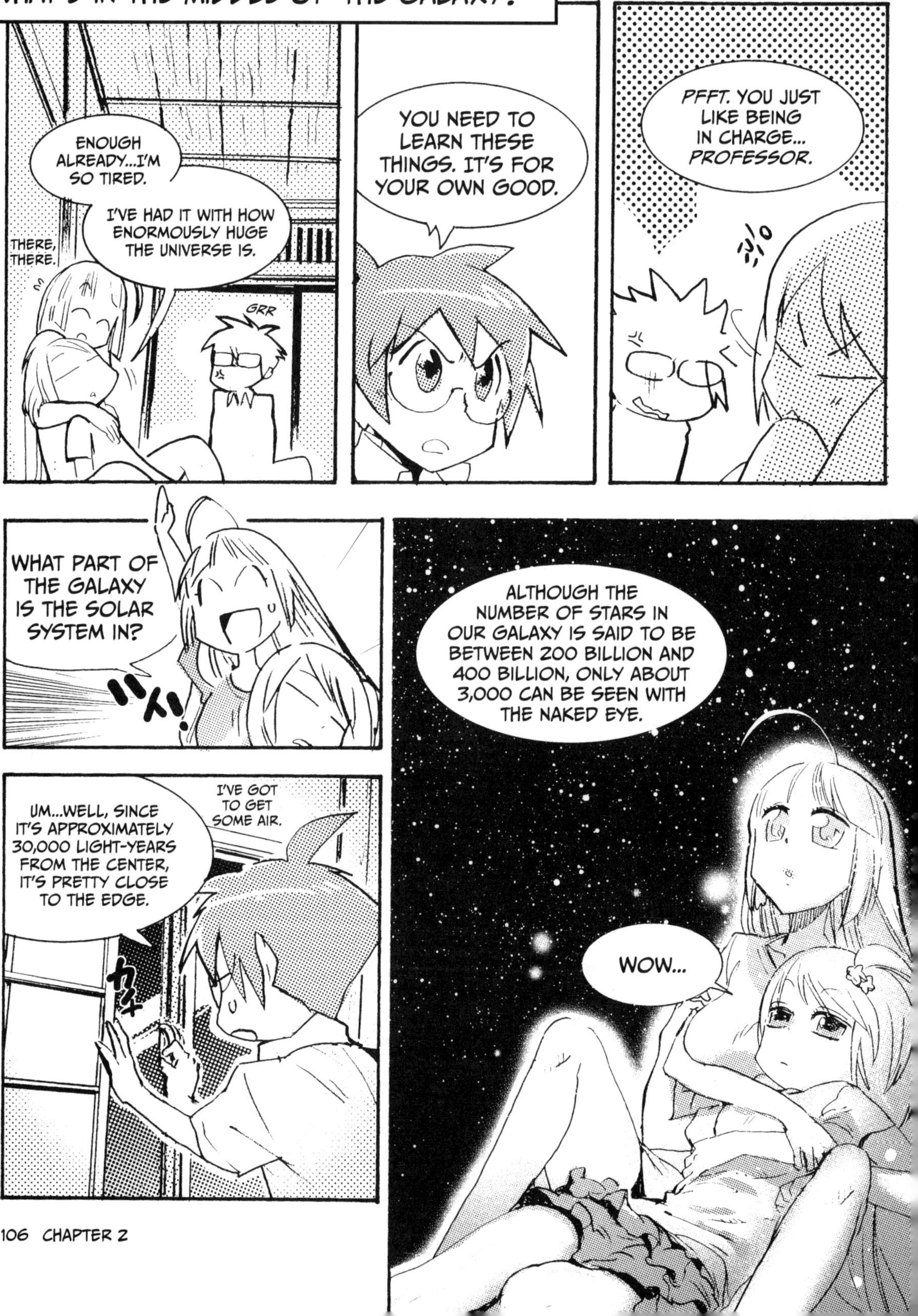
ENOUGH ALREADY...I'M SO TIRED.
THERE, THERE.
I'VE HAD IT WITH HOW ENORMOUSLY HUGE THE UNIVERSE IS.
GRR
YOU NEED TO LEARN THESE THINGS. IT'S FOR YOUR OWN GOOD.
PFFT. YOU JUST LIKE BEING IN CHARGE... PROFESSOR.
WHAT PART OF THE GALAXY IS THE SOLAR SYSTEM IN?
UM...WELL, SINCE IT'S APPROXIMATELY 30,000 LIGHT-YEARS FROM THE CENTER, IT'S PRETTY CLOSE TO THE EDGE.
I'VE GOT TO GET SOME AIR.
ALTHOUGH THE NUMBER OF STARS IN OUR GALAXY IS SAID TO BE BETWEEN 200 BILLION AND 400 BILLION, ONLY ABOUT 3,000 CAN BE SEEN WITH THE NAKED EYE.
WOW...

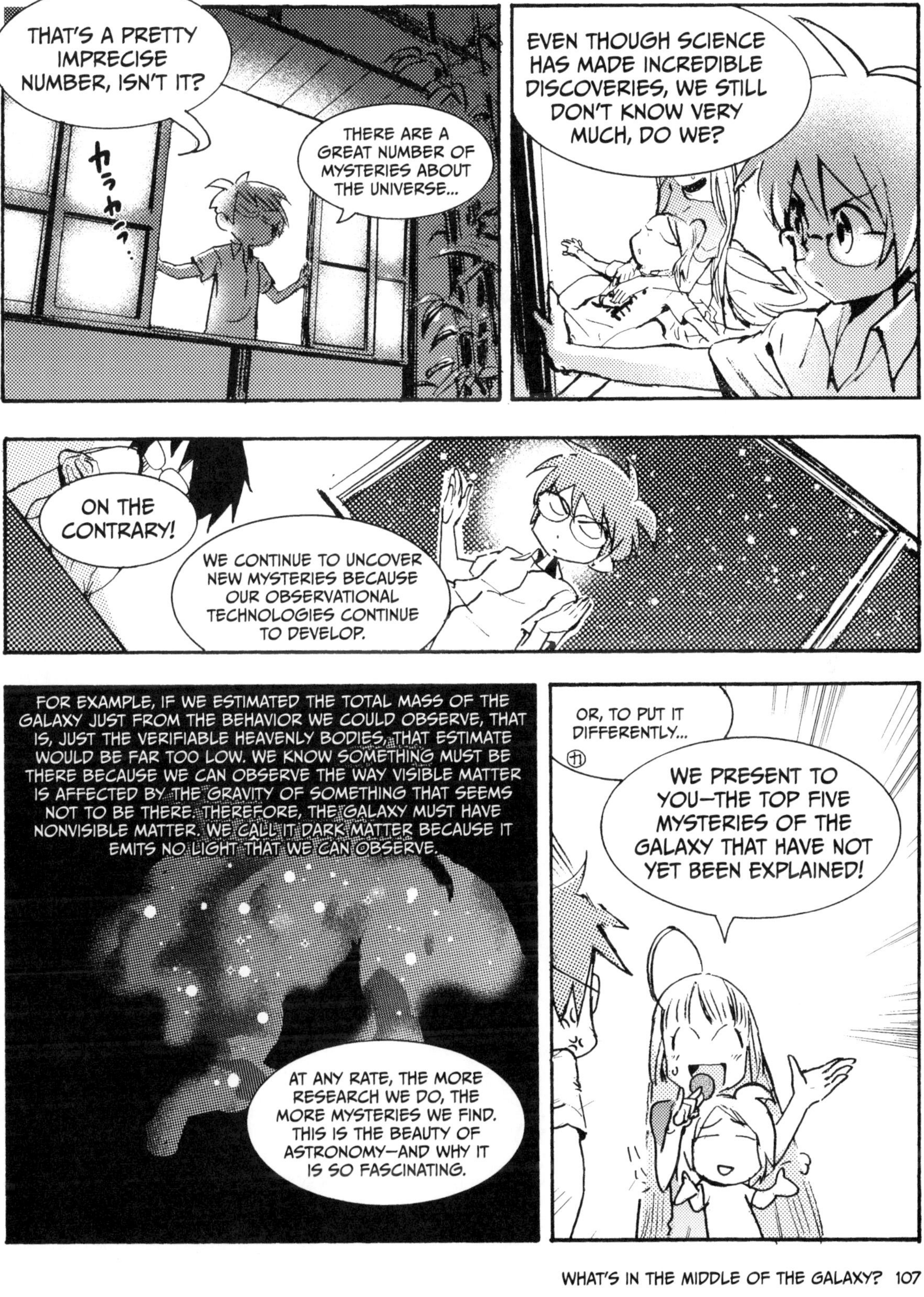
THAT'S A PRETTY IMPRECISE NUMBER, ISN'T IT?
THERE ARE A GREAT NUMBER OF MYSTERIES ABOUT THE UNIVERSE...
EVEN THOUGH SCIENCE HAS MADE INCREDIBLE DISCOVERIES, WE STILL DON'T KNOW VERY MUCH, DO WE?
ON THE CONTRARY!
WE CONTINUE TO UNCOVER NEW MYSTERIES BECAUSE OUR OBSERVATIONAL TECHNOLOGIES CONTINUE TO DEVELOP.
FOR EXAMPLE, IF WE ESTIMATED THE TOTAL MASS OF THE GALAXY JUST FROM THE BEHAVIOR WE COULD OBSERVE, THAT IS, JUST THE VERIFIABLE HEAVENLY BODIES, THAT ESTIMATE WOULD BE FAR TOO LOW. WE KNOW SOMETHING MUST BE THERE BECAUSE WE CAN OBSERVE THE WAY VISIBLE MATTER IS AFFECTED BY THE GRAVITY OF SOMETHING THAT SEEMS NOT TO BE THERE. THEREFORE, THE GALAXY MUST HAVE NONVISIBLE MATTER. WE CALL IT DARK MATTER BECAUSE IT EMITS NO LIGHT THAT WE CAN OBSERVE.
AT ANY RATE, THE MORE RESEARCH WE DO, THE MORE MYSTERIES WE FIND. THIS IS THE BEAUTY OF ASTRONOMY—AND WHY IT IS SO FASCINATING.
OR, TO PUT IT DIFFERENTLY...
WE PRESENT TO YOU—THE TOP FIVE MYSTERIES OF THE GALAXY THAT HAVE NOT YET BEEN EXPLAINED!

TOP FIVE MYSTERIES OF THE GALAXY THAT HAVE NOT YET BEEN EXPLAINED!

Astronomers and astrophysicists are still exploring our solar system and galaxy.

WHAT IS THE GALAXY'S SHAPE, AND HOW DID IT FORM?

Previously, the Milky Way was considered a spiral galaxy. However, American astronomers analyzed observational data from the Spitzer Space Telescope, which was launched by NASA in 2003, and concluded that a bar structure approximately 27,000 light-years long runs through the center of the Milky Way. A barred spiral galaxy is a type of spiral galaxy but one that looks as though the spiraling arms have been moved to the end of a bar of stellar and interstellar matter that goes right through the center of the galaxy.

Therefore, the theory that the Milky Way is a barred spiral galaxy is now dominant. Some research suggests that the barred shape may be an indicator of more mature galaxies; however, we still don't know why or how spiral galaxies and barred spiral galaxies are formed.

Spiral Galaxies
Elliptical Galaxies
Barred Spiral Galaxies

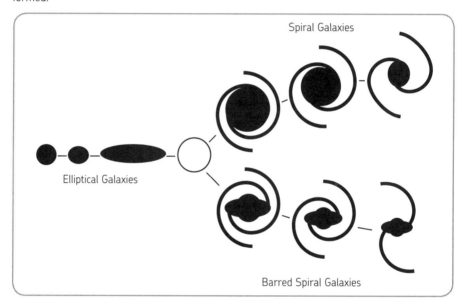

Types of galaxies

WHAT'S AT THE CENTER?

Our solar system is an aggregation of heavenly bodies gravitationally collected by the star known as the Sun. So what is at the center of the galaxy, the thing that gathered it all together? The answer is still not definitively known. The enormous mass of the galaxy, which is between 300 billion and 3 trillion times the mass of the Sun, cannot be accounted for by normal stars, so the galaxy must contain some kind of extremely heavy but small heavenly body at its center. Currently, the only plausible solution is a black hole.

A *black hole* is a region in which the force of gravity is so great that even light cannot escape from it. Some scientists propose that a supermassive black hole is at the center of the galaxy.

HOW ARE SUPERMASSIVE BLACK HOLES FORMED?

Although there are smaller black holes with approximately the same mass as a star, black holes like the one hypothesized to be in the center of the galaxy are enormous, having a mass from several million to several hundred million times the mass of the Sun. It is still not known how or why these supermassive black holes are formed.

A stellar-mass black hole is believed to be formed from the remnants of a star. When a supernova explosion of a large star at least 20 times the size of the Sun occurs, the core that remains continues to be compressed by gravitational collapse, creating a black hole. It seems reasonable that a coalescence of smaller-sized black holes or other heavenly bodies would lead to a larger black hole.

Since such intermediate-sized black holes (ISBH) have not been found so far, scientists aren't sure how they are created. However, an explanation may not be far off, since X-ray sources from nearby galaxies seem to be ISBHs. The existence of these ISBHs cannot be confirmed, though, until a mass measurement has been made using the gravitational effect the ISBH has on surrounding bodies.

WHAT IS THE GALAXY MADE OF?

Although the mass of the galaxy is thought to be equivalent to 600 billion to 1 trillion Suns (based on motion analysis), all of the heavenly bodies that can be observed by telescopes and radio telescopes combined do not account for more than 10 percent of it. This is the same for other galaxies and galaxy groups, and astronomers currently think that more than 90 percent of what forms the universe is dark matter that cannot be observed because it does not emit or reflect light.

Opinions on what dark matter is made of range from *neutrinos* (a type of elementary particle) or some unknown elementary particle to black holes. The discovery of the composition of dark matter will surely merit a Nobel Prize.

Incidentally, the actual condition of more than 70 percent of the energy that exists in the universe is unknown, and this is referred to as *dark energy*.

WHAT WILL HAPPEN WHEN WE COLLIDE WITH THE ANDROMEDA GALAXY?

The Andromeda galaxy is close to the Milky Way and is known to be approaching at a speed of 100 km per second (62 miles per second). Since its current distance is approximately 2.52 million light-years away, if it continues approaching at the current rate, the two galaxies should collide in 7 to 8 billion years. What will happen at that time is not well understood.

Of course, individual heavenly bodies are quite far apart even inside a galaxy,* and although it seems unlikely that stars will actually collide, various predictions have been made as to what might occur. These include that the two galaxies will combine to form one new galaxy and that there will be enormous gravitational effects accompanying the collision.

---

* The closest nearby star, Proxima Centauri, is 4.2 light-years away from Earth (approximately 25 trillion miles).

THE MILKY WAY GALAXY IS ONE OF MANY GALAXIES

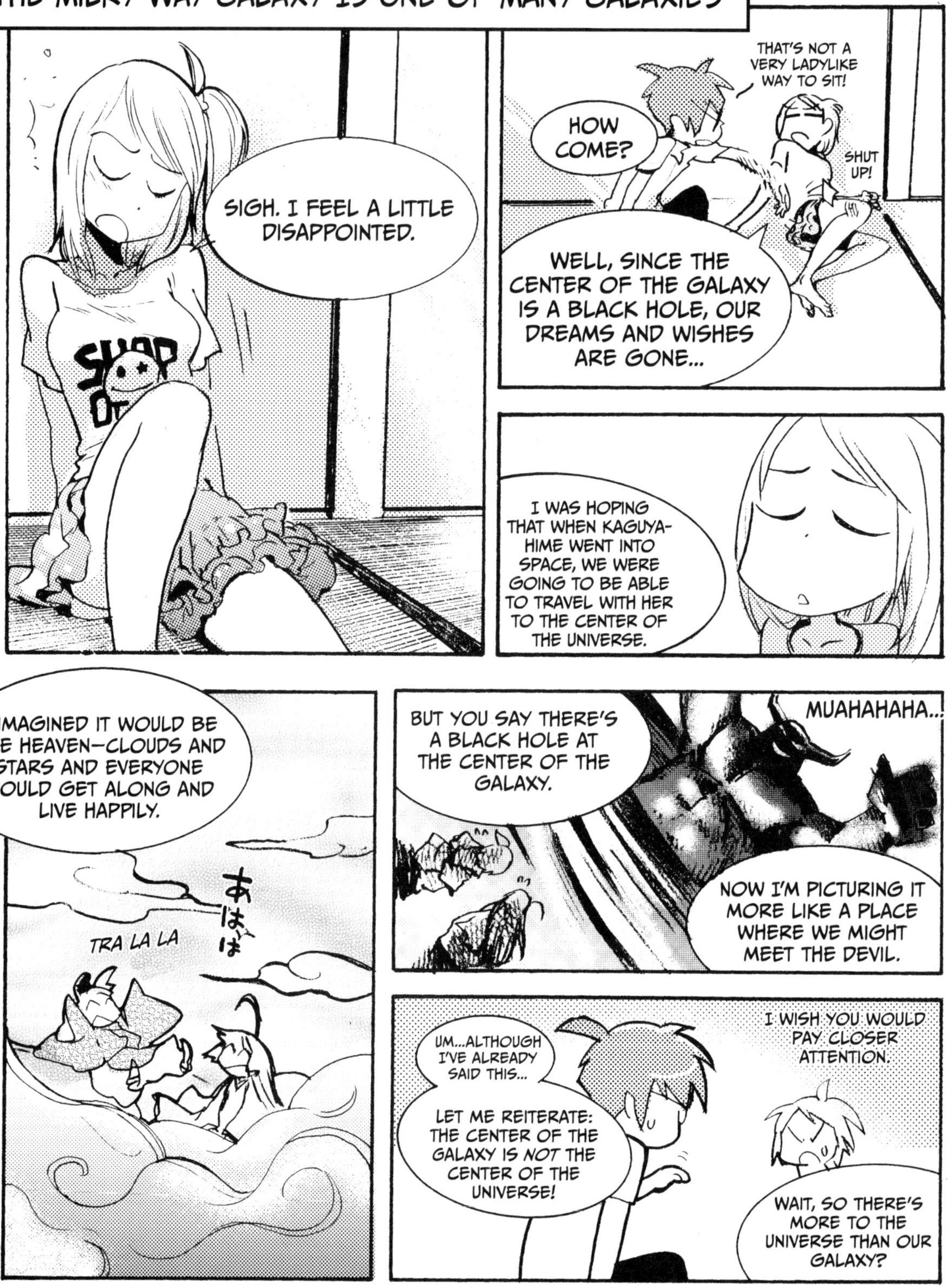
SIGH. I FEEL A LITTLE DISAPPOINTED.
THAT'S NOT A VERY LADYLIKE WAY TO SIT!
HOW COME?
SHUT UP!
WELL, SINCE THE CENTER OF THE GALAXY IS A BLACK HOLE, OUR DREAMS AND WISHES ARE GONE...
I WAS HOPING THAT WHEN KAGUYA-HIME WENT INTO SPACE, WE WERE GOING TO BE ABLE TO TRAVEL WITH HER TO THE CENTER OF THE UNIVERSE.
I IMAGINED IT WOULD BE LIKE HEAVEN—CLOUDS AND STARS AND EVERYONE WOULD GET ALONG AND LIVE HAPPILY.
あはは
TRA LA LA
BUT YOU SAY THERE'S A BLACK HOLE AT THE CENTER OF THE GALAXY.
MUAHAHAHA...
NOW I'M PICTURING IT MORE LIKE A PLACE WHERE WE MIGHT MEET THE DEVIL.
UM...ALTHOUGH I'VE ALREADY SAID THIS...
LET ME REITERATE: THE CENTER OF THE GALAXY IS *NOT* THE CENTER OF THE UNIVERSE!
I WISH YOU WOULD PAY CLOSER ATTENTION.
WAIT, SO THERE'S MORE TO THE UNIVERSE THAN OUR GALAXY?

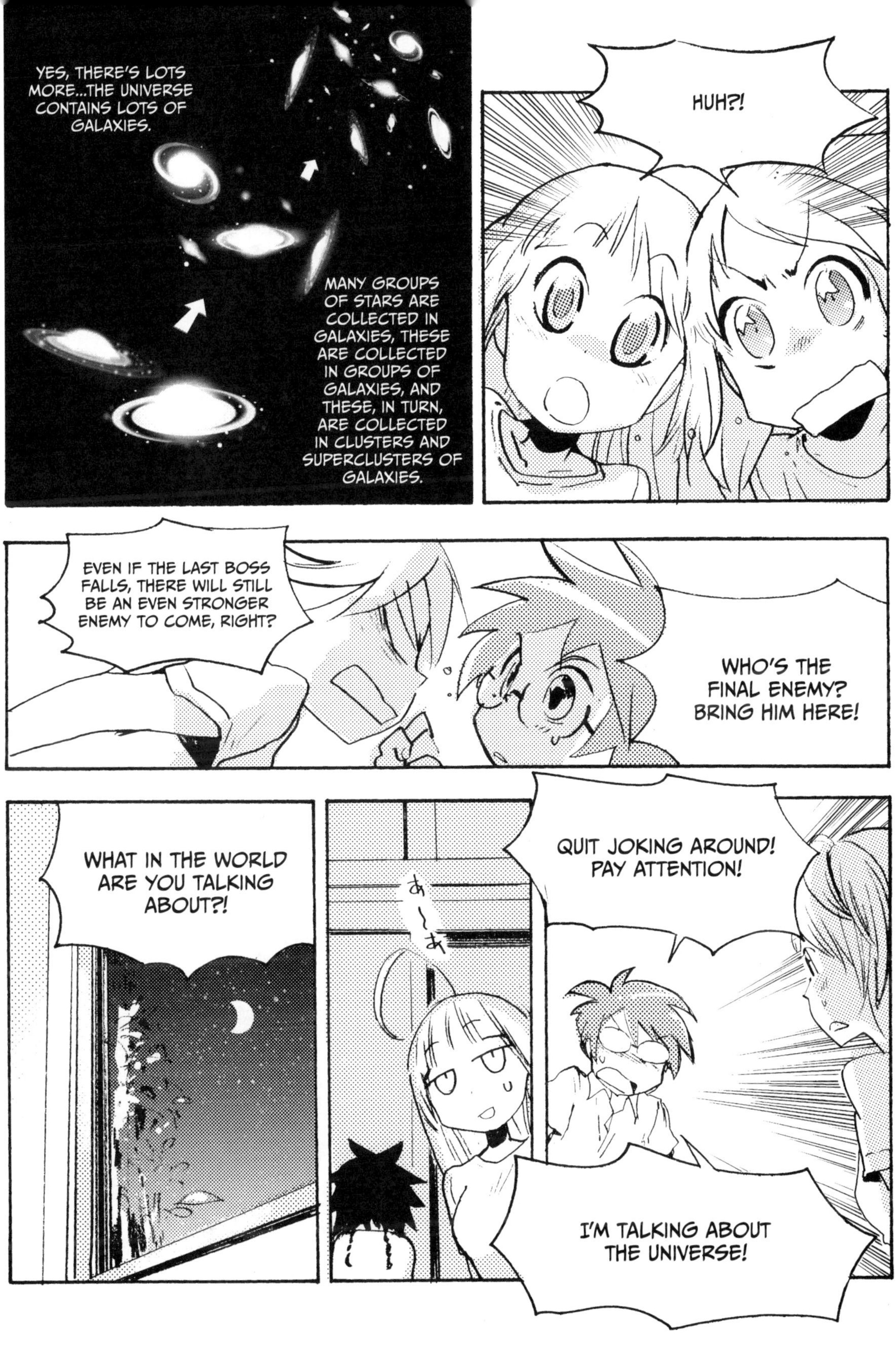
YES, THERE'S LOTS MORE...THE UNIVERSE CONTAINS LOTS OF GALAXIES.
MANY GROUPS OF STARS ARE COLLECTED IN GALAXIES, THESE ARE COLLECTED IN GROUPS OF GALAXIES, AND THESE, IN TURN, ARE COLLECTED IN CLUSTERS AND SUPERCLUSTERS OF GALAXIES.
HUH?!
EVEN IF THE LAST BOSS FALLS, THERE WILL STILL BE AN EVEN STRONGER ENEMY TO COME, RIGHT?
WHO'S THE FINAL ENEMY? BRING HIM HERE!
WHAT IN THE WORLD ARE YOU TALKING ABOUT?!
あ〜あ
QUIT JOKING AROUND! PAY ATTENTION!
I'M TALKING ABOUT THE UNIVERSE!

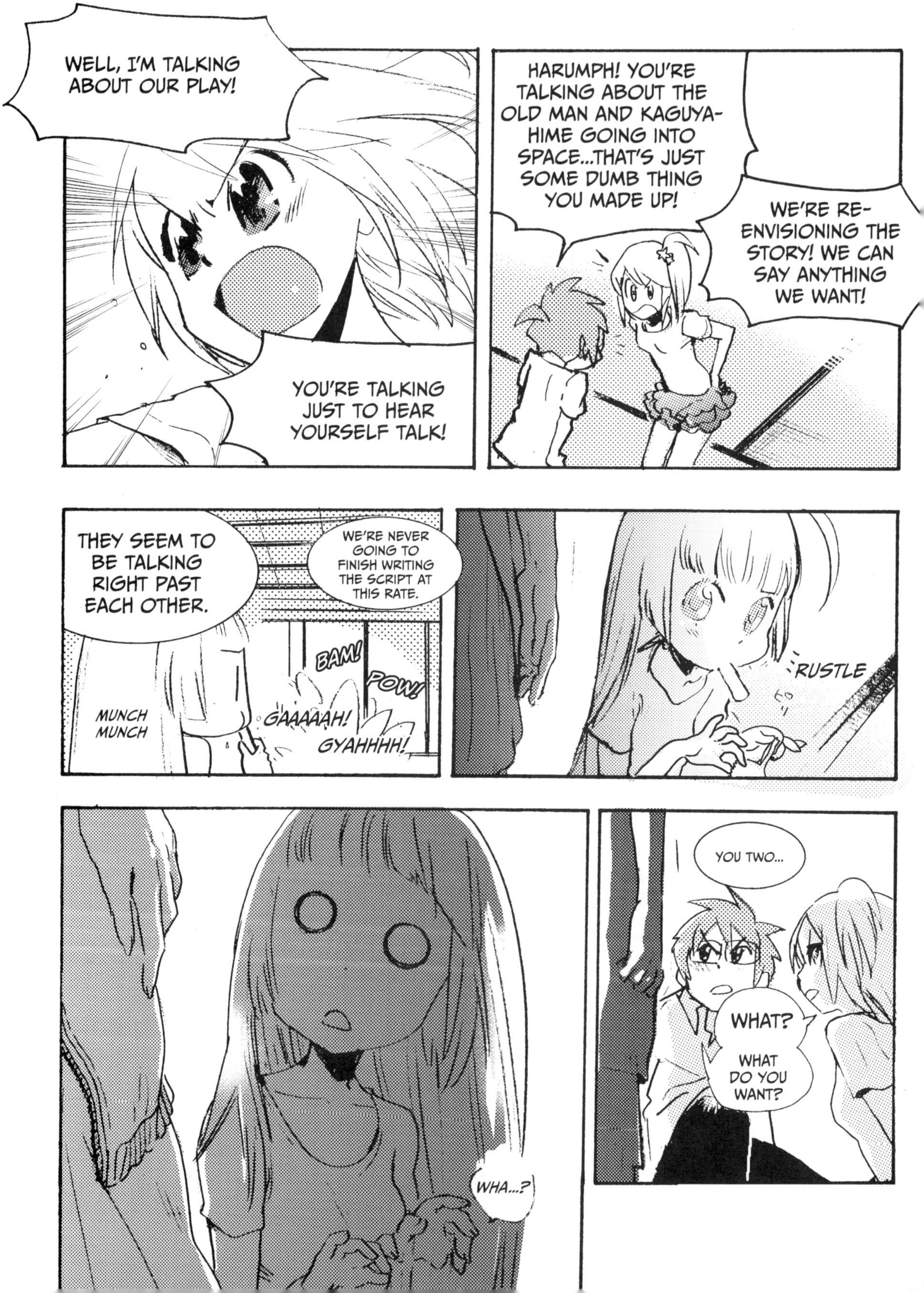
WELL, I'M TALKING ABOUT OUR PLAY!
YOU'RE TALKING JUST TO HEAR YOURSELF TALK!
HARUMPH! YOU'RE TALKING ABOUT THE OLD MAN AND KAGUYA-HIME GOING INTO SPACE...THAT'S JUST SOME DUMB THING YOU MADE UP!
WE'RE RE-ENVISIONING THE STORY! WE CAN SAY ANYTHING WE WANT!
THEY SEEM TO BE TALKING RIGHT PAST EACH OTHER.
WE'RE NEVER GOING TO FINISH WRITING THE SCRIPT AT THIS RATE.
BAM!
POW!
MUNCH MUNCH
GAAAAAH!
GYAHHHH!
RUSTLE
WHA...?
YOU TWO...
WHAT?
WHAT DO YOU WANT?

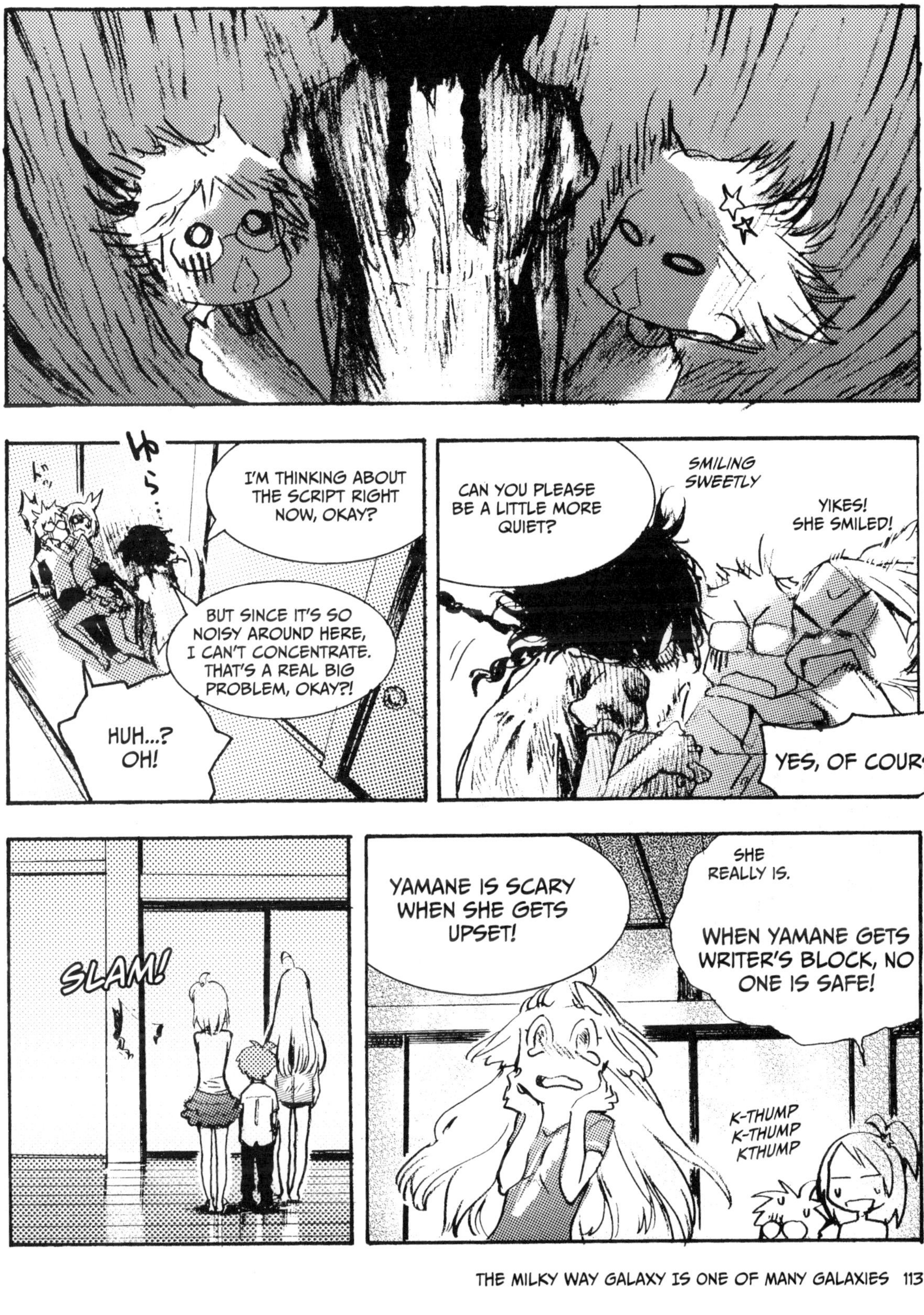
ゆら
I'M THINKING ABOUT THE SCRIPT RIGHT NOW, OKAY?
BUT SINCE IT'S SO NOISY AROUND HERE, I CAN'T CONCENTRATE. THAT'S A REAL BIG PROBLEM, OKAY?!
HUH...? OH!
CAN YOU PLEASE BE A LITTLE MORE QUIET?
SMILING SWEETLY
YIKES! SHE SMILED!
YES, OF COURSE!
SLAM!
YAMANE IS SCARY WHEN SHE GETS UPSET!
SHE REALLY IS.
WHEN YAMANE GETS WRITER'S BLOCK, NO ONE IS SAFE!
K-THUMP K-THUMP KTHUMP

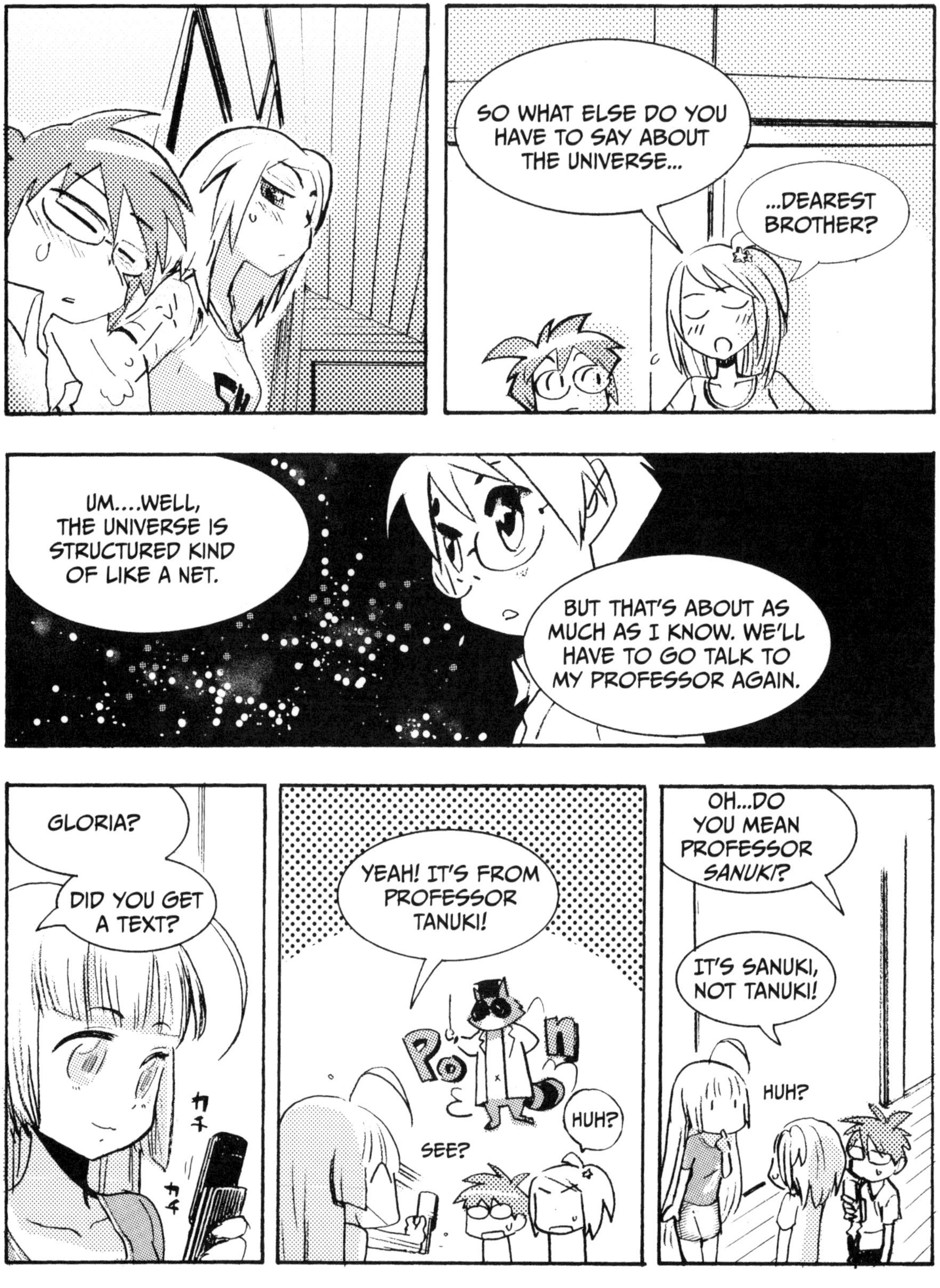
SO WHAT ELSE DO YOU HAVE TO SAY ABOUT THE UNIVERSE...
...DEAREST BROTHER?
UM....WELL, THE UNIVERSE IS STRUCTURED KIND OF LIKE A NET.
BUT THAT'S ABOUT AS MUCH AS I KNOW. WE'LL HAVE TO GO TALK TO MY PROFESSOR AGAIN.
GLORIA?
DID YOU GET A TEXT?
カチ
カチ
YEAH! IT'S FROM PROFESSOR TANUKI!
Pon
HUH?
SEE?
OH...DO YOU MEAN PROFESSOR SANUKI?
IT'S SANUKI, NOT TANUKI!
HUH?

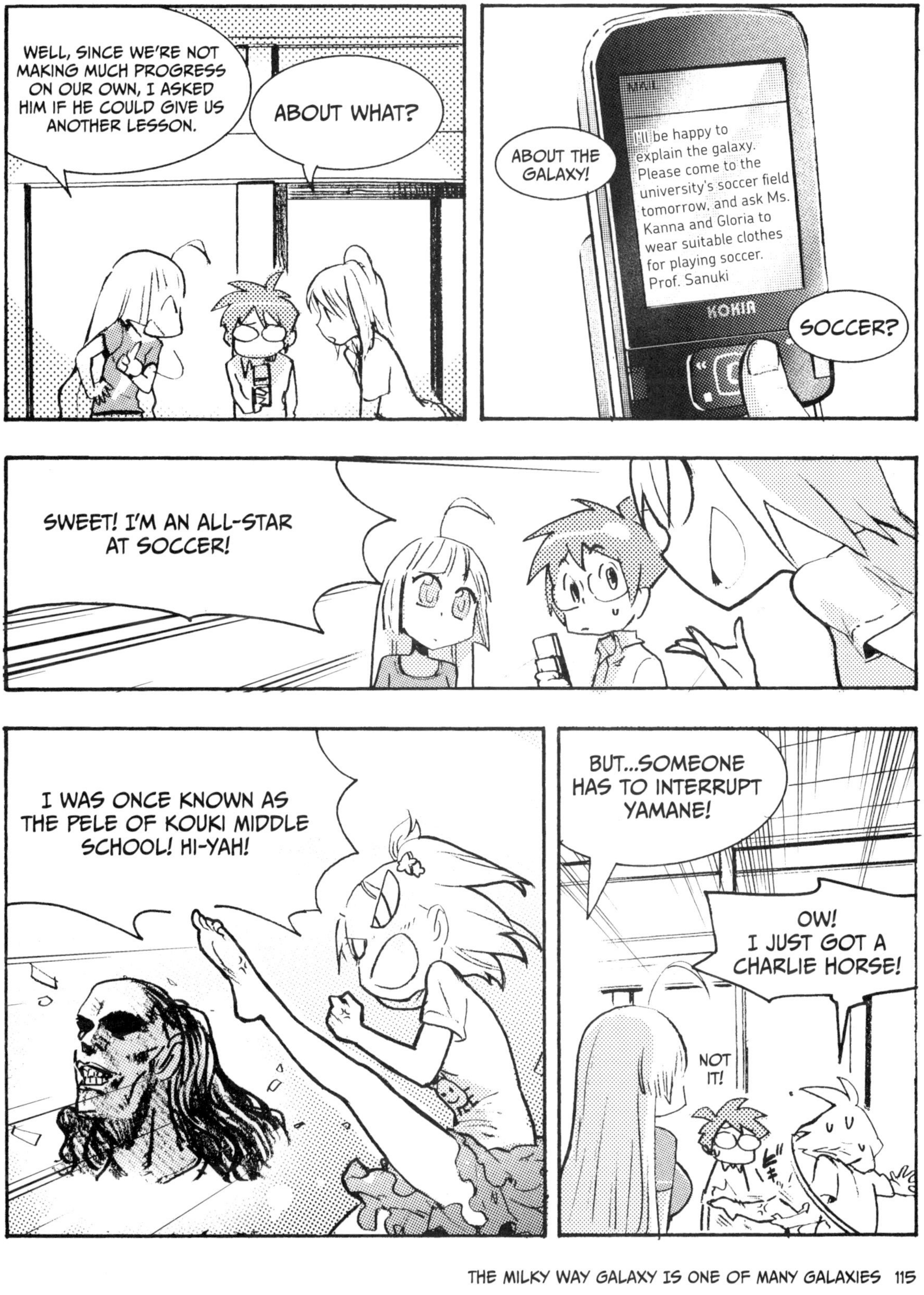
WELL, SINCE WE'RE NOT MAKING MUCH PROGRESS ON OUR OWN, I ASKED HIM IF HE COULD GIVE US ANOTHER LESSON.
ABOUT WHAT?
ABOUT THE GALAXY!
I'll be happy to explain the galaxy. Please come to the university's soccer field tomorrow, and ask Ms. Kanna and Gloria to wear suitable clothes for playing soccer. Prof. Sanuki
SOCCER?
SWEET! I'M AN ALL-STAR AT SOCCER!
I WAS ONCE KNOWN AS THE PELE OF KOUKI MIDDLE SCHOOL! HI-YAH!
BUT...SOMEONE HAS TO INTERRUPT YAMANE!
OW! I JUST GOT A CHARLIE HORSE!
NOT IT!

THE UNIVERSE IS STEADILY GETTING LARGER

Whether they believed in the geocentric theory or the heliocentric theory, people in the 16th century thought the universe consisted only of Earth, the Moon, the Sun, planets of the solar system, and many other stars. The only things moving independently in the sky were the planets—people thought the other stars were just affixed to the celestial sphere like lanterns hanging from a wall in the background, forming various constellations.

However, Galileo Galilei felt there were problems with that view of the universe. He used a telescope he had built himself to gaze at the sky every night, and in 1609, he discovered that the Milky Way was an uncountable aggregation of stars.

Before there were telescopes, people thought that the Milky Way was either a cloud or something that was flowing across the celestial sphere. The ancient Greek philosopher Democritus (about 460 BC–370 BC) advocated the theory that the Milky Way was a collection of faraway stars. Although it is unclear whether he arrived at this conclusion as the result of logical thought or whether he had extraordinarily good eyesight, we can guess what his reasoning must have been: If the Milky Way were like a gaseous cloud or river, its position or shape would vary with time. However, for as long as it has been observed, its state has not changed, just like the constellations. Therefore, it's natural to believe that it is also a collection of stars.

The Milky Way

Of course, making an inference is no way to prove a scientific theory, but Galileo was finally able to observe that this was the case, approximately 1,200 years after Democritus.

WHY CAN WE SEE THE MILKY WAY?

If the Milky Way were assumed to be an aggregation of many stars, the next obligation of scientists should have been to determine its structure. Both Galileo and Kepler must have been too busy advocating the heliocentric theory, or else they just never thought much about this point. So let's try to think about it a little bit ourselves. This will, of course, be an experiment with no prior knowledge required.

Here's what we know: Although the Milky Way is a collection of stars, the individual stars that comprise it don't shine brightly enough for us to be able to pick them out individually with the naked eye. So the Milky Way looks like a cloud to us. This means that the stars in the Milky Way are either smaller or farther away than the other stars that we can see from Earth.

From a common-sense point of view, the theory that small stars are collected together in one specific location seems improbable. It is hard to logically explain why stars of a certain size would be gathered together along a band at a certain location.

Therefore, we should probably consider the theory that the stars constituting the Milky Way are farther away from Earth than the other stars are. From there, we can come up with some ideas about what the structure of the Milky Way might look like.

We can start by thinking of the solar system as a sphere inside a larger sphere of stars. The Milky Way galaxy surrounds that sphere of stars near our Sun.

Sphere of Fixed Stars
Milky Way
Solar System

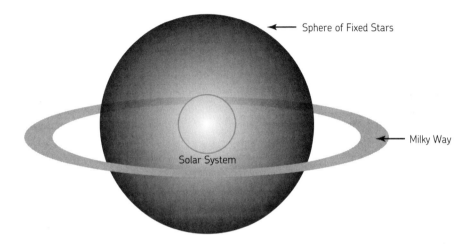

A faulty model of the structure of the Milky way, based only on what we can see from Earth

A DISC-SHAPED GALACTIC MODEL IS THE EASIEST TO UNDERSTAND

There were certainly people who believed in this kind of model of the universe, even during Galileo's time. However, if you look carefully, a better idea will reveal itself.

The biggest defect in this model is that the sphere of stars and the Milky Way are separated. It is difficult to explain why this would occur.

What happens if we try to attach these two kinds of stars to each other? If we actually bring the sphere of stars and the Milky Way together to make a disc-shaped structure, a belt-shaped flow of stars should be visible in the celestial sphere—that is, precisely the current model of the galaxy.

Although the structure of the galaxy shown in the following figure became clearer in the 19th and 20th centuries—nearly 200 years after Galileo lived—it seems that there were people who believed that the universe was disc shaped even before that.

Here are a few other hypotheses to consider:

- Since Earth and the solar system are revolving, perhaps the universe is also revolving.
- If matter spreads out while rotating, it is likely that the universe will become disc shaped.

This second hypothesis is actually easy to understand and not unreasonable—it's essentially the same phenomenon you see when someone tosses spinning pizza dough into the air.

Diameter is approximately 100,000 light-years
Bulge
Thickness is approximately 15,000 light-years
Position of the Sun
Length is approximately 28,000 light-years
Galactic disc
Central core of the Milky Way galaxy

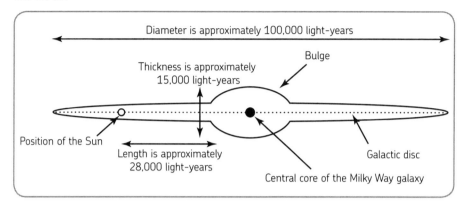

Model of the galaxy that is currently accepted

RESULTS OF SCIENTIFIC OBSERVATION ALSO PROVE A DISC-SHAPED UNIVERSE

Until now, we have guessed at the structure of the universe based on the beliefs of people from the 17th to 19th centuries. However, one person who attempted to elucidate the shape of the galaxy from actual observational results was the German-born astronomer Frederick William Herschel (1738–1822).

His method is extremely easy to understand. From the observable sky, he sampled areas, which he called "blocks," each of which was a section of sky equal to the area covered by approximately 1/4 of the full moon. He used a telescope to count the number of stars in 683 "blocks" in various locations around the night sky. Although this was only 0.1 percent of the area of the entire sky, this technique is believed to be statistically reliable.

The number of stars that humans can see at night with just the naked eye depends on a lot of factors—where you are on the planet, what season it is, whether or not you are in a city, whether or not the Moon is out, pollution, and more. Stars are measured on what's called a magnitude scale. Very dim stars have very high magnitudes (the dimmest stars are magnitude 20) and really bright stars have very low magnitudes (magnitude -26 being the brightest star we can see, our Sun).

Humans cannot see stars dimmer than magnitude 6. The total number of magnitude 6 or less stars is approximately 8,600. But that is for the entire planet! If you are in New York, you can't see stars that are over Japan! So the average night sky for Kanna, Yamane, and Gloria (or for you and your friends) seems to have approximately 2,000 stars visible to the naked eye. Herschel was using telescopes that he designed, though. With telescopes, we can see stars much dimmer than magnitude 6. Although no record remains of the total number of stars Herschel counted, it was certainly more than 10,000. That was an enormous amount of work.

When we look at the Milky Way, we see that the stars of the universe seem to have created a cluster that has some kind of structure. Herschel made a superhuman effort in counting stars in an attempt to identify this structure of our galaxy.

In most cases, the contents of clusters in the natural world are homogeneously scattered. Therefore, Herschel assumed the stars also would exist uniformly with almost the same density in the cluster.

Assuming in addition that all stars had the same brightness and were evenly scattered about the galaxy, Herschel postulated that the more stars he counted in a given block of space, the farther away those stars had to be.

Think of it this way: Van Gogh used many small dots to paint his famous self-portrait. If you stood really close to that painting and looked at only 1 square inch of it, you would only be able to see a few points. But if you backed up a few feet, then looked at 1 square inch of the painting, you would see more points. In fact, it would be hard to distinguish the separate points from that distance. So for Herschel, the more stars that were in his "1 square inch," the farther away he was from those stars.

After this enormous effort, Herschel created the model of the universe shown in the figure below.

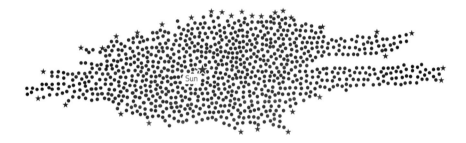

Model of the universe created by Herschel

Actually, since parts of the sky that were hidden by dust could not be verified with the naked eye, the shape Herschel came up with was somewhat odd. He ended up estimating that the size of the galaxy was quite small—less than 1/10 the size of the actual galaxy—but he did conclude that the galaxy was disc shaped. His results certainly contributed significantly to the development of cosmology.

AN IDEA FROM KANT ENLARGED THE PERCEIVED UNIVERSE IN A FLASH

The shape of the galaxy came to be vaguely understood as a result of Herschel's efforts. However, since he believed, without foundation, that the solar system was at the center of the galaxy and thought that the stars that could be observed from Earth (which we now know are our galaxy) made up the entire universe, he was unable to consider a more extensive universe. But then, since this was the established paradigm for astronomers at the time, Herschel alone should not take all the blame.

However, at the same time, someone did come close to the truth of the universe from a completely different viewpoint. This was the German-born, world-renowned philosopher Immanuel Kant (1728–1804).

Kant brought about a revolution in epistemology by providing a rebuttal to the existing assertion of empiricists at that time—that all knowledge and concepts held by mankind arise through experience. Kant asserted that cognitive contents provided through experience are processed intellectually, enabling people to continue to gain further knowledge or new concepts (this is considered to be a synthesis of rationalism and empiricism).

Kant also turned his thoughts toward the universe. He became aware of the disc-shaped model of the galaxy early on and wrote that the reason systems like the Milky Way galaxy can be seen is that stars are often organized in a lens-shaped pattern. Incidentally, Herschel may have read this and may have begun counting stars to try to scientifically prove Kant's idea, but we don't know for sure.

Kant's greatest insight related to understanding the universe is his hypothesis that the entire universe contains many "island universes," which are systems (or collections) of stars like the many islands in the ocean. The stars that mankind had observed until then were just one island universe called the Milky Way. Kant suggested that there were also countless other similar island universes, which together constituted the whole universe.

In the latter half of the 18th century, when Kant was active, many other nonstellar heavenly bodies came to be known because of advances in observation technology. These were named *nebulas* because they shone faintly like clouds.

For example, the "Great Andromeda Nebula" and the Large and Small Magellanic Clouds, which are classified today as galaxies, exist in ancient records, since they could be seen even by the naked eye. However, soon after telescopes were available, it became apparent that these objects that were thought to be cosmic clouds were actually uncountable collections of stars.

If the Milky Way is considered to be a collection of stars and the galactic structure is explained on that basis, then a nebula is also probably a cluster of stars. Kant's theory was really quite insightful.*

HOW DID TECHNOLOGY FOR OBSERVING THE UNIVERSE PROGRESS?

Mankind's conception of the universe has changed and developed over time based on observational science as well as logical reasoning. The results of astronomy throughout the 19th and 20th centuries, after Herschel, are presented in Chapter 3. This section gives a simple overview of the technology we use today to generate new observational data.

We've already mentioned that telescopes were invented at the beginning of the 17th century and that Galileo built one himself and made many discoveries with it. Amazingly, given how crude such early telescopes were, Galileo was able to observe some of the faint stars composing the Milky Way galaxy and thus prove Democritus's theory that the Milky Way was an aggregation of stars. In 1612, the German astronomer Simon Marius observed the Andromeda galaxy (in those days, the "Great Andromeda Nebula"), which appears next to the Milky Way in the night sky, but he still did not recognize that it was an aggregation of stars. If he had discovered at that time that Andromeda was the same as the Milky Way, the structure of the universe might have become clearer much sooner.

---

* However, we should note that Kant expanded upon the unique ideas found in Thomas Wright's *An Original Theory or New Hypothesis of the Universe* (1750). In this work, which precedes Kant's hypothesis, Wright proposes that we are immersed within a "flat layer of stars."

Kepler, who lived in the same era as Galileo, invented and used a slightly different kind of telescope. The Galilean telescope produces an upright image, but increasing the magnification of the image is fairly difficult. However, the Keplerian telescope has the advantage that, although it produces an inverted image (that is, the image is upside down), the amount of sky observable through the telescope at one time, the *field of view*, is not likely to be narrowed, even for high magnifications. Therefore, the Keplerian telescope became the optimal type of telescope for viewing the heavens, while the Galilean telescope is better suited to viewing terrestrial objects.

Objective lens
Eyepiece
Galilean Type
Objective lens
Eyepiece
Keplerian Type

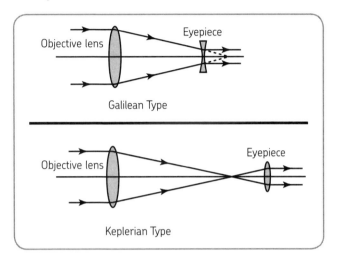

Two types of telescopes

However, regardless of the type of telescope, since a large lens could not be created using the technology of those times, the resolution was limited. *Resolution*, which is the capability of the telescope to distinguish two points, is one of the most important performance characteristics of an astronomical telescope. Resolution can be increased by increasing the telescope's *aperture*, that is, increasing the diameter of the opening housing the objective lens of the telescope (the lens that gathers incoming light), thereby enabling it to collect more light. However, unless the lens is manufactured to extremely precise specifications, increasing the telescope's aperture is useless. This is because lenses, especially very thick ones, separate colors like a prism. In an attempt to get rid of this effect, astronomers in the 17th and 18th century developed extremely long telescopes, some up to 200 feet long, that required massive structures to hold them in place.

Therefore, a mirror is a good alternative to a lens. Telescopes using a mirror to collect light instead of a wide objective lens are known as Newtonian telescopes. The first Newtonian telescope was invented in 1688, and while several improvements have been made since, the basic concept is still used in astronomical (optical) telescopes today.

Mirror
Concave mirror
Convex
lens
Eye

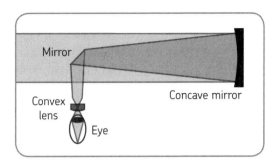

Newtonian telescope

FAMOUS TELESCOPES

Several famous telescopes that aided scientists in making many important discoveries about the universe are introduced below.

Mount Wilson Observatory 100-inch (2.54 m) Hooker Telescope

This telescope was used by Edwin Hubble (who will be introduced in Chapter 3) to make his great discovery linked to the riddle of the galaxies and the birth of the universe, now known as Hubble's law. It was completed in 1917 and was known as the world's largest telescope for approximately 30 years.

Mount Palomar Observatory 200-inch (5.08 m) Hale Telescope

The Hale telescope snatched the title of "world's largest telescope" from the Mount Wilson Observatory when it was completed in 1948 and retained this distinction for 27 years. Its high performance helped 20th-century astronomers discover over 100 asteroids.

Hubble Space Telescope

Launched by the space shuttle *Discovery* in 1990, this man-made satellite telescope orbits Earth at an altitude of approximately 600 km (373 miles). Although its aperture of nearly 2.4 m (8 feet) is less than half that of the Mount Palomar Observatory telescope, it can make high-precision astronomical observations, since it is unaffected by the atmosphere or weather. It continues to make major discoveries such as proving the existence of planets outside the solar system and clarifying the nature of dark matter.

National Astronomical Observatory of Japan Hawaii Observatory Subaru Telescope

This is the world's largest optical infrared telescope. It was completed by the National Astronomical Observatory of Japan in 1999, and the diameter of its primary reflecting mirror is 8.2 m (27 feet)!

When a mirror is this large, it usually ends up being distorted by its own weight. However, technology for correcting this distortion was refined by the Mitsubishi Electric Corporation, which enabled this telescope to be fully operational. The Subaru telescope continues to produce magnificent results, such as discovering an extremely distant cluster of galaxies, nearly 12.88 million light-years from Earth.

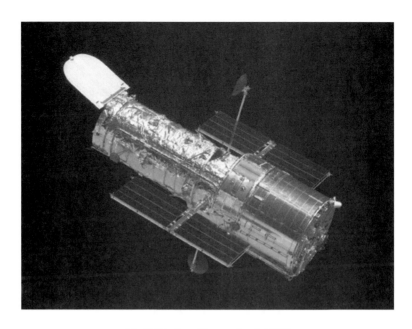

The Hubble Space Telescope (Credit: NASA)

Galaxy IOK-1
Suprime-Cam (NB973, R, B)
September 13, 2006
Subaru Telescope, National Astronomical Observatory of Japan
Copyright © 2006 National Astronomical Observatory of Japan. All rights reserved.

A galaxy 12.88 billion light-years from Earth

WHAT CAN A RADIO TELESCOPE OBSERVE?

A *radio telescope* is another type of astronomical observation tool. But what in the world can radio waves tell us?

Radio waves are electromagnetic waves, just like light rays or infrared rays. However, since the wavelength of radio waves is so long, obstacles in their paths don't cause much obstruction. This is why you can use your cell phone indoors, whereas light from outside is obstructed by walls.

Space contains various kinds of interstellar matter, and if some of this matter consists of heavenly bodies that absorb light such as dark nebulas, it is impossible to use an optical telescope to observe what is there. Therefore, radio waves are used instead.

Radio telescopes, which were developed in the middle of the 20th century, led to major advances in astronomy. For example, one of the results that will go down in history is the discovery of what is believed to be evidence of the Big Bang (this will be explained in Chapter 3).

A famous radio observatory in the United States is the Very Large Array (VLA), located near the town of Soccoro in southern New Mexico. Completed in 1980, it consists of 27 radio telescopes, each of which has a diameter of 25 meters (82 feet). The telescopes are affixed to train tracks, and they can be repositioned and rearranged to act as a single antenna with a maximum diameter of 36 km (22.3 miles). Astronomers use the VLA to study objects like radio-emitting stars, black holes, and supernova remnants. Observations from the VLA have been used to discover water on Mercury and microquasars in the Milky Way galaxy.

The Very Large Array in New Mexico (Credit: NASA)

ANOTHER WAY TO MEASURE THE SIZE OF THE UNIVERSE: A TRIANGULATION TRICK

The distance to the Moon can be calculated by making a baseline between two spots on Earth and using triangulation (see pages 68–69). However, this method cannot be used to calculate the distance to the Sun, since that distance is too great. The mean distance to the Sun is approximately 150 million km (93 million miles), which is approximately 12,000 times the diameter of Earth. A slight measurement error will have a significant effect, and the length of the baseline that can be taken on Earth is limited. To calculate the distance to the Sun, we would need a spot on Earth and a spot somewhere off of Earth to create the baseline.

Luckily, we can use the distance to the Moon, which we discovered by triangulating from a baseline on Earth, as the baseline for triangulating the distance to the Sun. If this were ancient Greece, Aristarchus would help us out here because he already told us the Moon-Earth-Sun angle is 87°. But this isn't ancient Greece.

Since the angle created by the Moon and Sun was incorrectly measured to be 87°, according to the surveying technology available at the time, Aristarchus concluded that the distance to the Sun was approximately 20 times farther than the distance to the Moon. With the help of current technology, we now know it is approximately 390 times further. Nevertheless, this method thought up by Aristarchus more than 2,300 years ago is really quite remarkable.

Moon
Equatorial radius of Earth:
6,378 km (3,963 miles)
87° (Actually 89.85°)
Sun
Mean distance between Earth and the Sun: 150 million km (93 million miles)
Earth

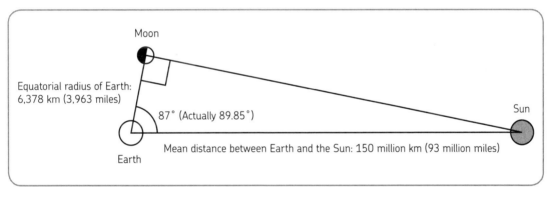

Method of finding the distance between Earth and the Sun

TRIANGULATION CAN GIVE US THE DISTANCE TO STARS BEYOND THE SOLAR SYSTEM

Using the Moon to create a baseline is one way of using Aristarchus's method to calculate long distances in space. But there is an even better method that uses Earth's radius of revolution.

It goes without saying that Earth revolves once around the Sun in one year. Therefore, if we observe a heavenly body outside of our solar system at some time and then observe it six months later, we can see how much the star's position has changed compared to the stars around it. Because measuring the change of position gives us the change in the angle (angle of inclination) required to look up at it, we can use that value to determine the distance to that heavenly body.

Apparent position changes
Star
Annual parallax
Earth (summer)
Sun
Earth (winter)
Radius of revolution
(1 AU)

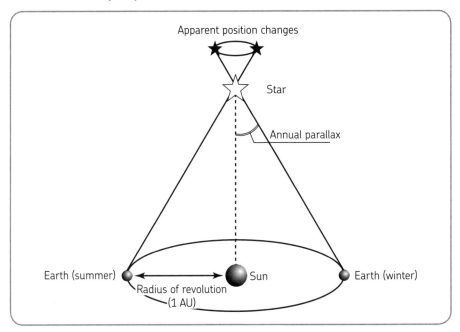

Triangulation using Earth's radius of revolution

Incidentally, half of the difference in those angles is called the *annual parallax*, and the distance to a heavenly body for which this parallax is 1 arc second (1 arc second is 1/3600 of a degree) is a unit called a *parsec* (short for *parallax per one arc second*). This unit is related to other units of distance as follows.

1 pc (parsec) = approximately 3.26 light-years = approximately 206,265 AU = approximately 3.08568×10^{16} m = approximately 31 trillion km = approximately 1.9×10^{13} miles.

For heavenly bodies that can be observed from Earth by this method, the maximum parallax is approximately 0.033 arc seconds (approximately 1/100,000 of a degree), and the maximum distance is approximately 30 parsecs or 100 light-years. In 1989, the European Space Agency launched the Hipparcos High Precision Parallax Collecting Satellite, which can accurately measure the distance to stars up to 500 light-years away or up to 1,000 light-years away within a reasonable margin of error.

HOW BIG IS THE SOLAR SYSTEM?

Although Pluto was removed from the category of planet in 2006, this did not change the fact that it is one of the heavenly bodies that form the solar system. So what indicates the extent of the solar system, and how large is it?

First, the distance (semimajor axis) to the outermost planet Neptune in terms of astronomical units is approximately 30 AU. Beyond that, in the range from 30 to 100 AU, is the *Kuiper Belt*, where many comets orbit the Sun. Unlike asteroids, these *Kuiper Belt objects (KBOs)* consist mainly of different types of ices like water, methane, and ammonia. The Kuiper Belt is believed to contain more than 70,000 objects that have a diameter of at least 50 km (30 miles), and Pluto is also considered to be part of it.

In addition, there is believed to be a group of heavenly bodies consisting of ice and rock in a contiguous area more distant than the Kuiper Belt. This is called the *Oort cloud*, which ranges from 50 AU to 100,000 AU away from the Sun! This is approximately 3,300 times the orbital radius of Neptune.

Generally, the solar system is said to extend as far as the Oort cloud, which is the gravitational boundary of the Sun (the farthest extent to which heavenly bodies are affected by the Sun's gravity). The radius of this region is approximately 1.6 light-years, so it would take that many years to escape from the solar system if your spaceship were traveling at the speed of light—and even longer if it were traveling more slowly.

The gravitational effects of the Sun extend 3,300 times the distance of the furthest planet!

3
THE UNIVERSE WAS BORN WITH A BIG BANG

GALAXIES ARE ISLANDS OF LIGHT IN THE VOID OF SPACE

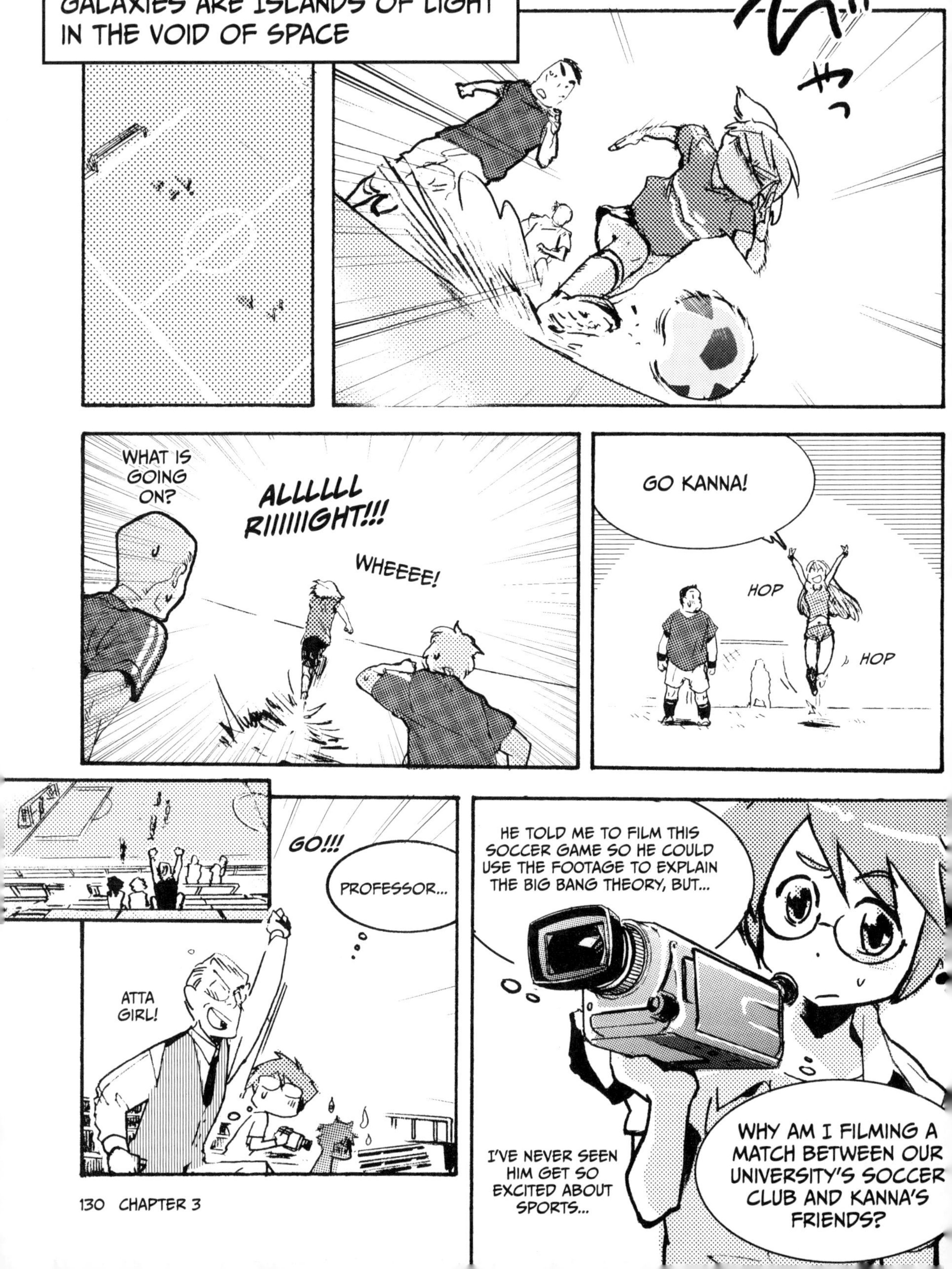

WHAT IS GOING ON?
ALLLLLL RIIIIIIGHT!!!
WHEEEEE!
GO KANNA!
HOP
HOP
GO!!!
PROFESSOR...
ATTA GIRL!
HE TOLD ME TO FILM THIS SOCCER GAME SO HE COULD USE THE FOOTAGE TO EXPLAIN THE BIG BANG THEORY, BUT...
I'VE NEVER SEEN HIM GET SO EXCITED ABOUT SPORTS...
WHY AM I FILMING A MATCH BETWEEN OUR UNIVERSITY'S SOCCER CLUB AND KANNA'S FRIENDS?

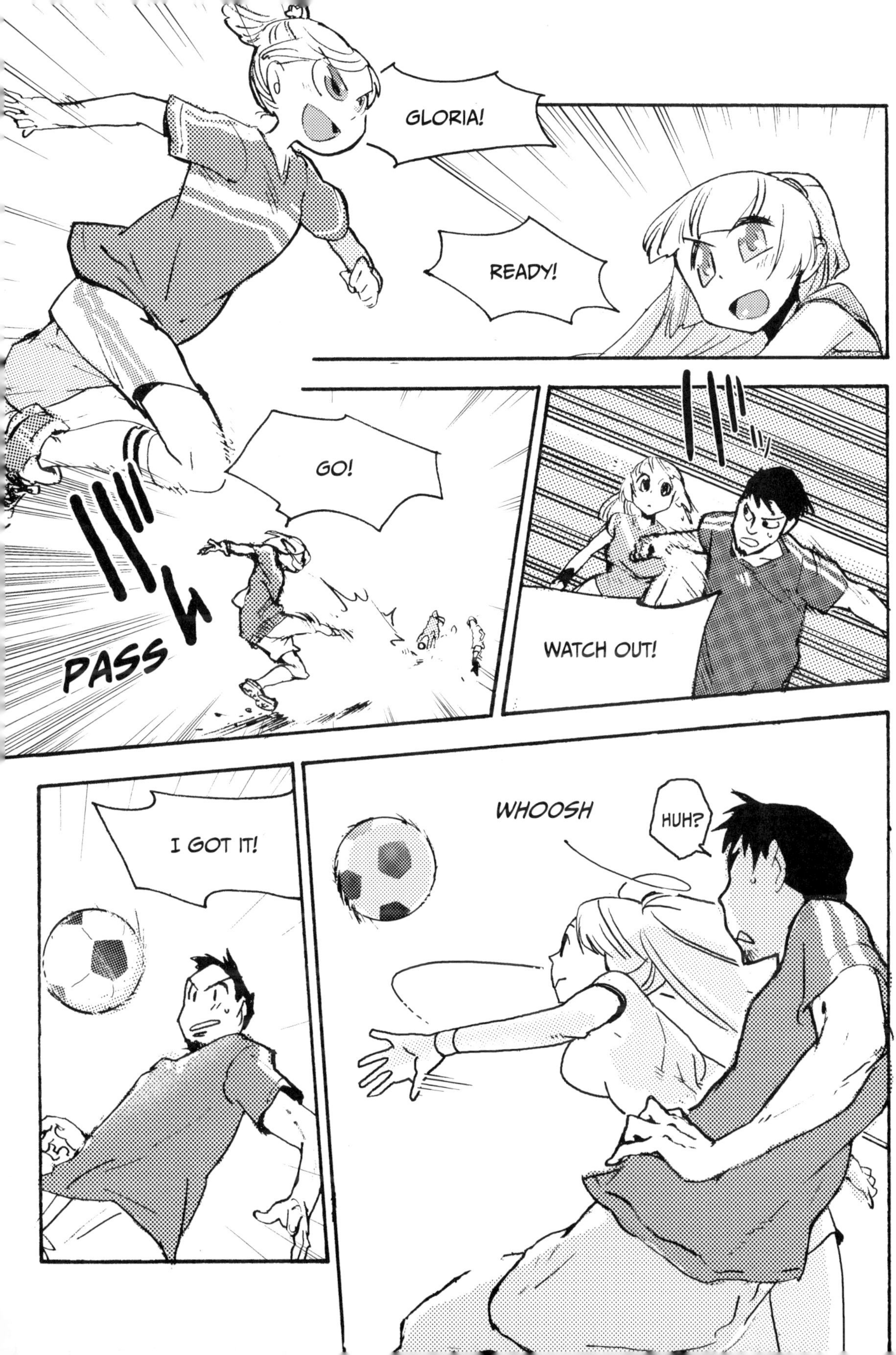
GLORIA!
READY!
GO!
PASS
WATCH OUT!
I GOT IT!
WHOOSH
HUH?

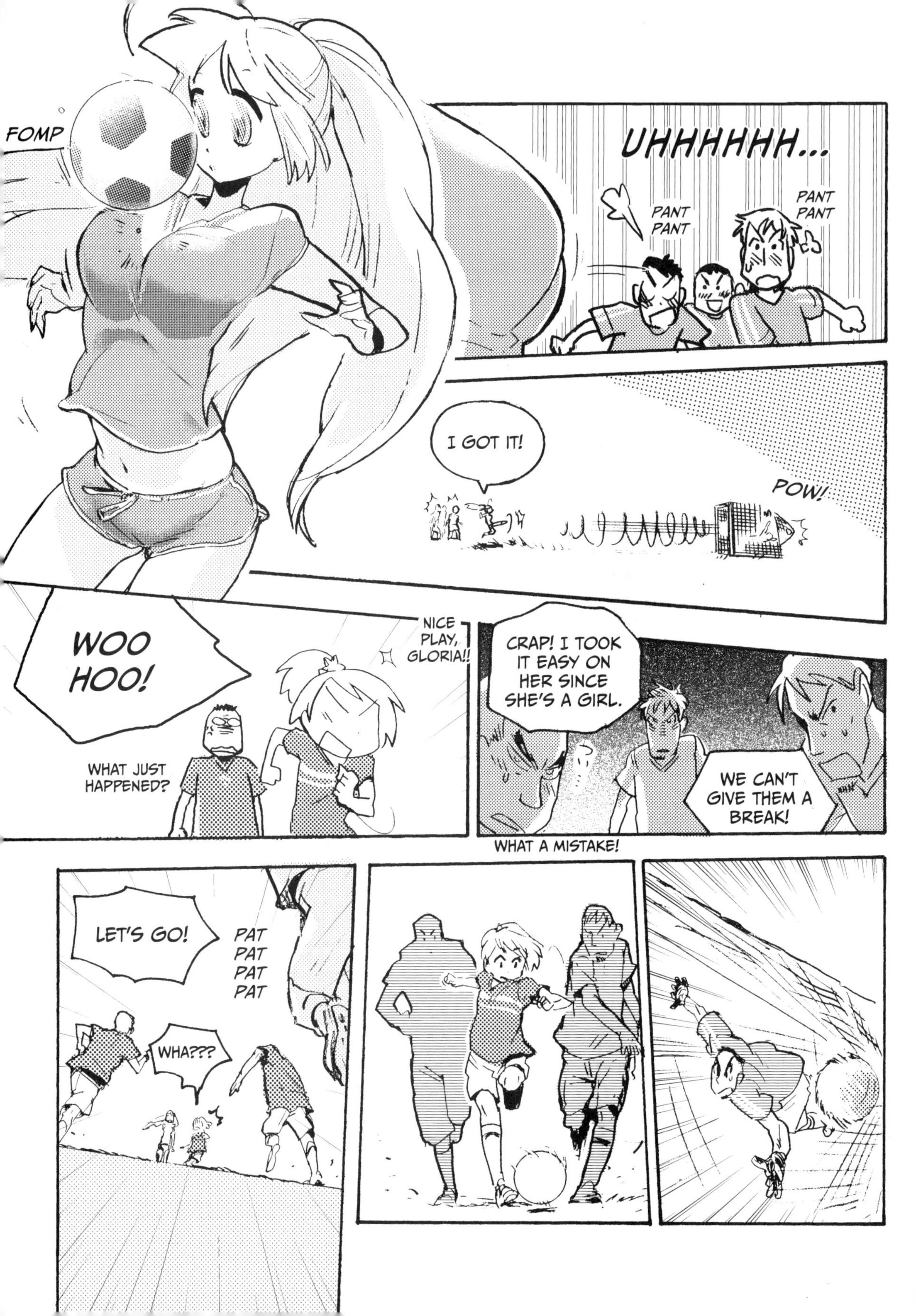
FOMP
UHHHHHH...
PANT PANT
PANT PANT
I GOT IT!
POW!
WOO HOO!
WHAT JUST HAPPENED?
NICE PLAY, GLORIA!!
CRAP! I TOOK IT EASY ON HER SINCE SHE'S A GIRL.
WE CAN'T GIVE THEM A BREAK!
WHAT A MISTAKE!
LET'S GO!
PAT PAT PAT PAT
WHA???

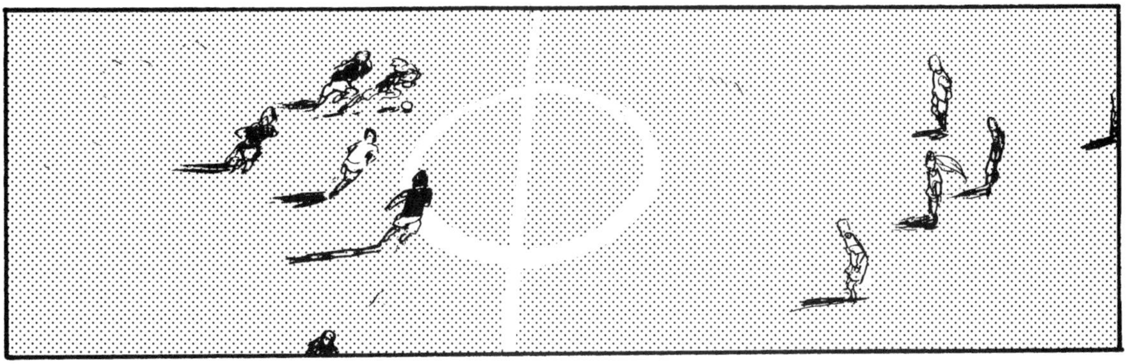

THE WINNING TEAM LEARNS A LESSON

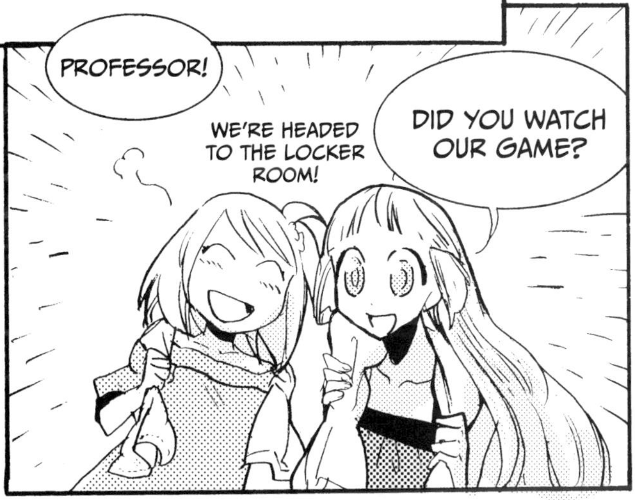
PROFESSOR!
WE'RE HEADED TO THE LOCKER ROOM!
DID YOU WATCH OUR GAME?

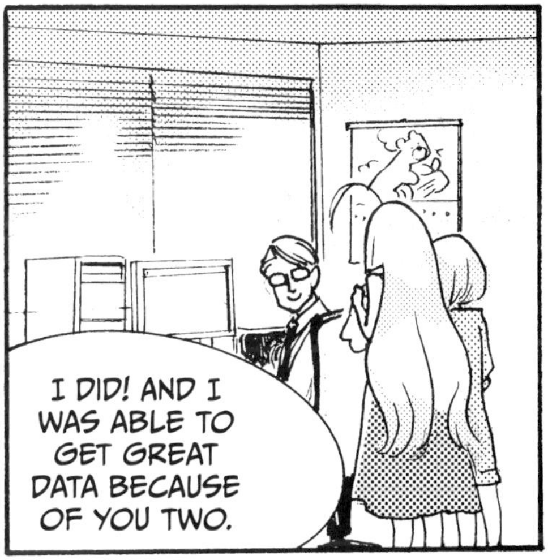
I DID! AND I WAS ABLE TO GET GREAT DATA BECAUSE OF YOU TWO.

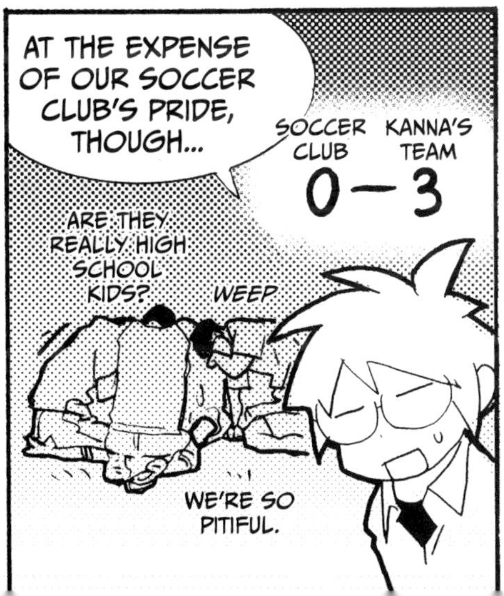
AT THE EXPENSE OF OUR SOCCER CLUB'S PRIDE, THOUGH...
SOCCER CLUB
KANNA'S TEAM
0—3
ARE THEY REALLY HIGH SCHOOL KIDS?
WEEP
WE'RE SO PITIFUL.

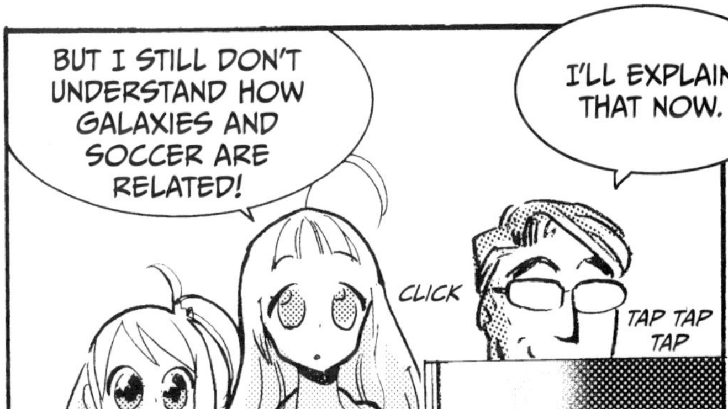
BUT I STILL DON'T UNDERSTAND HOW GALAXIES AND SOCCER ARE RELATED!
I'LL EXPLAIN THAT NOW.
CLICK
TAP TAP TAP

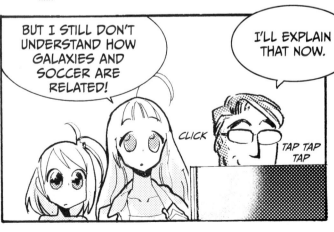

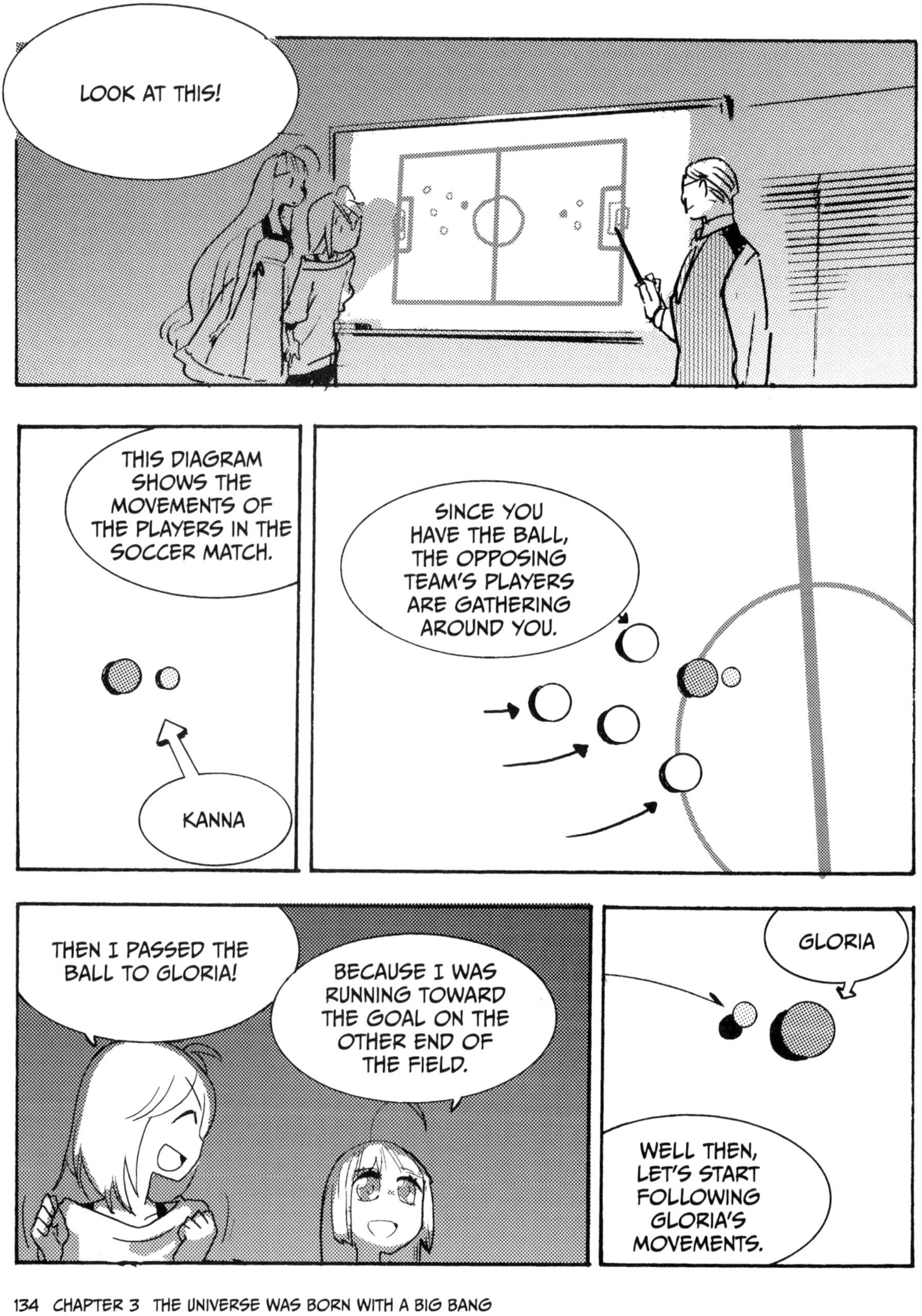
LOOK AT THIS!
THIS DIAGRAM SHOWS THE MOVEMENTS OF THE PLAYERS IN THE SOCCER MATCH.
KANNA
SINCE YOU HAVE THE BALL, THE OPPOSING TEAM'S PLAYERS ARE GATHERING AROUND YOU.
THEN I PASSED THE BALL TO GLORIA!
BECAUSE I WAS RUNNING TOWARD THE GOAL ON THE OTHER END OF THE FIELD.
GLORIA
WELL THEN, LET'S START FOLLOWING GLORIA'S MOVEMENTS.

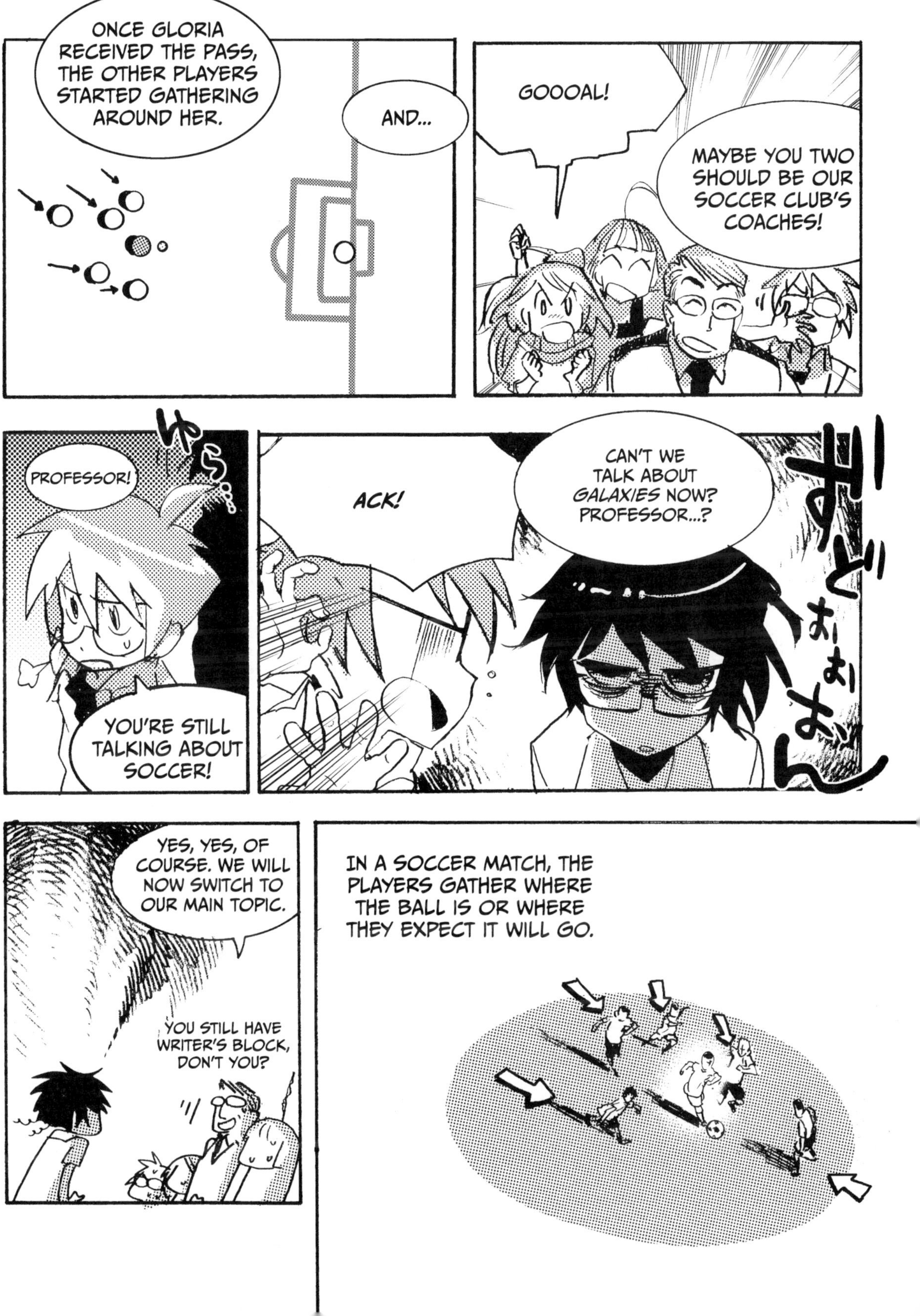

ONCE GLORIA RECEIVED THE PASS, THE OTHER PLAYERS STARTED GATHERING AROUND HER.
AND...
GOOOAL!
MAYBE YOU TWO SHOULD BE OUR SOCCER CLUB'S COACHES!
PROFESSOR!
ゆら…
YOU'RE STILL TALKING ABOUT SOCCER!
ACK!
CAN'T WE TALK ABOUT GALAXIES NOW? PROFESSOR...?
ずどぉぉぉん
YES, YES, OF COURSE. WE WILL NOW SWITCH TO OUR MAIN TOPIC.
YOU STILL HAVE WRITER'S BLOCK, DON'T YOU?
IN A SOCCER MATCH, THE PLAYERS GATHER WHERE THE BALL IS OR WHERE THEY EXPECT IT WILL GO.

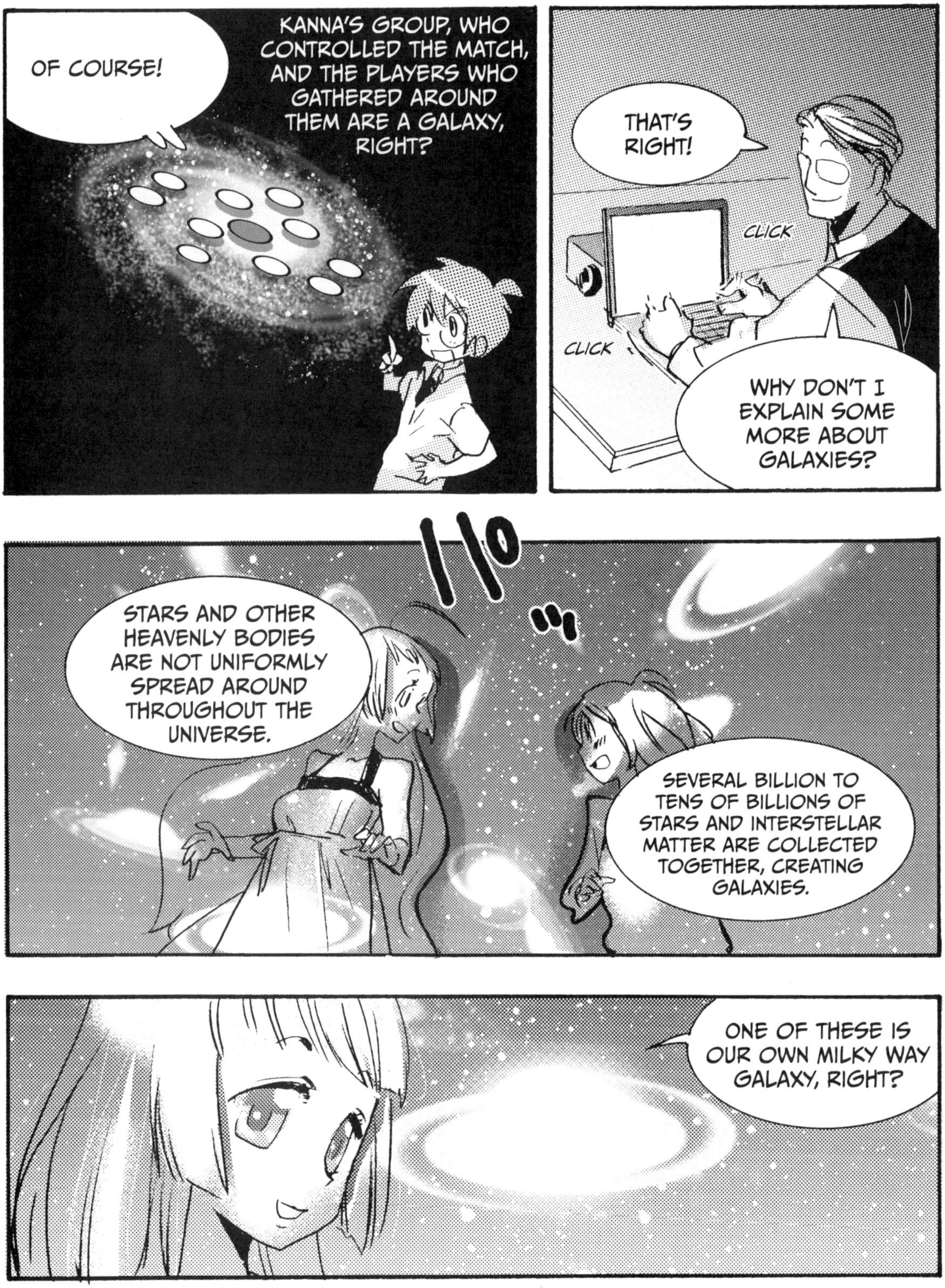

OF COURSE!
KANNA'S GROUP, WHO CONTROLLED THE MATCH, AND THE PLAYERS WHO GATHERED AROUND THEM ARE A GALAXY, RIGHT?
THAT'S RIGHT!
CLICK
CLICK
WHY DON'T I EXPLAIN SOME MORE ABOUT GALAXIES?
STARS AND OTHER HEAVENLY BODIES ARE NOT UNIFORMLY SPREAD AROUND THROUGHOUT THE UNIVERSE.
SEVERAL BILLION TO TENS OF BILLIONS OF STARS AND INTERSTELLAR MATTER ARE COLLECTED TOGETHER, CREATING GALAXIES.
ONE OF THESE IS OUR OWN MILKY WAY GALAXY, RIGHT?

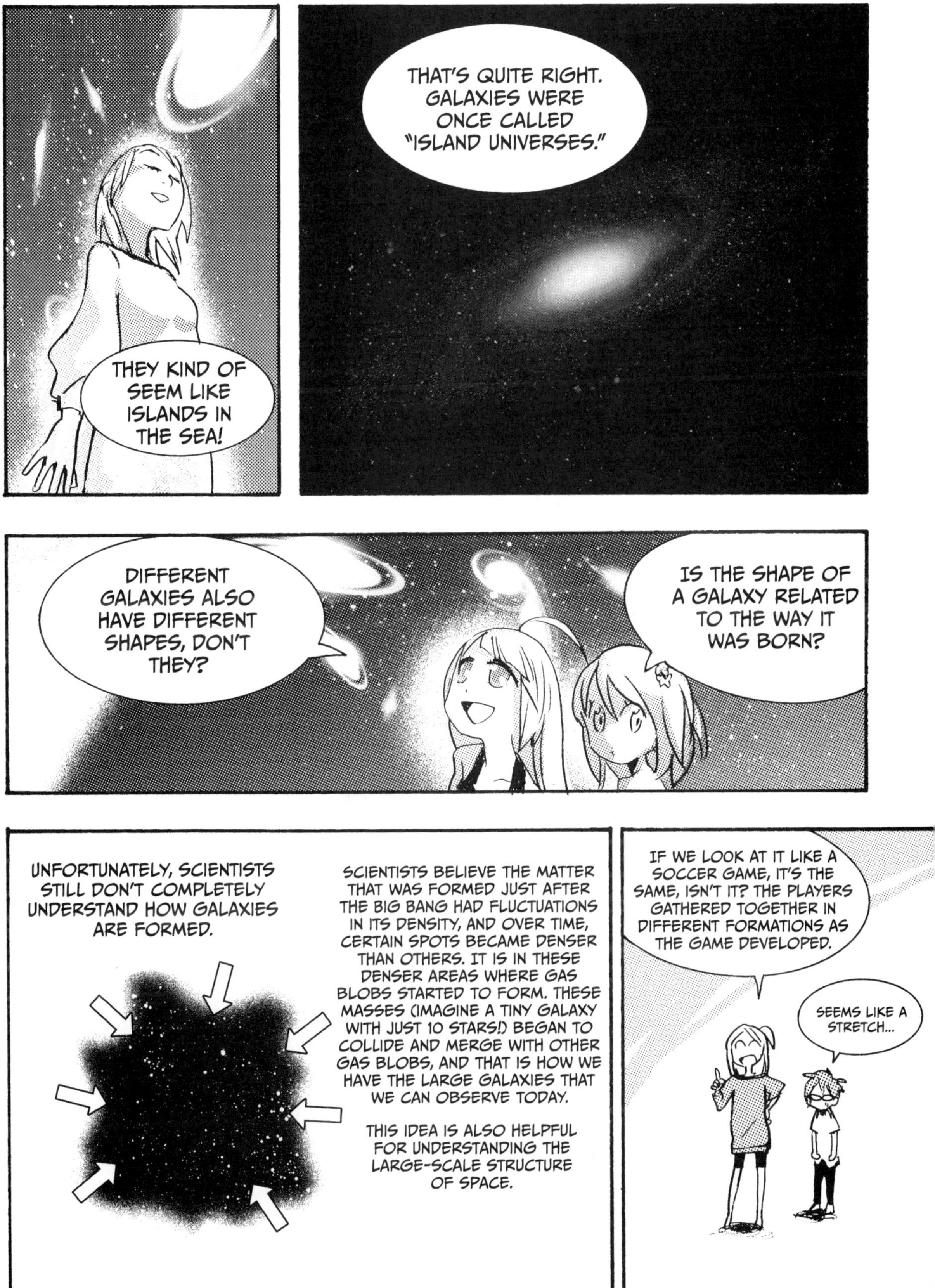
THEY KIND OF SEEM LIKE ISLANDS IN THE SEA!
THAT'S QUITE RIGHT. GALAXIES WERE ONCE CALLED "ISLAND UNIVERSES."
DIFFERENT GALAXIES ALSO HAVE DIFFERENT SHAPES, DON'T THEY?
IS THE SHAPE OF A GALAXY RELATED TO THE WAY IT WAS BORN?
UNFORTUNATELY, SCIENTISTS STILL DON'T COMPLETELY UNDERSTAND HOW GALAXIES ARE FORMED.
SCIENTISTS BELIEVE THE MATTER THAT WAS FORMED JUST AFTER THE BIG BANG HAD FLUCTUATIONS IN ITS DENSITY, AND OVER TIME, CERTAIN SPOTS BECAME DENSER THAN OTHERS. IT IS IN THESE DENSER AREAS WHERE GAS BLOBS STARTED TO FORM. THESE MASSES (IMAGINE A TINY GALAXY WITH JUST 10 STARS!) BEGAN TO COLLIDE AND MERGE WITH OTHER GAS BLOBS, AND THAT IS HOW WE HAVE THE LARGE GALAXIES THAT WE CAN OBSERVE TODAY.
THIS IDEA IS ALSO HELPFUL FOR UNDERSTANDING THE LARGE-SCALE STRUCTURE OF SPACE.
IF WE LOOK AT IT LIKE A SOCCER GAME, IT'S THE SAME, ISN'T IT? THE PLAYERS GATHERED TOGETHER IN DIFFERENT FORMATIONS AS THE GAME DEVELOPED.
SEEMS LIKE A STRETCH...

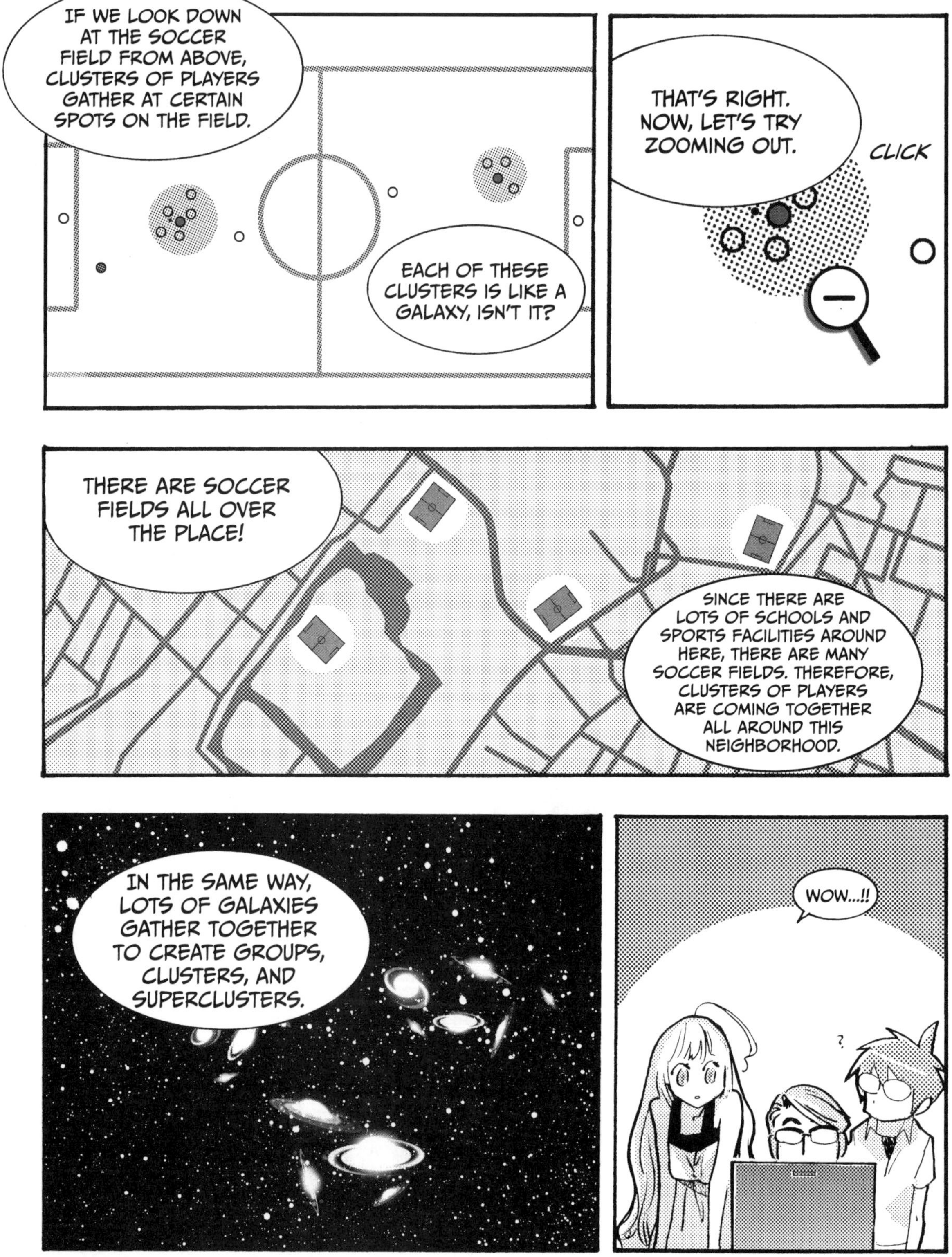
IF WE LOOK DOWN AT THE SOCCER FIELD FROM ABOVE, CLUSTERS OF PLAYERS GATHER AT CERTAIN SPOTS ON THE FIELD.
EACH OF THESE CLUSTERS IS LIKE A GALAXY, ISN'T IT?
THAT'S RIGHT. NOW, LET'S TRY ZOOMING OUT.
CLICK
THERE ARE SOCCER FIELDS ALL OVER THE PLACE!
SINCE THERE ARE LOTS OF SCHOOLS AND SPORTS FACILITIES AROUND HERE, THERE ARE MANY SOCCER FIELDS. THEREFORE, CLUSTERS OF PLAYERS ARE COMING TOGETHER ALL AROUND THIS NEIGHBORHOOD.
IN THE SAME WAY, LOTS OF GALAXIES GATHER TOGETHER TO CREATE GROUPS, CLUSTERS, AND SUPERCLUSTERS.
WOW...!!
?

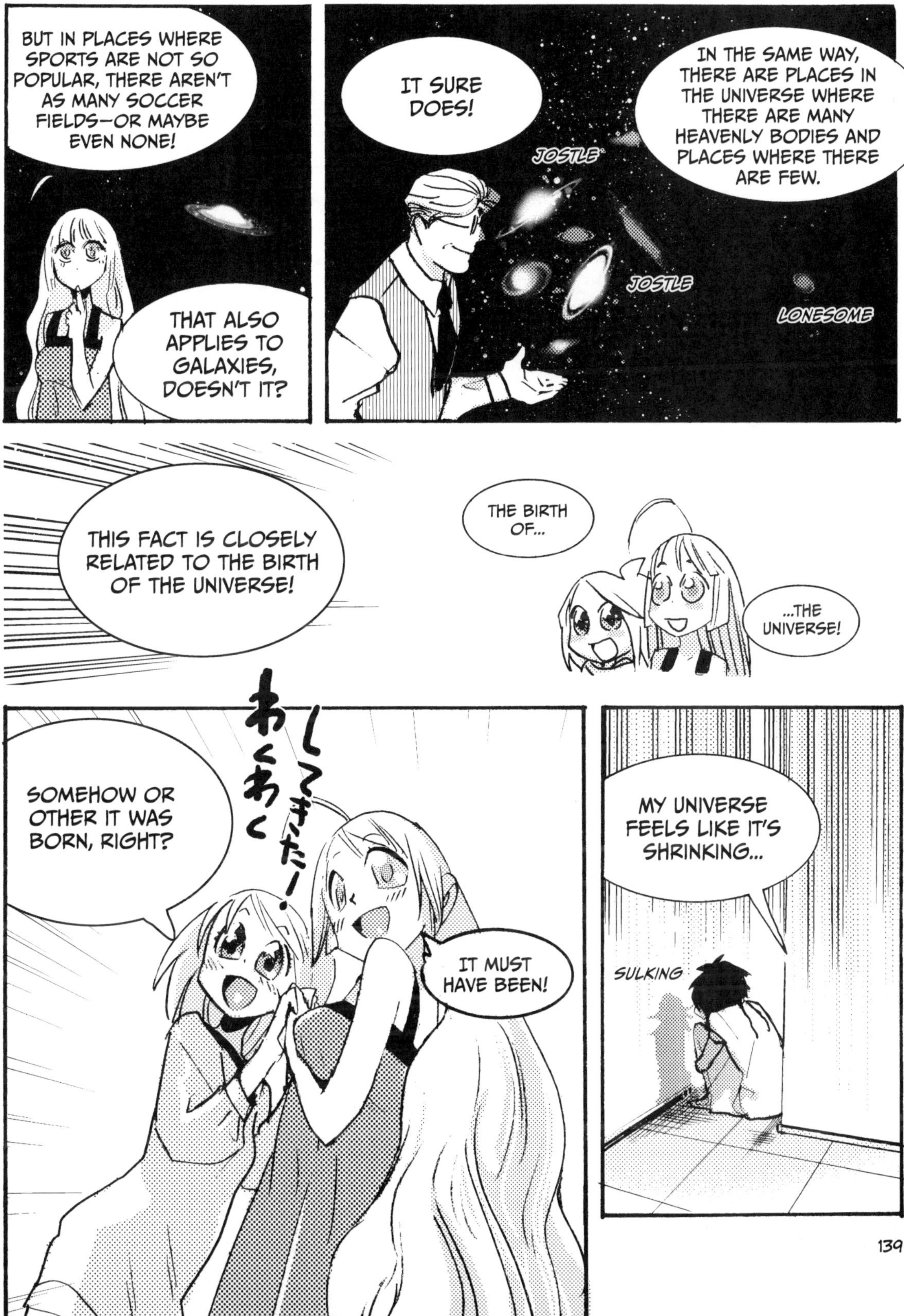
BUT IN PLACES WHERE SPORTS ARE NOT SO POPULAR, THERE AREN'T AS MANY SOCCER FIELDS—OR MAYBE EVEN NONE!
THAT ALSO APPLIES TO GALAXIES, DOESN'T IT?
IT SURE DOES!
IN THE SAME WAY, THERE ARE PLACES IN THE UNIVERSE WHERE THERE ARE MANY HEAVENLY BODIES AND PLACES WHERE THERE ARE FEW.
JOSTLE
JOSTLE
LONESOME
THIS FACT IS CLOSELY RELATED TO THE BIRTH OF THE UNIVERSE!
THE BIRTH OF...
...THE UNIVERSE!
SOMEHOW OR OTHER IT WAS BORN, RIGHT?
わくわくしてきたー！
IT MUST HAVE BEEN!
MY UNIVERSE FEELS LIKE IT'S SHRINKING...
SULKING

WHAT IS THE LARGE-SCALE STRUCTURE OF THE COSMOS?

The philosopher Kant and many others reasoned that there must be a hierarchical, large-scale structure in the universe, just like the hierarchical group structures in the world we live in—that is, just as houses make up neighborhoods, neighborhoods make up cities, many cities make up states, and states make up a country, so too must the heavenly bodies be grouped into patterns. However, this really became clear when the existence of galaxies outside of our own galaxy was verified. Later observations and research also made it apparent that many galaxies collect together to create a group, and these in turn form aggregations at even higher levels of the cosmic hierarchy.

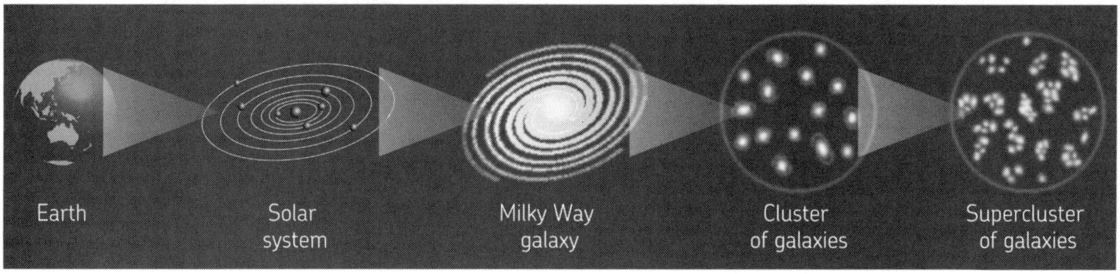
Earth
Solar system
Milky Way galaxy
Cluster of galaxies
Supercluster of galaxies

Large-scale structure of the cosmos

PLANETARY SYSTEM

A *planetary system* is a system like our solar system, in which planets, asteroids, satellites, comets, and other matter form a single system orbiting around a star.

GALAXY

A *galaxy* is a heavenly body that is formed by the gravitational attraction of several tens of billions to several hundreds of billions of stars and interstellar matter (including dark matter). Since galaxies exist in space like islands in the sea, they used to be referred to as *island universes*. The galaxy to which our solar system belongs is called the Milky Way galaxy.

The phrase "island universe" fell out of use once astronomers realized that there were many "island universes" (that is, galaxies) besides the Milky Way.

GROUP OF GALAXIES OR CLUSTER OF GALAXIES

Groups and *clusters* are multiple galaxies that are gravitationally unified. When the number of galaxies is fewer than 50, we call the aggregation a group of galaxies; when the number is greater (up to several thousand), we call it a cluster. Our Milky Way galaxy belongs to the Local Group, which is made up of 30 to 40 galaxies, including the Andromeda galaxy and the Large and Small Magellanic Clouds. The closest cluster of galaxies to the Local Group is the Virgo Cluster, which is approximately 60 million light-years away. It appears in the constellation Virgo, and it has a diameter of approximately 12 million light-years.

SUPERCLUSTER OF GALAXIES

A *supercluster* is an aggregation of several hundred groups or clusters of galaxies. A supercluster has a diameter of several hundred million light-years.

Astronomers formerly thought that since this kind of large-scale structure existed, the galaxies would be distributed uniformly throughout the universe. However, in the 1980s, regions were discovered in space in which no galaxies could be observed at all. The diameter of each of these regions was more than 100 million light-years. When they were further investigated, these voids were found to be lined up like bubbles, with groups of galaxies or clusters of galaxies distributed on their surfaces.

In 1989, Margaret Geller and John Huchra of the Harvard-Smithsonian Center for Astrophysics named a grouping of galaxies the *Great Wall*, since it appears to be a long, continuous entity like the Great Wall of China. The Great Wall is a gigantic structure 500 million light-years long and 200 million light-years wide. At the time of its discovery, the Great Wall was the largest known structure in the universe.

However, on October 20, 2003, a new Great Wall was found. This structure is approximately 1 billion light-years from Earth and has a length of 1.4 billion light-years, which is approximately three times the scale of the previous discovery. In other words, this structure currently holds the record for being the largest thing in the known universe.

These structures are distinguished from each other by calling the one discovered in 1989 the *CfA2 Great Wall* and the one discovered in 2003 the *Sloan Great Wall*.

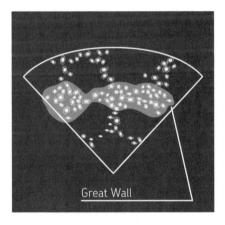

Galaxies form galactic walls.

HUBBLE'S GREAT DISCOVERY

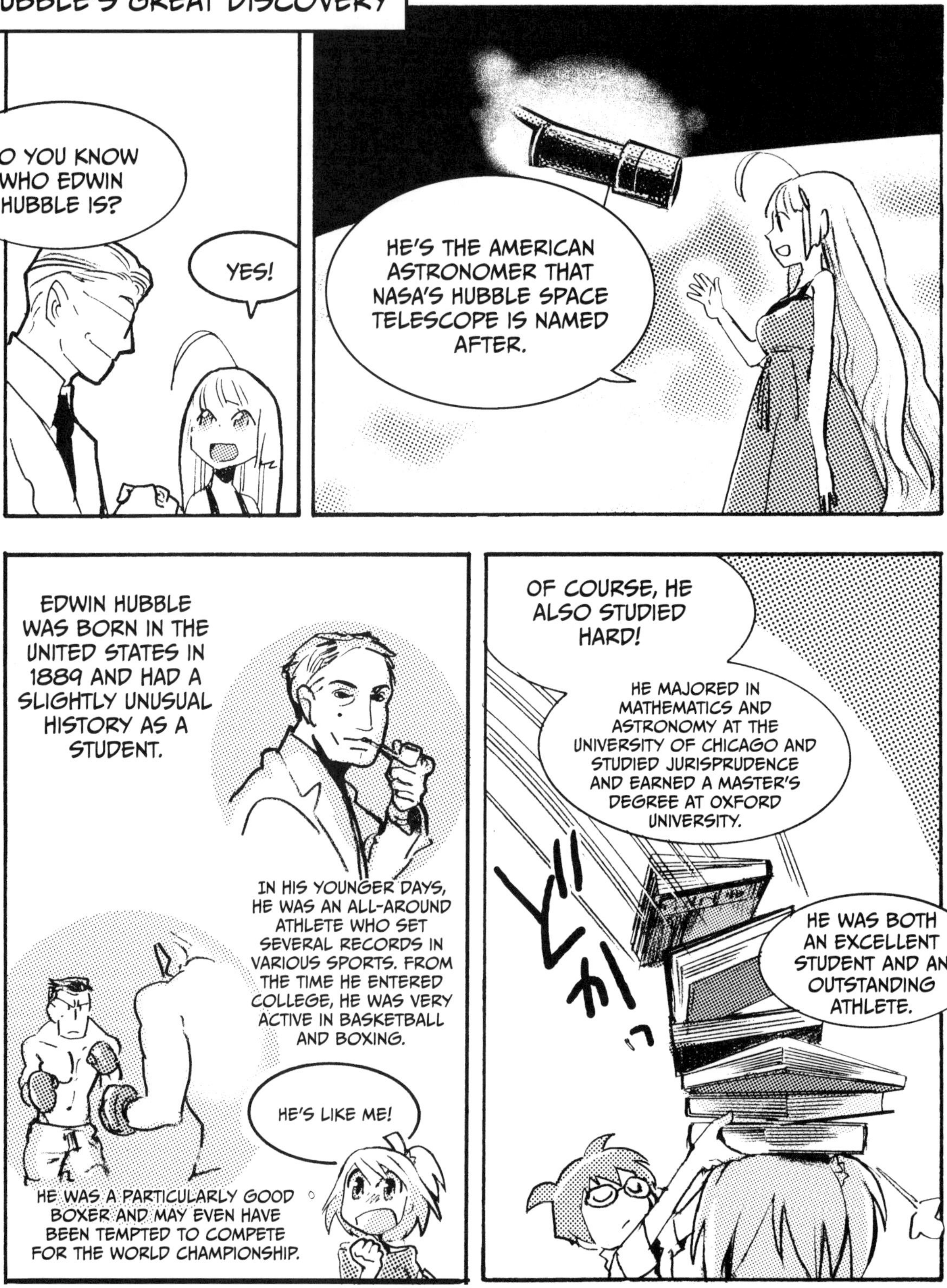
DO YOU KNOW WHO EDWIN HUBBLE IS?
YES!
HE'S THE AMERICAN ASTRONOMER THAT NASA'S HUBBLE SPACE TELESCOPE IS NAMED AFTER.
EDWIN HUBBLE WAS BORN IN THE UNITED STATES IN 1889 AND HAD A SLIGHTLY UNUSUAL HISTORY AS A STUDENT.
IN HIS YOUNGER DAYS, HE WAS AN ALL-AROUND ATHLETE WHO SET SEVERAL RECORDS IN VARIOUS SPORTS. FROM THE TIME HE ENTERED COLLEGE, HE WAS VERY ACTIVE IN BASKETBALL AND BOXING.
HE WAS A PARTICULARLY GOOD BOXER AND MAY EVEN HAVE BEEN TEMPTED TO COMPETE FOR THE WORLD CHAMPIONSHIP.
HE'S LIKE ME!
OF COURSE, HE ALSO STUDIED HARD!
HE MAJORED IN MATHEMATICS AND ASTRONOMY AT THE UNIVERSITY OF CHICAGO AND STUDIED JURISPRUDENCE AND EARNED A MASTER'S DEGREE AT OXFORD UNIVERSITY.
HE WAS BOTH AN EXCELLENT STUDENT AND AN OUTSTANDING ATHLETE.

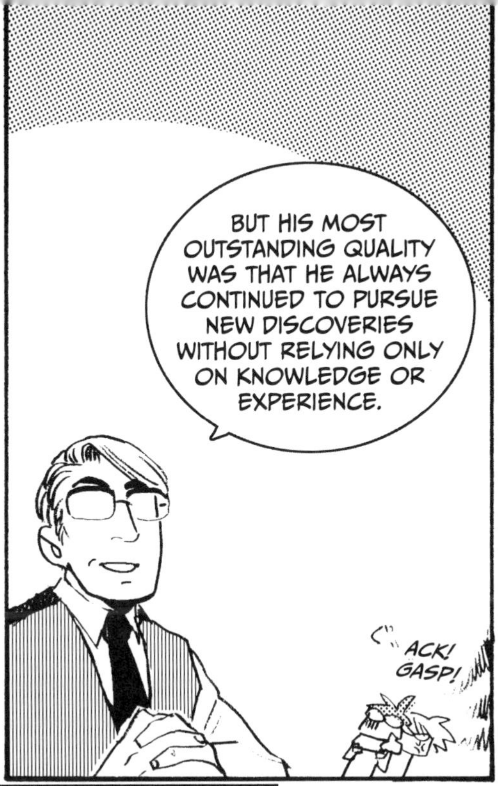
BUT HIS MOST OUTSTANDING QUALITY WAS THAT HE ALWAYS CONTINUED TO PURSUE NEW DISCOVERIES WITHOUT RELYING ONLY ON KNOWLEDGE OR EXPERIENCE.
ACK! GASP!

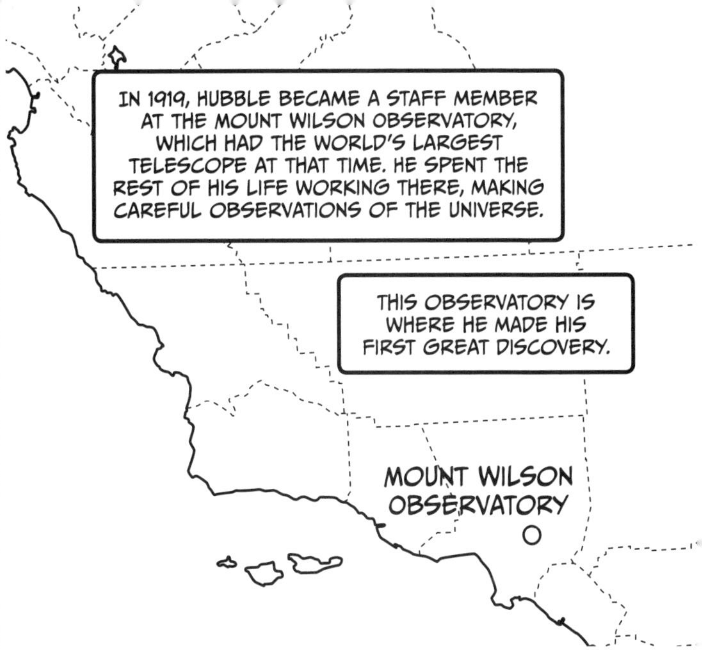
IN 1919, HUBBLE BECAME A STAFF MEMBER AT THE MOUNT WILSON OBSERVATORY, WHICH HAD THE WORLD'S LARGEST TELESCOPE AT THAT TIME. HE SPENT THE REST OF HIS LIFE WORKING THERE, MAKING CAREFUL OBSERVATIONS OF THE UNIVERSE.
THIS OBSERVATORY IS WHERE HE MADE HIS FIRST GREAT DISCOVERY.
MOUNT WILSON OBSERVATORY

THE MOUNT WILSON OBSERVATORY (MWO) IN LOS ANGELES COUNTY, CALIFORNIA, IS LOCATED ON MOUNT WILSON, WHICH HAS AN ELEVATION OF 1,742 METERS (5,715 FEET). IT IS SAID THAT MOUNT WILSON IS ONE OF THE PLACES IN NORTH AMERICA WITH THE MOST STABLE ATMOSPHERE. THE OBSERVATORY WAS BUILT IN 1902.

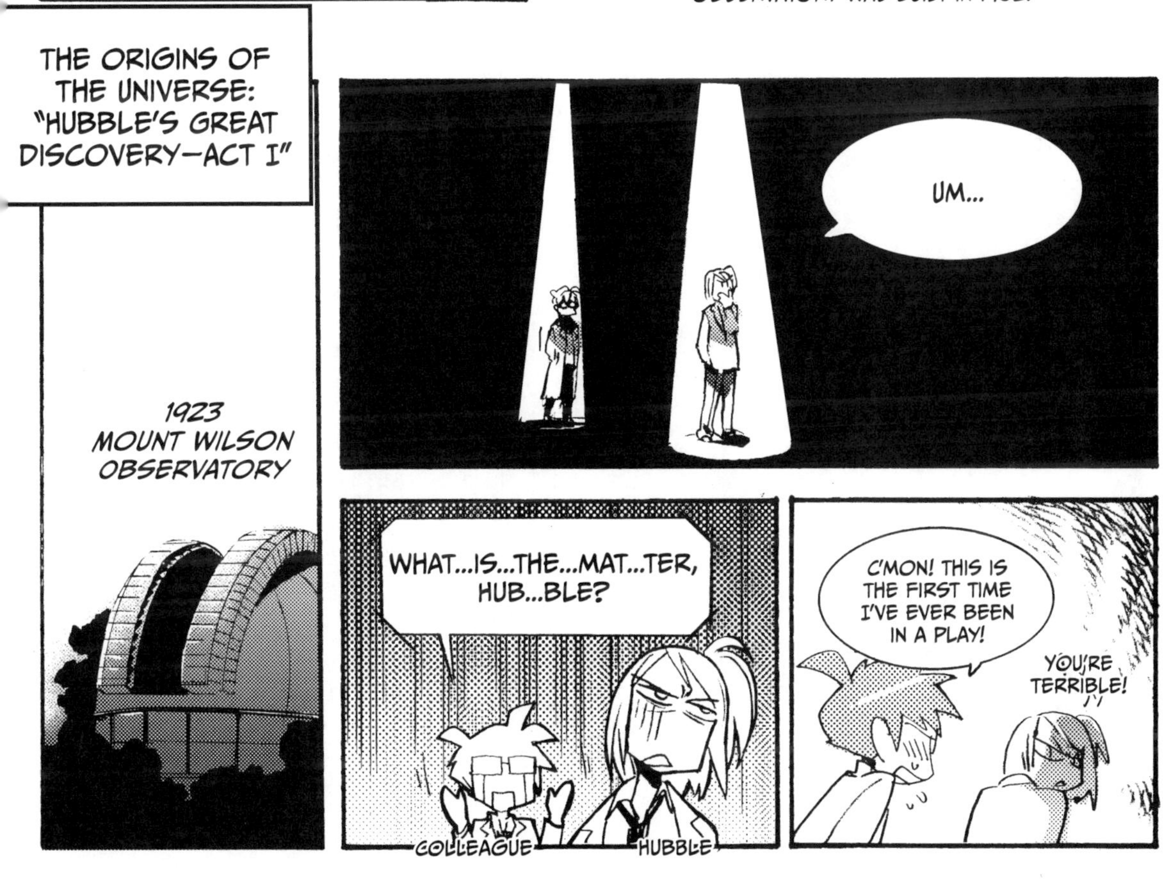
THE ORIGINS OF THE UNIVERSE: "HUBBLE'S GREAT DISCOVERY—ACT I"
1923 MOUNT WILSON OBSERVATORY
UM...
WHAT...IS...THE...MAT...TER, HUB...BLE?
COLLEAGUE
HUBBLE
C'MON! THIS IS THE FIRST TIME I'VE EVER BEEN IN A PLAY!
YOU'RE TERRIBLE!

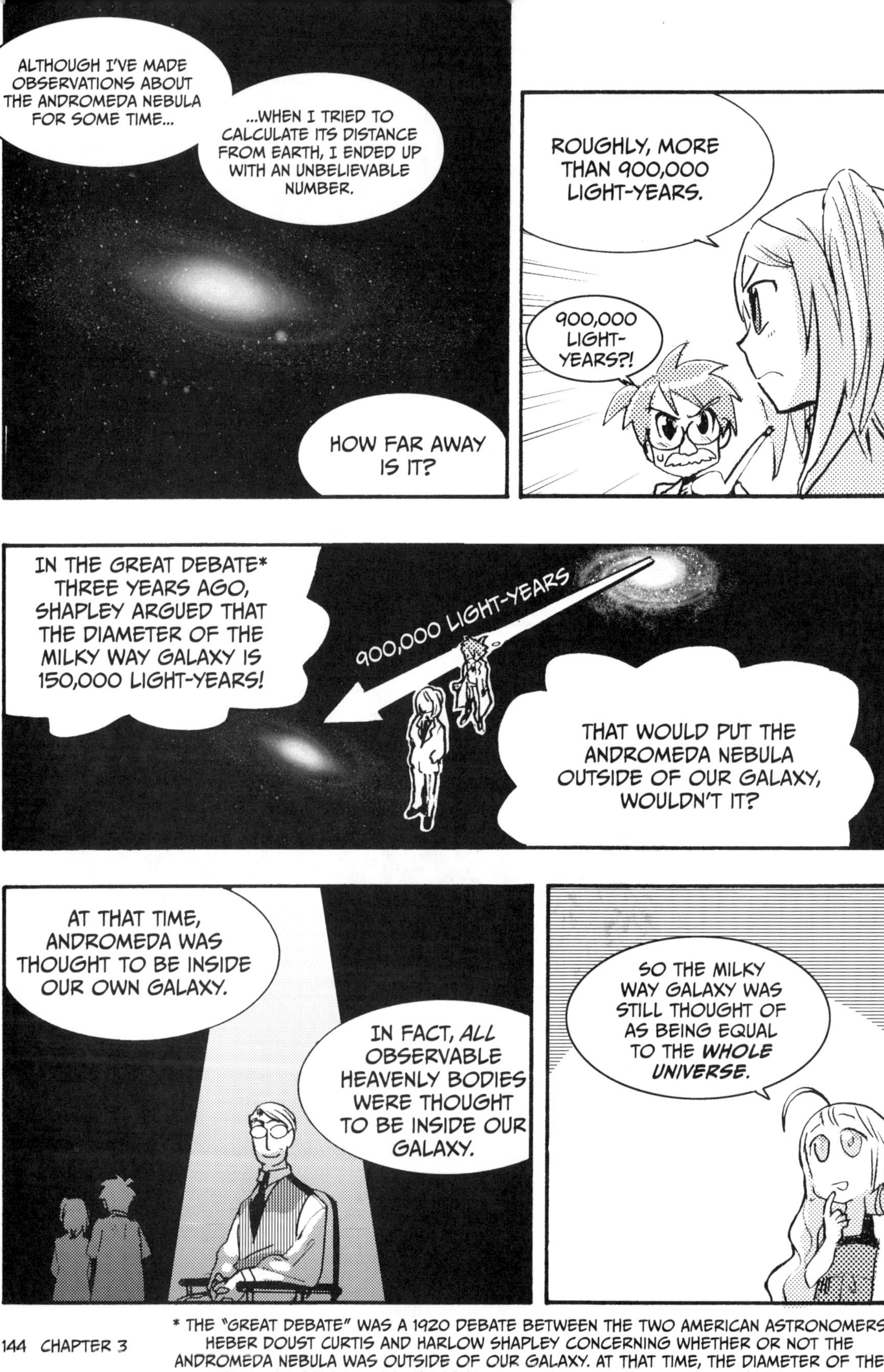
ALTHOUGH I'VE MADE OBSERVATIONS ABOUT THE ANDROMEDA NEBULA FOR SOME TIME...
...WHEN I TRIED TO CALCULATE ITS DISTANCE FROM EARTH, I ENDED UP WITH AN UNBELIEVABLE NUMBER.
HOW FAR AWAY IS IT?
ROUGHLY, MORE THAN 900,000 LIGHT-YEARS.
900,000 LIGHT-YEARS?!
IN THE GREAT DEBATE* THREE YEARS AGO, SHAPLEY ARGUED THAT THE DIAMETER OF THE MILKY WAY GALAXY IS 150,000 LIGHT-YEARS!
900,000 LIGHT-YEARS
THAT WOULD PUT THE ANDROMEDA NEBULA OUTSIDE OF OUR GALAXY, WOULDN'T IT?
AT THAT TIME, ANDROMEDA WAS THOUGHT TO BE INSIDE OUR OWN GALAXY.
IN FACT, *ALL* OBSERVABLE HEAVENLY BODIES WERE THOUGHT TO BE INSIDE OUR GALAXY.
SO THE MILKY WAY GALAXY WAS STILL THOUGHT OF AS BEING EQUAL TO THE ***WHOLE UNIVERSE.***

* THE "GREAT DEBATE" WAS A 1920 DEBATE BETWEEN THE TWO AMERICAN ASTRONOMERS HEBER DOUST CURTIS AND HARLOW SHAPLEY CONCERNING WHETHER OR NOT THE ANDROMEDA NEBULA WAS OUTSIDE OF OUR GALAXY. AT THAT TIME, THE DIAMETER OF THE MILKY WAY GALAXY WAS THOUGHT TO BE APPROXIMATELY 150,000 LIGHT-YEARS.

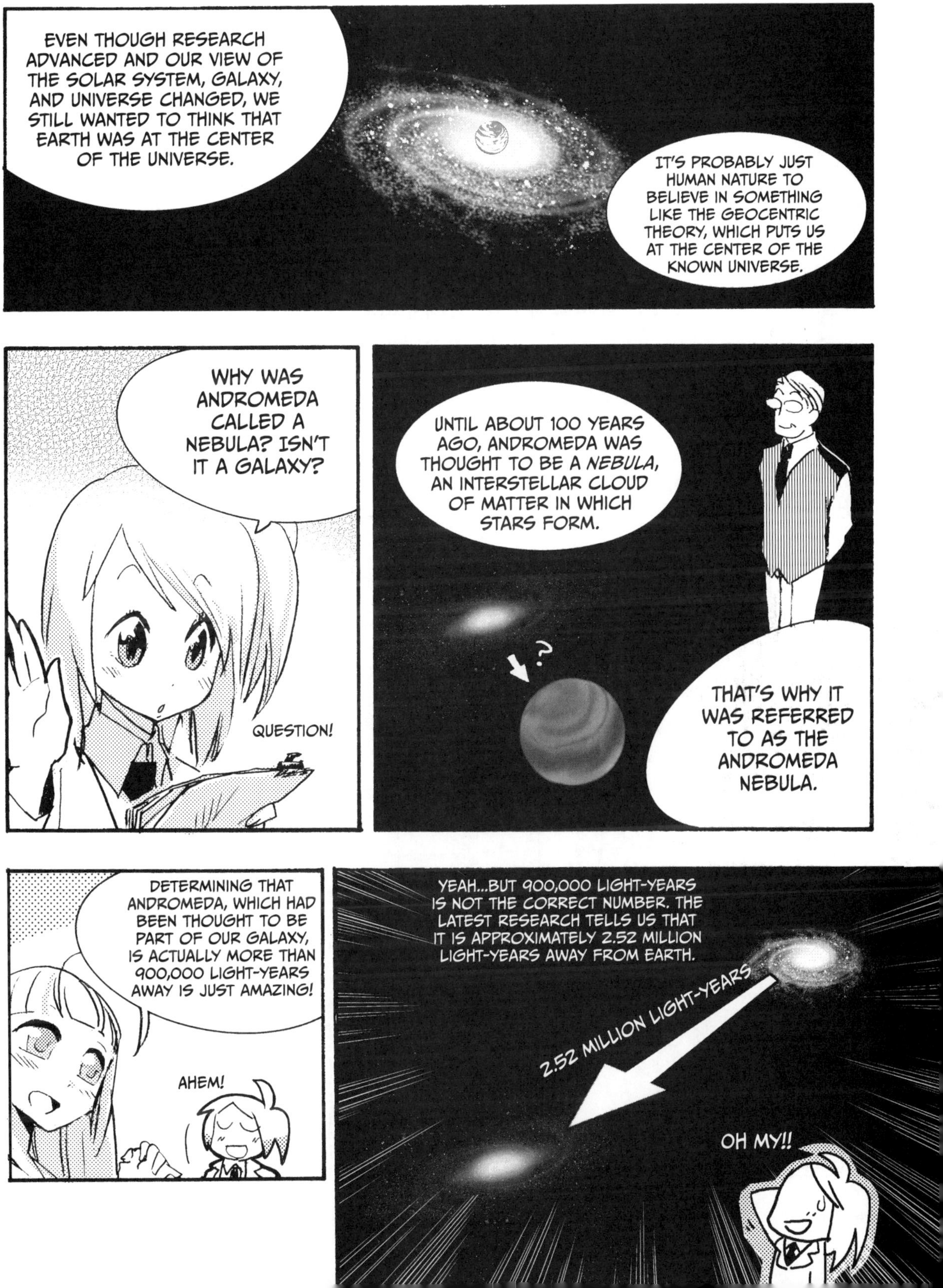
EVEN THOUGH RESEARCH ADVANCED AND OUR VIEW OF THE SOLAR SYSTEM, GALAXY, AND UNIVERSE CHANGED, WE STILL WANTED TO THINK THAT EARTH WAS AT THE CENTER OF THE UNIVERSE.
IT'S PROBABLY JUST HUMAN NATURE TO BELIEVE IN SOMETHING LIKE THE GEOCENTRIC THEORY, WHICH PUTS US AT THE CENTER OF THE KNOWN UNIVERSE.
WHY WAS ANDROMEDA CALLED A NEBULA? ISN'T IT A GALAXY?
QUESTION!
UNTIL ABOUT 100 YEARS AGO, ANDROMEDA WAS THOUGHT TO BE A *NEBULA*, AN INTERSTELLAR CLOUD OF MATTER IN WHICH STARS FORM.
?
THAT'S WHY IT WAS REFERRED TO AS THE ANDROMEDA NEBULA.
DETERMINING THAT ANDROMEDA, WHICH HAD BEEN THOUGHT TO BE PART OF OUR GALAXY, IS ACTUALLY MORE THAN 900,000 LIGHT-YEARS AWAY IS JUST AMAZING!
AHEM!
YEAH...BUT 900,000 LIGHT-YEARS IS NOT THE CORRECT NUMBER. THE LATEST RESEARCH TELLS US THAT IT IS APPROXIMATELY 2.52 MILLION LIGHT-YEARS AWAY FROM EARTH.
2.52 MILLION LIGHT-YEARS
OH MY!!

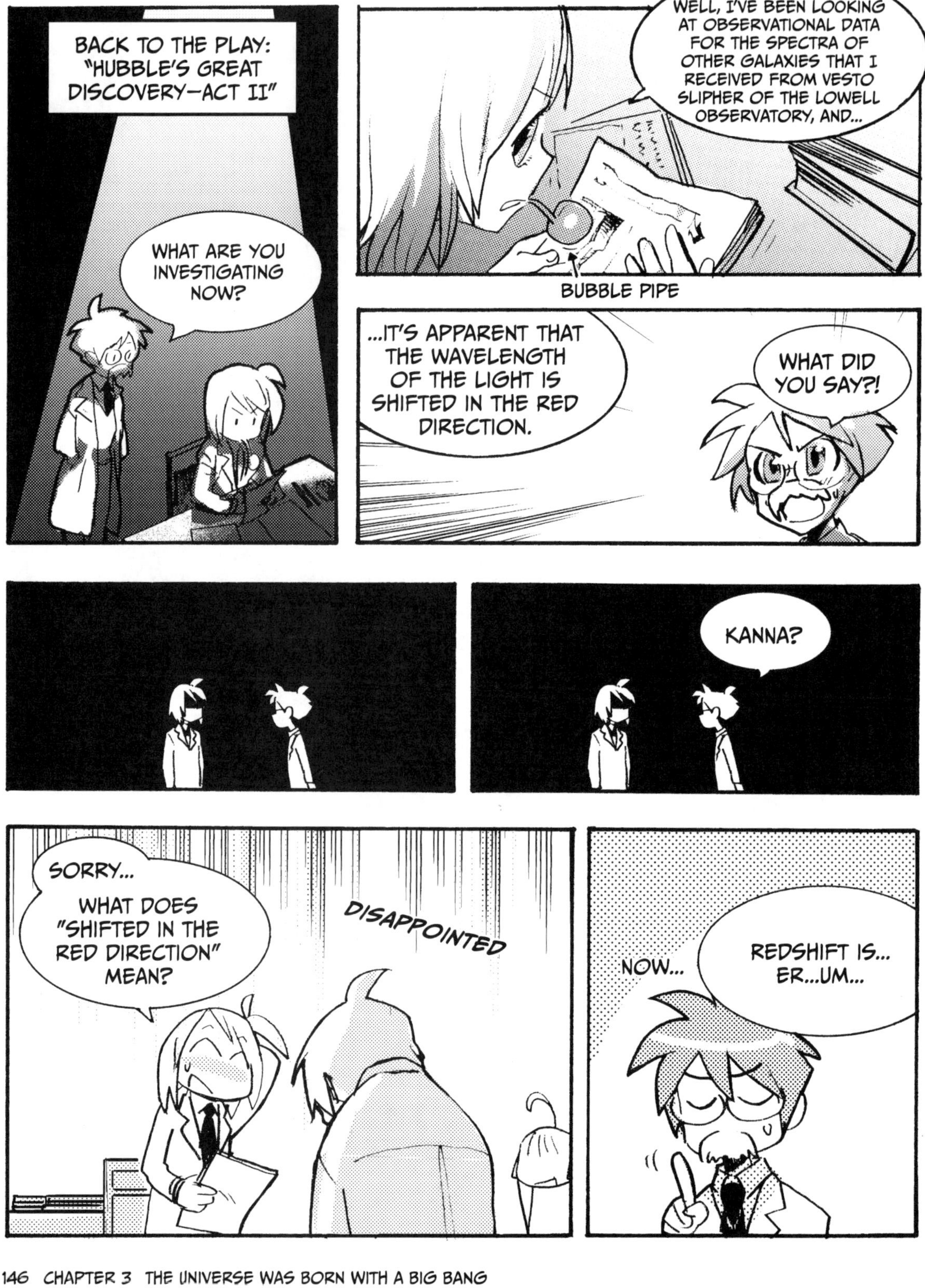
BACK TO THE PLAY: "HUBBLE'S GREAT DISCOVERY—ACT II"
WHAT ARE YOU INVESTIGATING NOW?
WELL, I'VE BEEN LOOKING AT OBSERVATIONAL DATA FOR THE SPECTRA OF OTHER GALAXIES THAT I RECEIVED FROM VESTO SLIPHER OF THE LOWELL OBSERVATORY, AND...
BUBBLE PIPE
...IT'S APPARENT THAT THE WAVELENGTH OF THE LIGHT IS SHIFTED IN THE RED DIRECTION.
WHAT DID YOU SAY?!
KANNA?
SORRY... WHAT DOES "SHIFTED IN THE RED DIRECTION" MEAN?
DISAPPOINTED
NOW...
REDSHIFT IS... ER...UM...

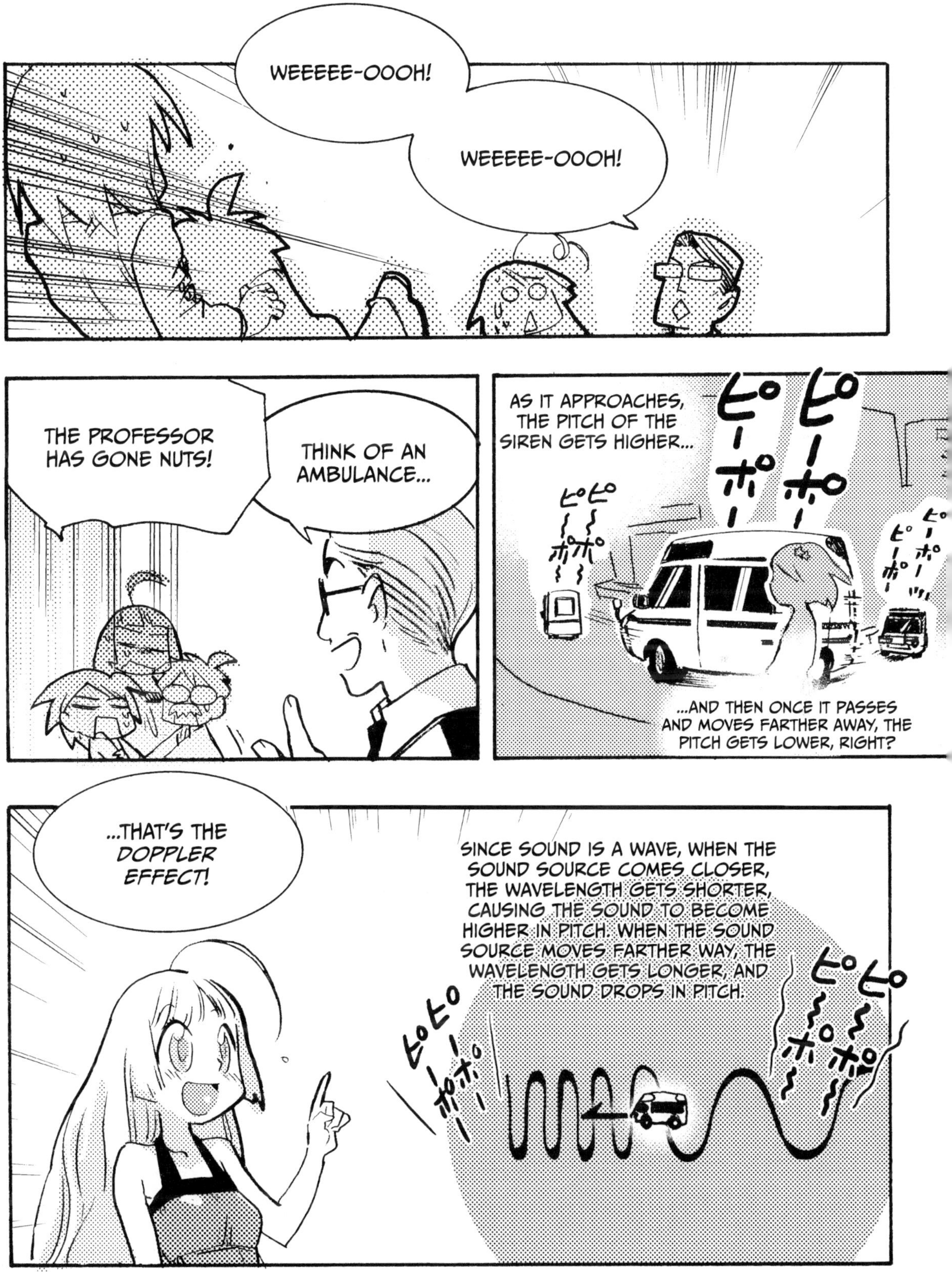
WEEEEE-OOOH!
WEEEEE-OOOH!
THE PROFESSOR HAS GONE NUTS!
THINK OF AN AMBULANCE...
AS IT APPROACHES, THE PITCH OF THE SIREN GETS HIGHER...
ピーポー ピーポー
...AND THEN ONCE IT PASSES AND MOVES FARTHER AWAY, THE PITCH GETS LOWER, RIGHT?
...THAT'S THE DOPPLER EFFECT!
SINCE SOUND IS A WAVE, WHEN THE SOUND SOURCE COMES CLOSER, THE WAVELENGTH GETS SHORTER, CAUSING THE SOUND TO BECOME HIGHER IN PITCH. WHEN THE SOUND SOURCE MOVES FARTHER WAY, THE WAVELENGTH GETS LONGER, AND THE SOUND DROPS IN PITCH.

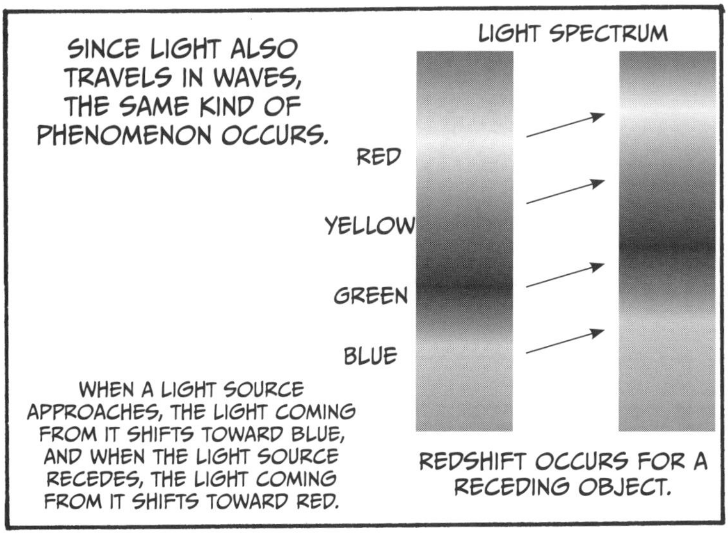
SINCE LIGHT ALSO TRAVELS IN WAVES, THE SAME KIND OF PHENOMENON OCCURS.
LIGHT SPECTRUM
RED
YELLOW
GREEN
BLUE
WHEN A LIGHT SOURCE APPROACHES, THE LIGHT COMING FROM IT SHIFTS TOWARD BLUE, AND WHEN THE LIGHT SOURCE RECEDES, THE LIGHT COMING FROM IT SHIFTS TOWARD RED.
REDSHIFT OCCURS FOR A RECEDING OBJECT.

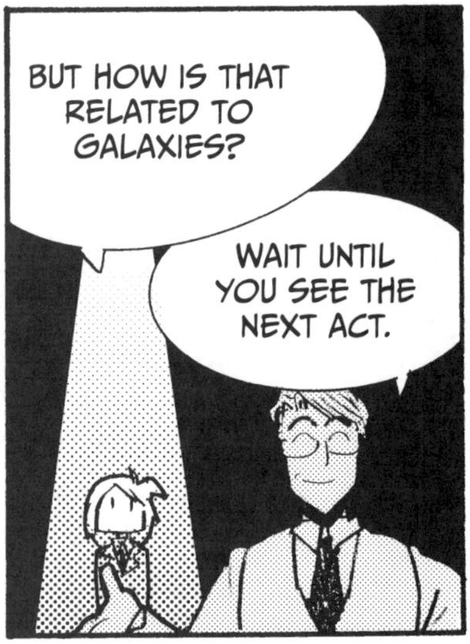
BUT HOW IS THAT RELATED TO GALAXIES?
WAIT UNTIL YOU SEE THE NEXT ACT.

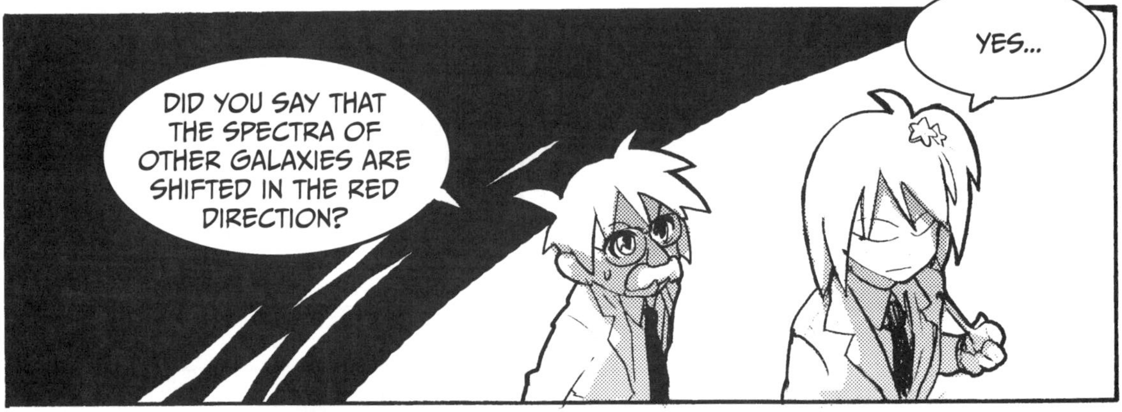
DID YOU SAY THAT THE SPECTRA OF OTHER GALAXIES ARE SHIFTED IN THE RED DIRECTION?
YES...

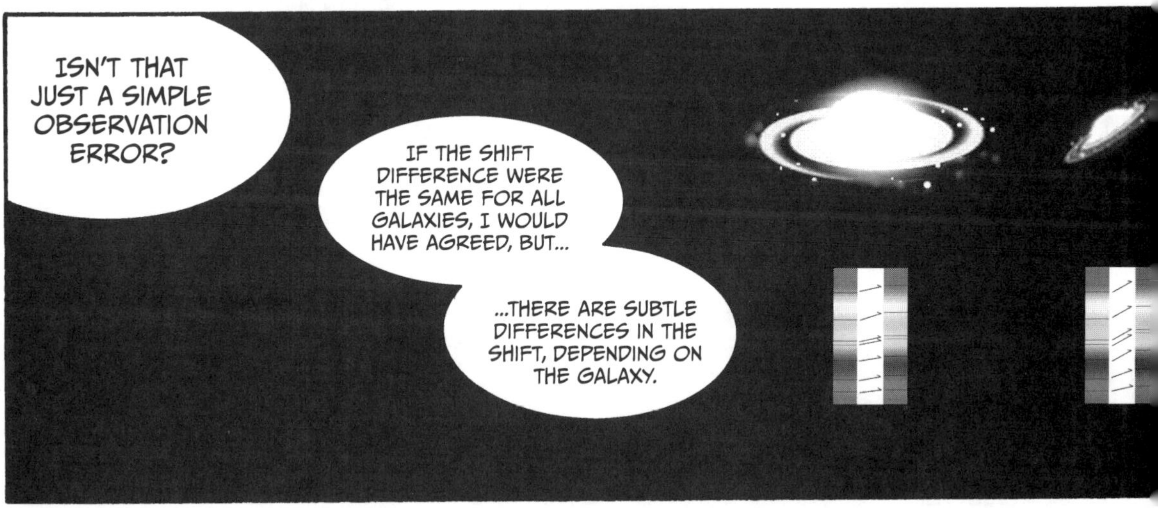
ISN'T THAT JUST A SIMPLE OBSERVATION ERROR?
IF THE SHIFT DIFFERENCE WERE THE SAME FOR ALL GALAXIES, I WOULD HAVE AGREED, BUT...
...THERE ARE SUBTLE DIFFERENCES IN THE SHIFT, DEPENDING ON THE GALAXY.

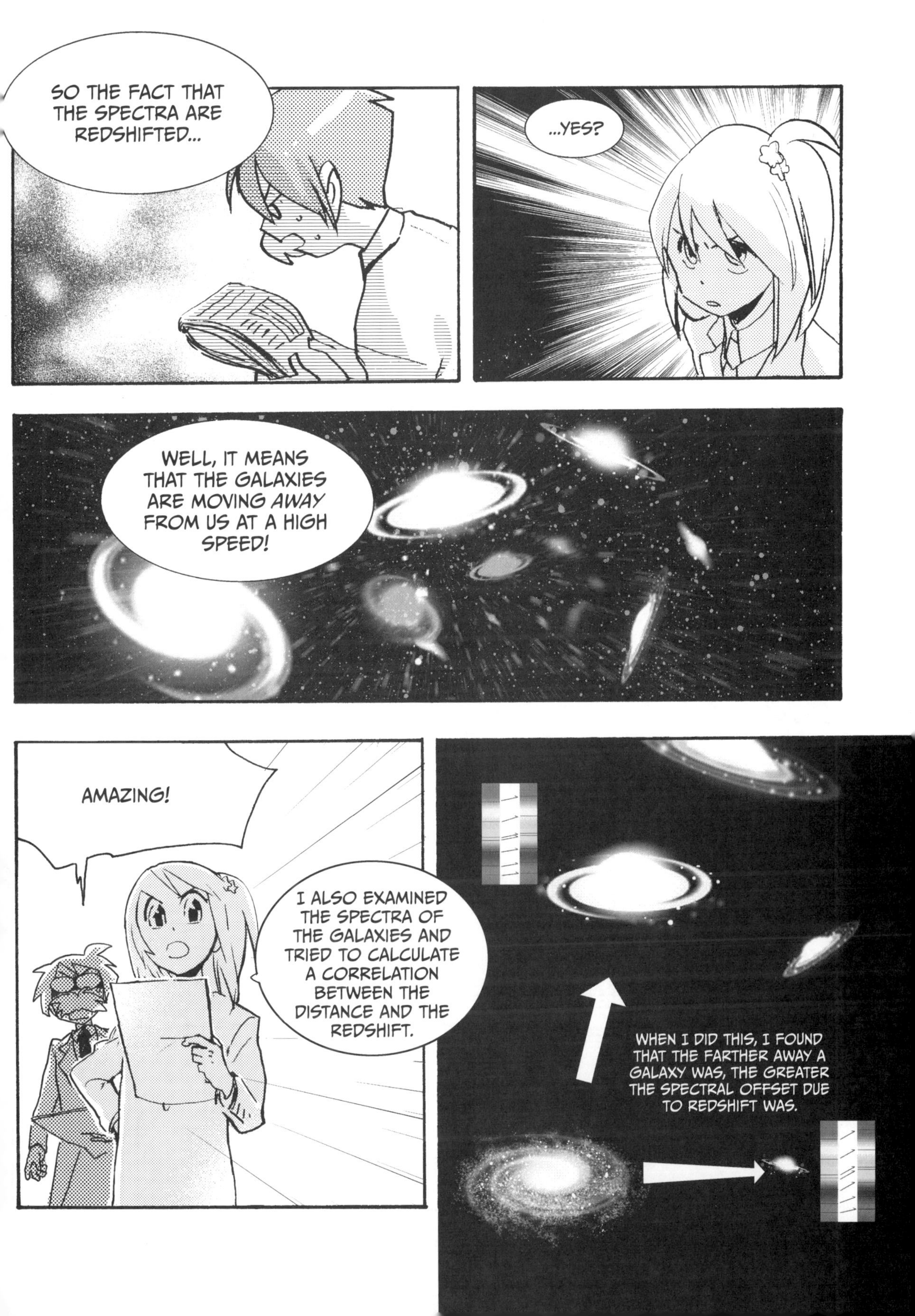
SO THE FACT THAT THE SPECTRA ARE REDSHIFTED...
...YES?
WELL, IT MEANS THAT THE GALAXIES ARE MOVING *AWAY* FROM US AT A HIGH SPEED!
AMAZING!
I ALSO EXAMINED THE SPECTRA OF THE GALAXIES AND TRIED TO CALCULATE A CORRELATION BETWEEN THE DISTANCE AND THE REDSHIFT.
WHEN I DID THIS, I FOUND THAT THE FARTHER AWAY A GALAXY WAS, THE GREATER THE SPECTRAL OFFSET DUE TO REDSHIFT WAS.

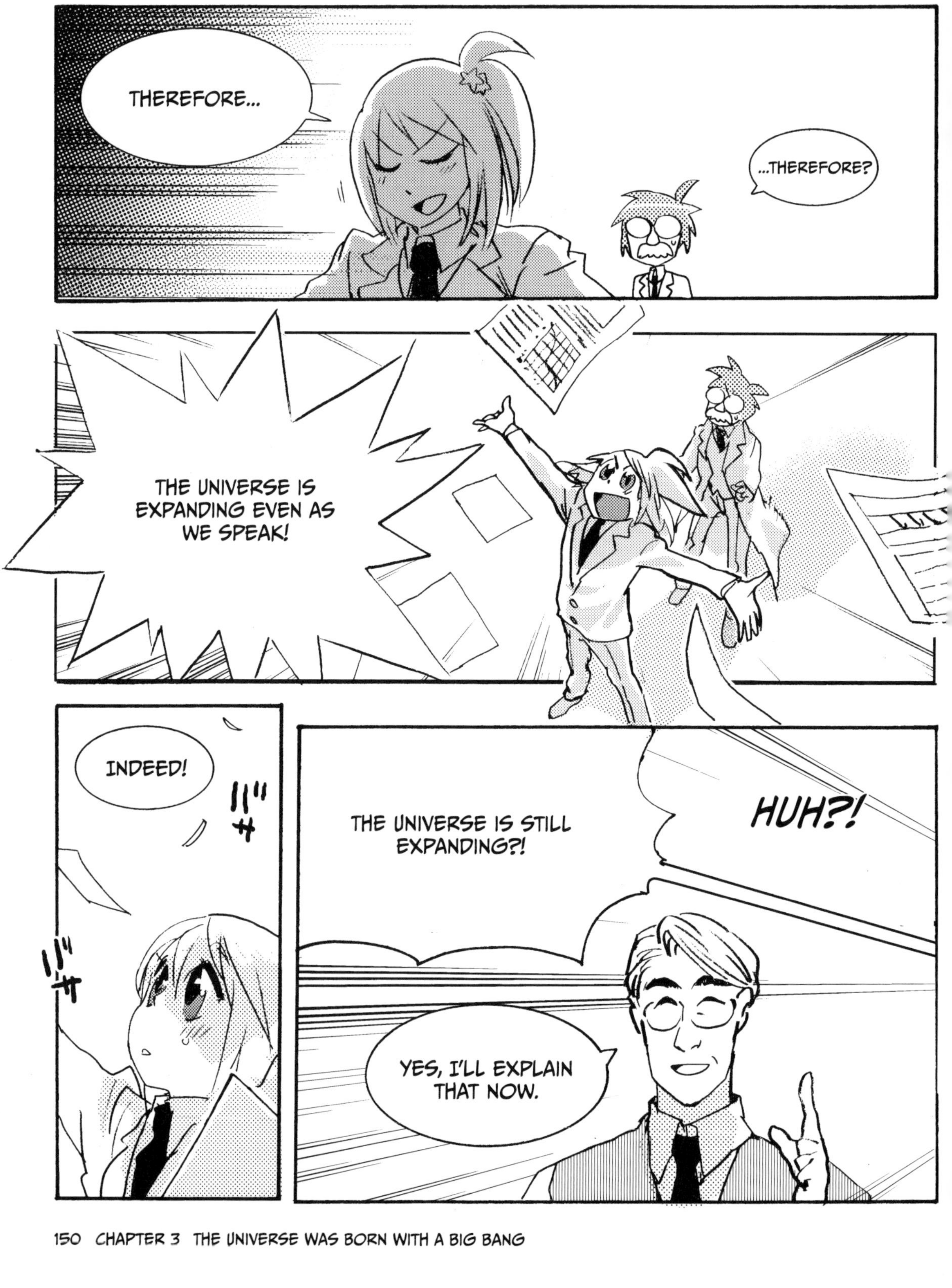
THEREFORE...
...THEREFORE?
THE UNIVERSE IS EXPANDING EVEN AS WE SPEAK!
INDEED!
THE UNIVERSE IS STILL EXPANDING?!
HUH?!
YES, I'LL EXPLAIN THAT NOW.

IF THE UNIVERSE IS EXPANDING...

Let me explain this in a little more detail.

Please do!

Have you all done a flame reaction test in chemistry class?

Hmm, did I do that?

This is the phenomenon that occurs when a substance is placed in a flame and the color varies according to the chemical elements that are contained in the substance.

Salt is yellow since it contains sodium, and copper is blue-green.

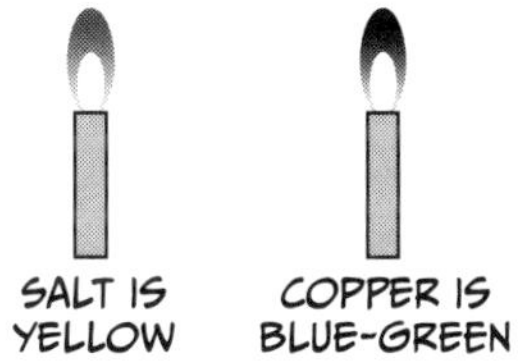
SALT IS YELLOW
COPPER IS BLUE-GREEN

That's right. The chemical elements emit or absorb light with specific wavelengths according to their structures. Therefore, if you use a prism to separate the light emitted by a heavenly body according to wavelengths, the prism creates a rainbow that you can analyze to find out the substances that are contained in that heavenly body.

Does the spectrum differ much depending on the star?

Yes and no. Since most stars are composed of very similar substances,* one would assume that they would emit spectra that are also very similar. Stars are mainly composed of hydrogen and helium plus smaller amounts of a few heavier elements. However, different stars actually emit very different spectra, and that is because different stars vary in temperature.

* The chemical composition of stars in terms of mass ratio is predominantly hydrogen:helium = 3:1 with a very small percentage of other heavier elements, usually less than 2%. This ratio barely differs from star to star.

Very hot stars emit spectra that have prominent lines for helium and the ionized heavy elements (that is, atoms that have gained or lost an electron or electrons). On the other hand, very cool stars emit spectra with no visible helium lines but with lines for neutral atoms and molecules. Nevertheless, if temperature differences cause differences in spectra, then similar stars (that is, stars with similar mass and temperature) should have practically identical spectra. However, Slipher (see page 146) discovered that a shift toward red occurred in the wavelengths.

 If they were receding, table salt would become more orange, and copper would become yellowish green.

 Well, then Slipher, who gave the data to Hubble, seems to be the person who noticed that the universe is expanding—not Hubble, right?

 But Slipher isn't very famous in the United States.

 He believed that the redshift meant that many galaxies were receding from us, but he thought it was due to the basic motion of heavenly bodies. Hubble, however, conducted a thorough investigation of the correlation between distance and redshift. He discovered that the farther away a galaxy is, the faster it is receding—a fact that supports the theory of an expanding universe.

 But why does that mean the universe itself is expanding?

 You'll understand it if we write three letters on a balloon and then inflate it.

A—B—C
NOW...
EACH LETTER IS EVENLY SPACED.
BUT WATCH THE A-B AND A-C INTERVALS CAREFULLY.
THIS IS HARDER THAN IT LOOKS!
A-B-C
A—B—C
AS THE BALLOON GETS LARGER, THE DISTANCE BETWEEN ANY TWO POINTS STEADILY INCREASES.
HOWEVER, THE A-C INTERVAL INCREASES FASTER THAN THE A-B INTERVAL.
A-B—C

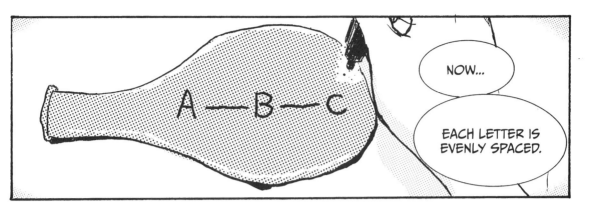

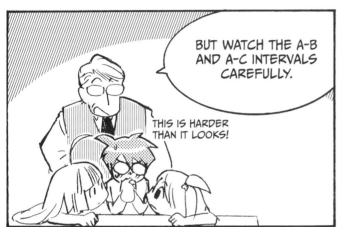

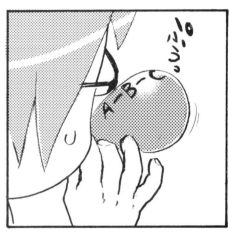

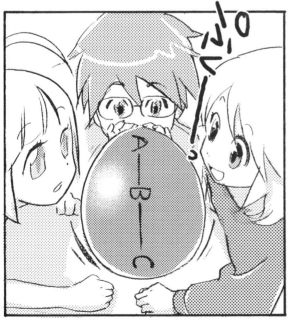

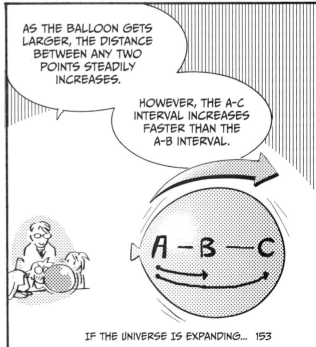

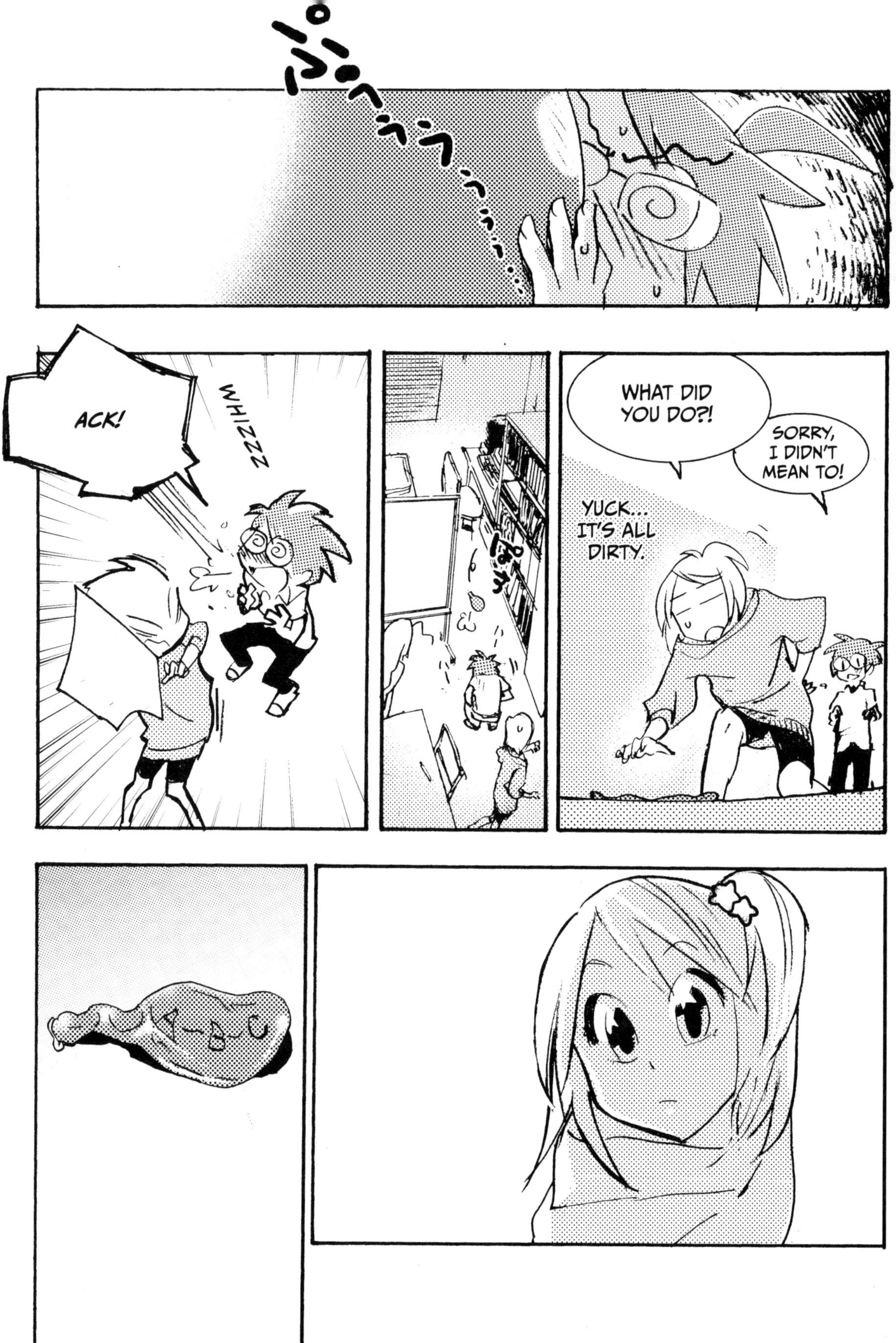

ACK!
WHIZZZ
WHAT DID YOU DO?!
SORRY, I DIDN'T MEAN TO!
YUCK... IT'S ALL DIRTY.

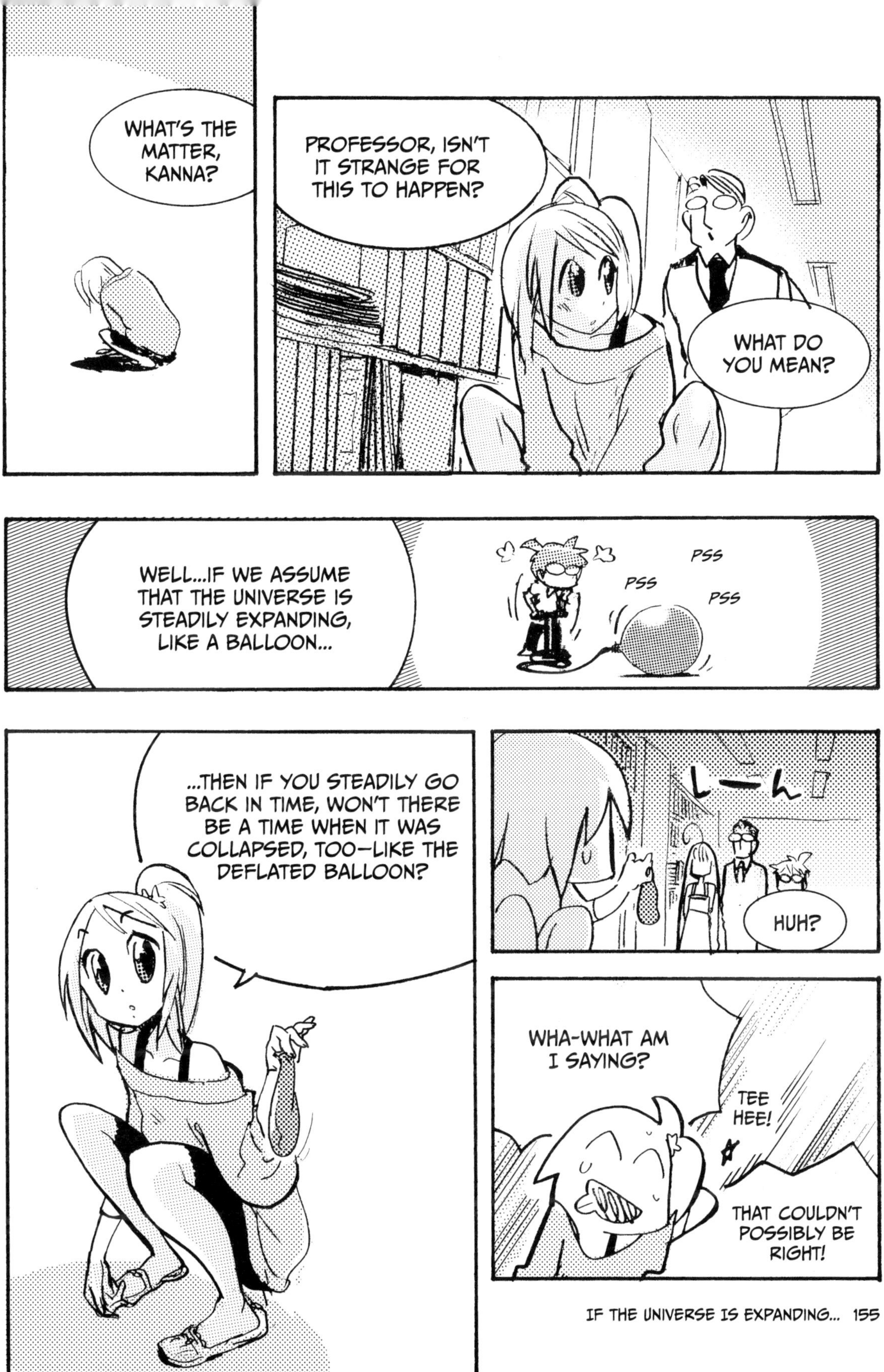
WHAT'S THE MATTER, KANNA?
PROFESSOR, ISN'T IT STRANGE FOR THIS TO HAPPEN?
WHAT DO YOU MEAN?
WELL...IF WE ASSUME THAT THE UNIVERSE IS STEADILY EXPANDING, LIKE A BALLOON...
PSS
PSS
PSS
...THEN IF YOU STEADILY GO BACK IN TIME, WON'T THERE BE A TIME WHEN IT WAS COLLAPSED, TOO—LIKE THE DEFLATED BALLOON?
しーん
HUH?
WHA-WHAT AM I SAYING?
TEE HEE!
THAT COULDN'T POSSIBLY BE RIGHT!

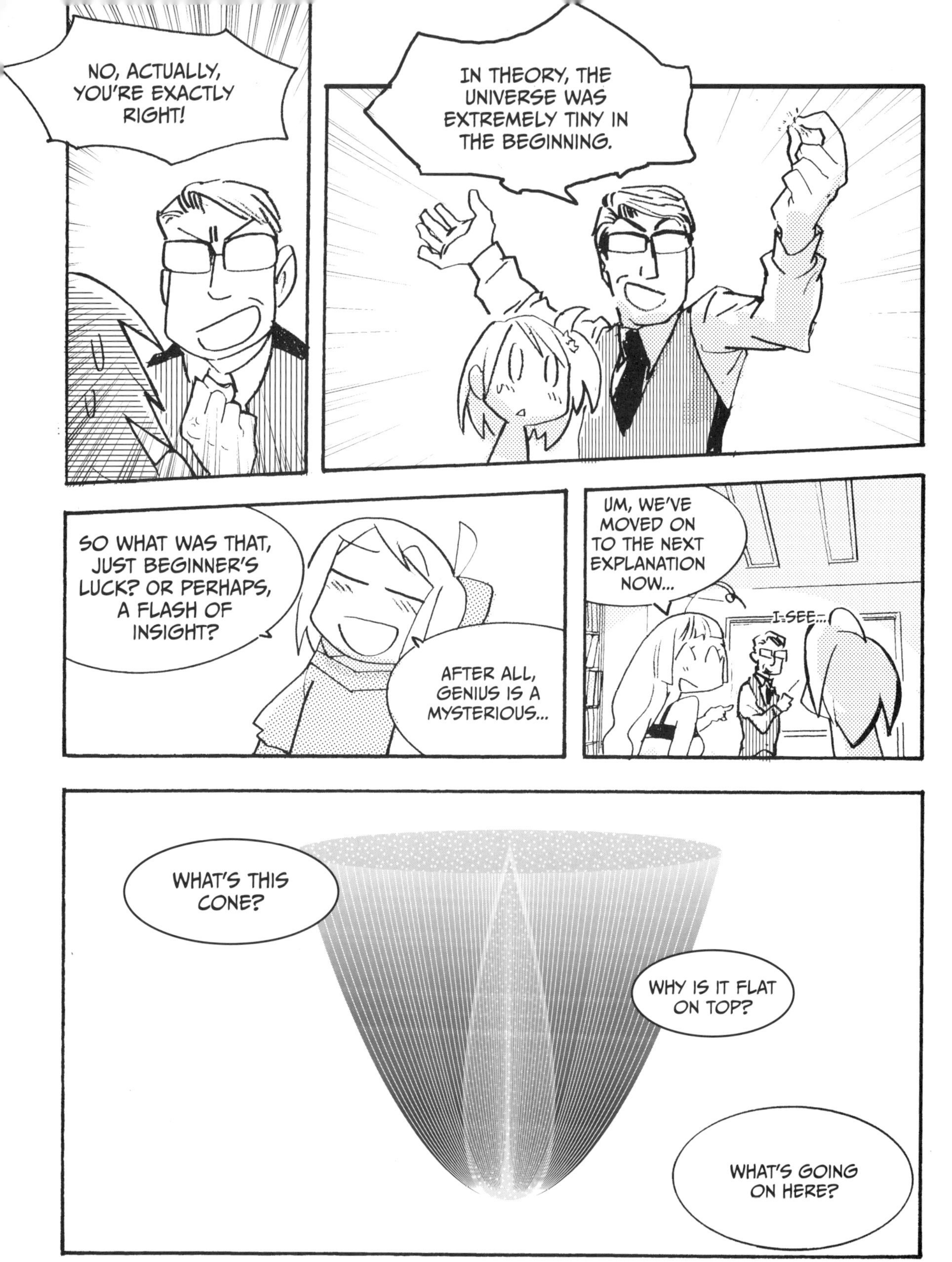
NO, ACTUALLY, YOU'RE EXACTLY RIGHT!
IN THEORY, THE UNIVERSE WAS EXTREMELY TINY IN THE BEGINNING.
SO WHAT WAS THAT, JUST BEGINNER'S LUCK? OR PERHAPS, A FLASH OF INSIGHT?
AFTER ALL, GENIUS IS A MYSTERIOUS...
UM, WE'VE MOVED ON TO THE NEXT EXPLANATION NOW...
I SEE...
WHAT'S THIS CONE?
WHY IS IT FLAT ON TOP?
WHAT'S GOING ON HERE?

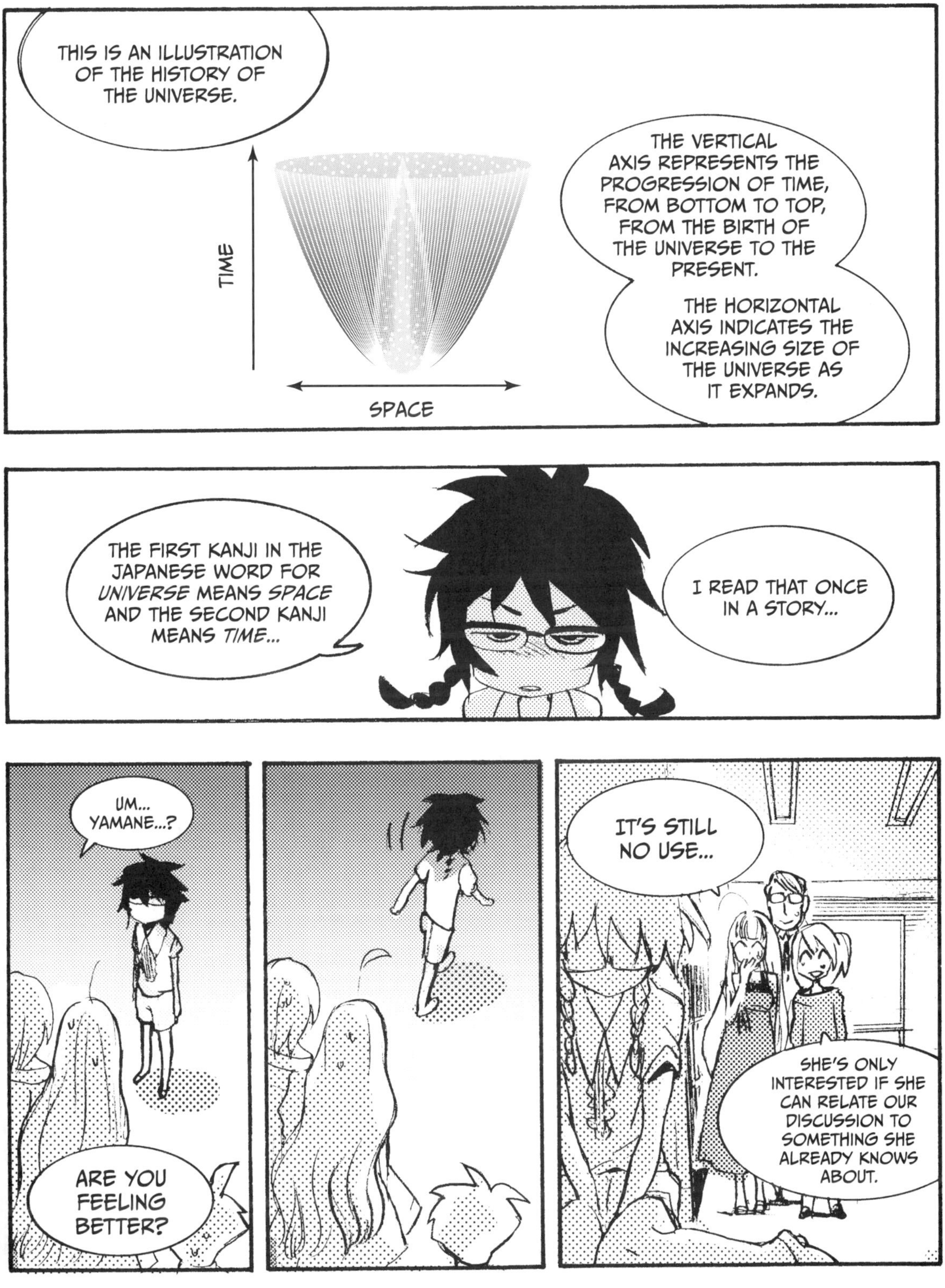
THIS IS AN ILLUSTRATION OF THE HISTORY OF THE UNIVERSE.
TIME
SPACE
THE VERTICAL AXIS REPRESENTS THE PROGRESSION OF TIME, FROM BOTTOM TO TOP, FROM THE BIRTH OF THE UNIVERSE TO THE PRESENT.
THE HORIZONTAL AXIS INDICATES THE INCREASING SIZE OF THE UNIVERSE AS IT EXPANDS.
THE FIRST KANJI IN THE JAPANESE WORD FOR *UNIVERSE* MEANS *SPACE* AND THE SECOND KANJI MEANS *TIME*...
I READ THAT ONCE IN A STORY...
UM... YAMANE...?
ARE YOU FEELING BETTER?
IT'S STILL NO USE...
SHE'S ONLY INTERESTED IF SHE CAN RELATE OUR DISCUSSION TO SOMETHING SHE ALREADY KNOWS ABOUT.

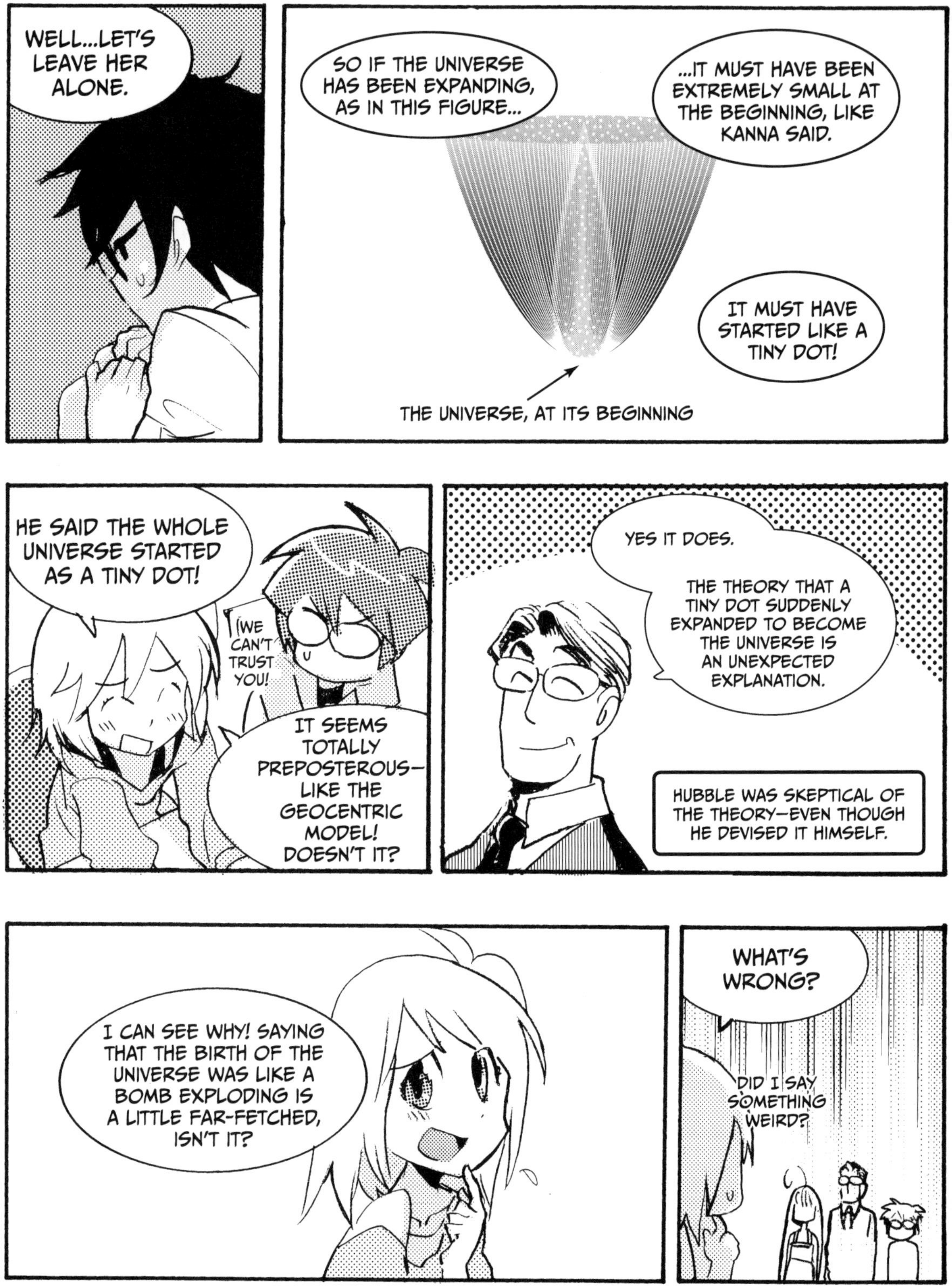
WELL...LET'S LEAVE HER ALONE.
SO IF THE UNIVERSE HAS BEEN EXPANDING, AS IN THIS FIGURE...
...IT MUST HAVE BEEN EXTREMELY SMALL AT THE BEGINNING, LIKE KANNA SAID.
IT MUST HAVE STARTED LIKE A TINY DOT!
THE UNIVERSE, AT ITS BEGINNING
HE SAID THE WHOLE UNIVERSE STARTED AS A TINY DOT!
WE CAN'T TRUST YOU!
IT SEEMS TOTALLY PREPOSTEROUS—LIKE THE GEOCENTRIC MODEL! DOESN'T IT?
YES IT DOES.
THE THEORY THAT A TINY DOT SUDDENLY EXPANDED TO BECOME THE UNIVERSE IS AN UNEXPECTED EXPLANATION.
HUBBLE WAS SKEPTICAL OF THE THEORY—EVEN THOUGH HE DEVISED IT HIMSELF.
I CAN SEE WHY! SAYING THAT THE BIRTH OF THE UNIVERSE WAS LIKE A BOMB EXPLODING IS A LITTLE FAR-FETCHED, ISN'T IT?
WHAT'S WRONG?
DID I SAY SOMETHING WEIRD?

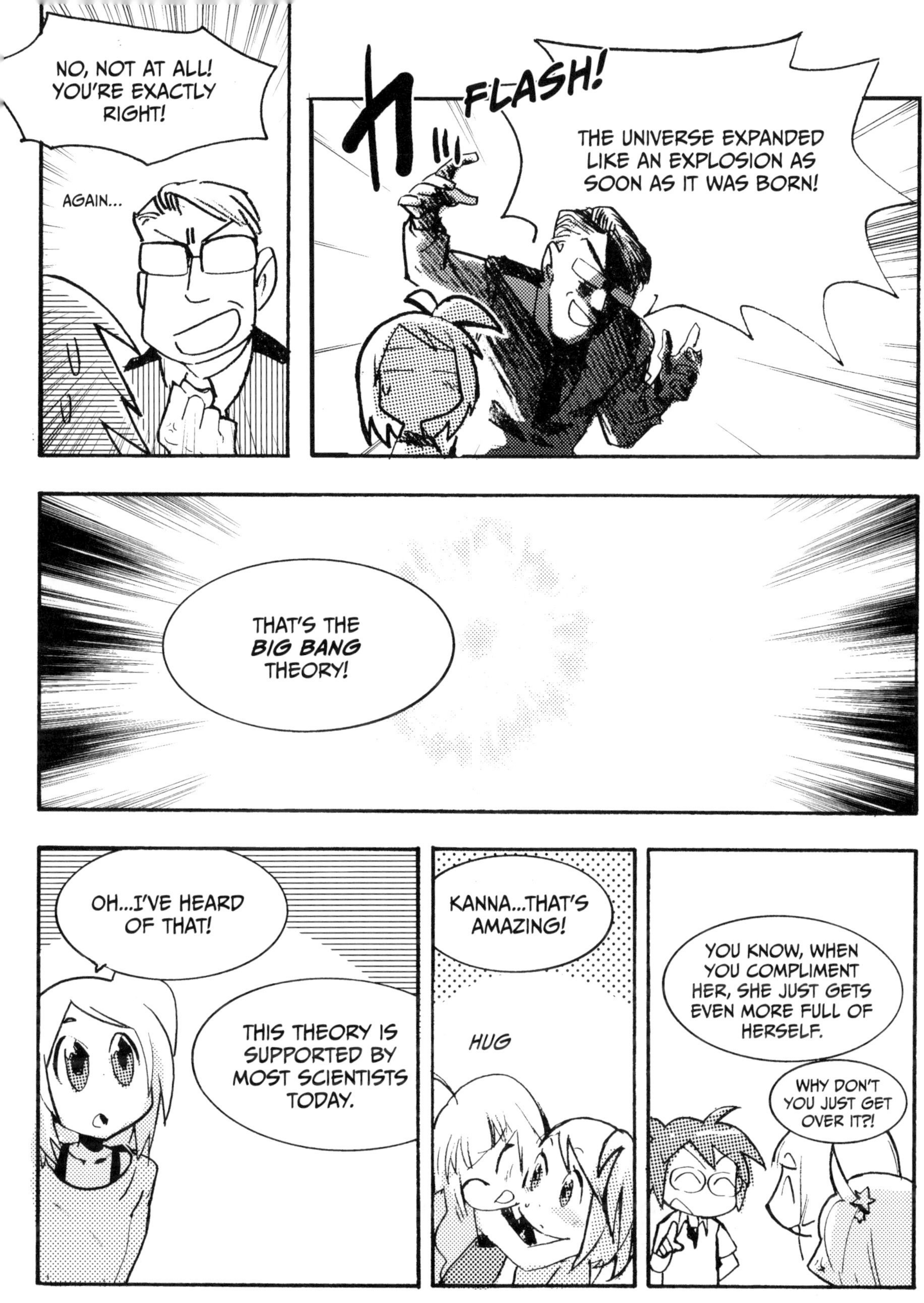
NO, NOT AT ALL! YOU'RE EXACTLY RIGHT!
AGAIN...
FLASH!
THE UNIVERSE EXPANDED LIKE AN EXPLOSION AS SOON AS IT WAS BORN!
THAT'S THE *BIG BANG* THEORY!
OH...I'VE HEARD OF THAT!
THIS THEORY IS SUPPORTED BY MOST SCIENTISTS TODAY.
KANNA...THAT'S AMAZING!
HUG
YOU KNOW, WHEN YOU COMPLIMENT HER, SHE JUST GETS EVEN MORE FULL OF HERSELF.
WHY DON'T YOU JUST GET OVER IT?!

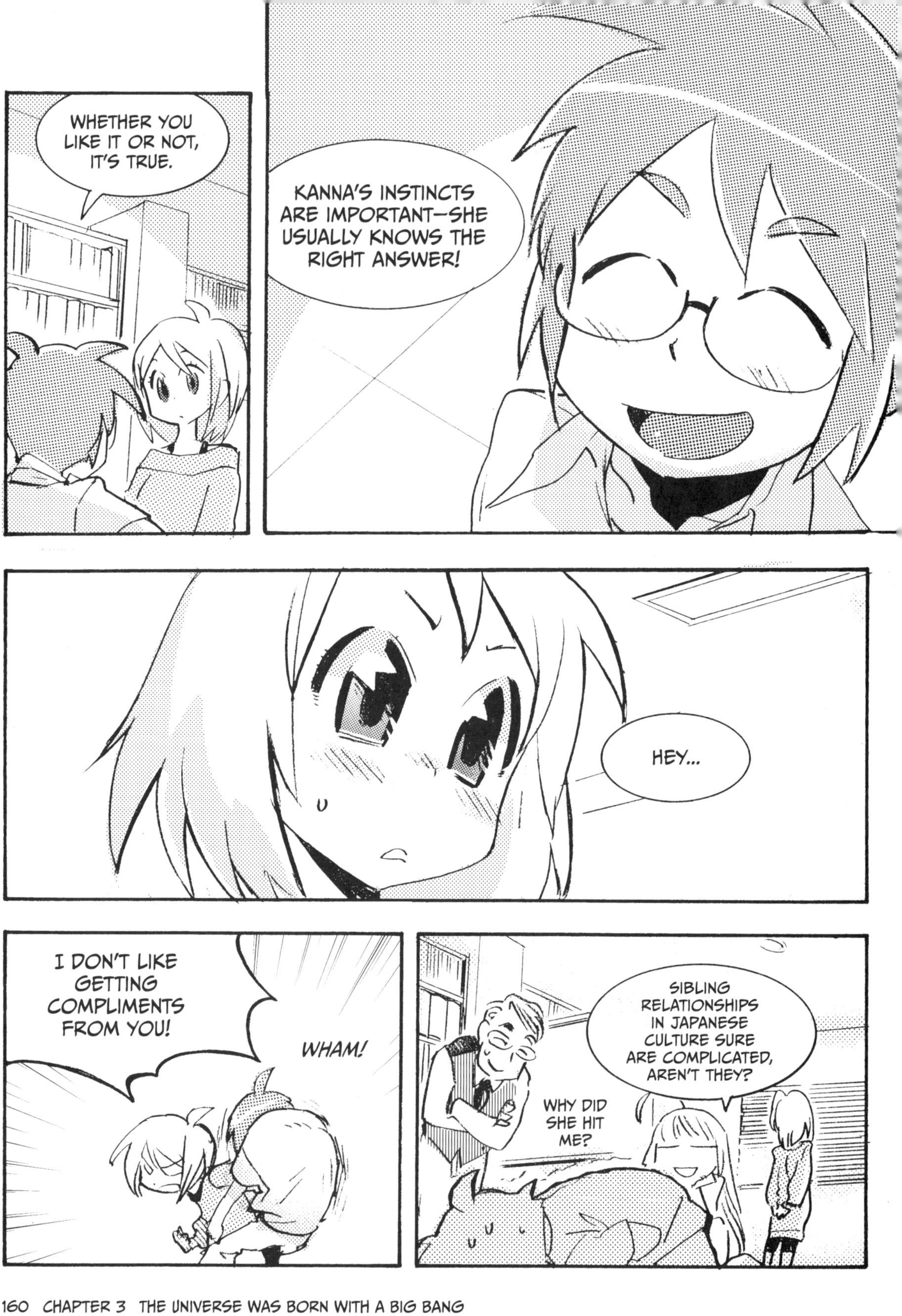
WHETHER YOU LIKE IT OR NOT, IT'S TRUE.
KANNA'S INSTINCTS ARE IMPORTANT—SHE USUALLY KNOWS THE RIGHT ANSWER!
HEY...
I DON'T LIKE GETTING COMPLIMENTS FROM YOU!
WHAM!
SIBLING RELATIONSHIPS IN JAPANESE CULTURE SURE ARE COMPLICATED, AREN'T THEY?
WHY DID SHE HIT ME?

EVERYTHING STARTED WITH THE BIG BANG

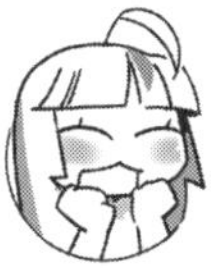

So approximately when did the Big Bang occur?

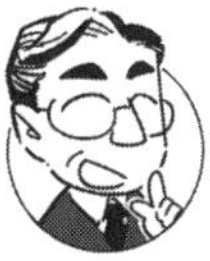

Calculating backward using the velocity at which the universe is expanding, it seems to have been about 13.7 billion years ago—plus or minus about 1 billion years.

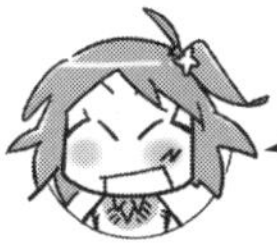

Wait a minute! What about the location? Professor, where did the Big Bang occur?

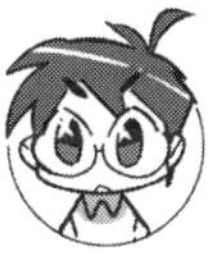

What are you talking about? Since the "location" called space came into being after the Big Bang, there couldn't be any concept of "where" before it occurred.

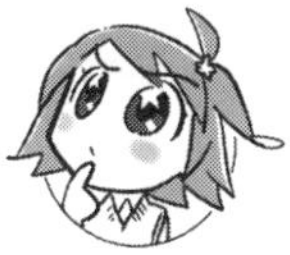

There was no location?

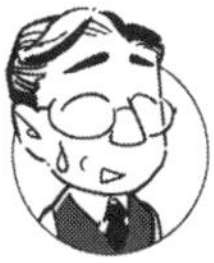

This is hard to understand, isn't it? An explosive phenomenon did not occur somewhere in space...space was created by the Big Bang. Not only that, but all substances as well as time originated from the Big Bang. Therefore, if it's a question of where, then everywhere in our entire universe is the spot where the Big Bang occurred.

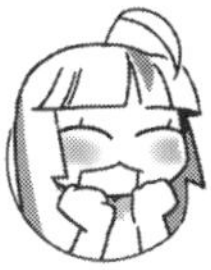

I think I get it. Remember the experiment when we inflated the balloon? After the balloon is inflated, there is no way to answer a question like "What part of the inflated balloon corresponds to the balloon before we inflated it?" It's the same thing with the universe.

HUBBLE'S THEORY OF THE EXPANSION OF THE UNIVERSE WAS IMPERFECT

Although Hubble discovered that the universe is expanding, it seems highly likely that he would have encountered fierce opposition from people around him, especially from his fellow researchers in the astronomical community. In fact, many accounts say that Hubble very prudently kept his ideas to himself.

The value that determines the expansion velocity of the universe is called *Hubble's constant*, and it is represented by H_0. Hubble obtained a value of 500 km/sec/Mpc for this velocity; this is approximately seven times the currently accepted value (74 km/sec/Mpc). But if Hubble's value is used to calculate how long ago the birth of the universe (that is, the Big Bang) occurred, it would indicate that the universe began 2 billion years ago, at most. Since scientists have found the age of Earth to be approximately 4.6 billion years (based on observations of rocks and fossils), then if we use Hubble's initial value, the universe would be younger than Earth, which it clearly cannot be.

Even Hubble himself did not obtain the correct value for Hubble's constant. It would take a very long time for the theory of the expansion of the universe to be widely accepted.

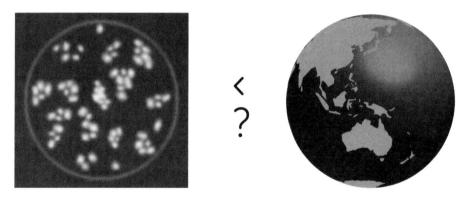

Hubble's experiments indicated that the entire universe was only 2 billion years old...
but that didn't make sense, as scientists had estimated that Earth itself was 4.6 billion years old!

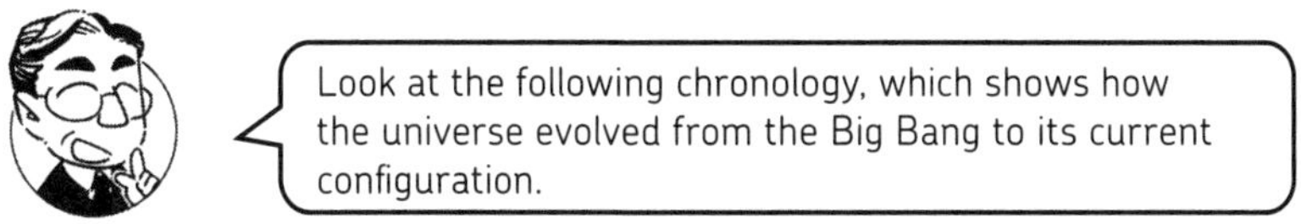
Look at the following chronology, which shows how the universe evolved from the Big Bang to its current configuration.

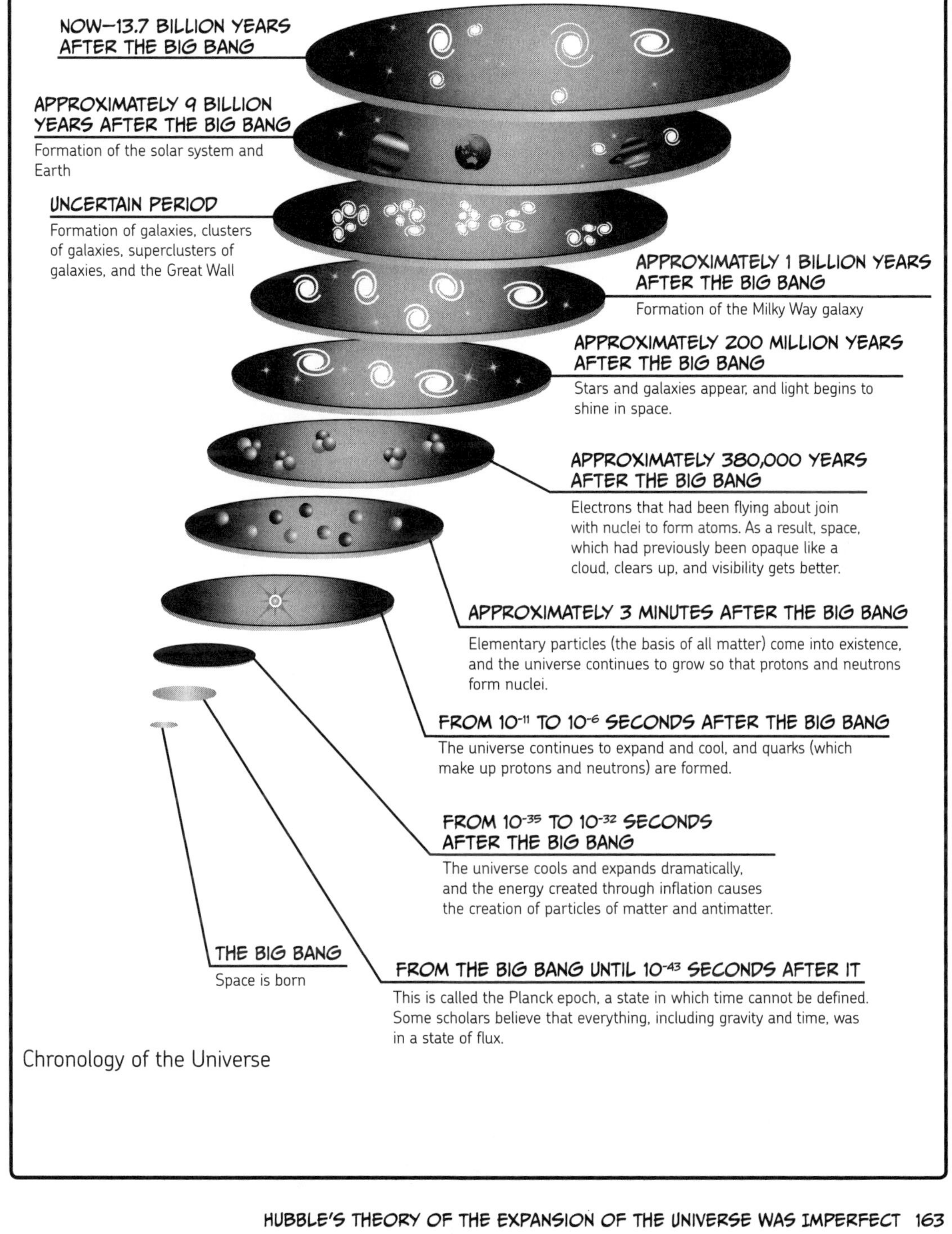
NOW—13.7 BILLION YEARS AFTER THE BIG BANG
APPROXIMATELY 9 BILLION YEARS AFTER THE BIG BANG
Formation of the solar system and Earth
UNCERTAIN PERIOD
Formation of galaxies, clusters of galaxies, superclusters of galaxies, and the Great Wall
APPROXIMATELY 1 BILLION YEARS AFTER THE BIG BANG
Formation of the Milky Way galaxy
APPROXIMATELY 200 MILLION YEARS AFTER THE BIG BANG
Stars and galaxies appear, and light begins to shine in space.
APPROXIMATELY 380,000 YEARS AFTER THE BIG BANG
Electrons that had been flying about join with nuclei to form atoms. As a result, space, which had previously been opaque like a cloud, clears up, and visibility gets better.
APPROXIMATELY 3 MINUTES AFTER THE BIG BANG
Elementary particles (the basis of all matter) come into existence, and the universe continues to grow so that protons and neutrons form nuclei.
FROM 10^{-11} TO 10^{-6} SECONDS AFTER THE BIG BANG
The universe continues to expand and cool, and quarks (which make up protons and neutrons) are formed.
FROM 10^{-35} TO 10^{-32} SECONDS AFTER THE BIG BANG
The universe cools and expands dramatically, and the energy created through inflation causes the creation of particles of matter and antimatter.
THE BIG BANG
Space is born
FROM THE BIG BANG UNTIL 10^{-43} SECONDS AFTER IT
This is called the Planck epoch, a state in which time cannot be defined. Some scholars believe that everything, including gravity and time, was in a state of flux.

Chronology of the Universe

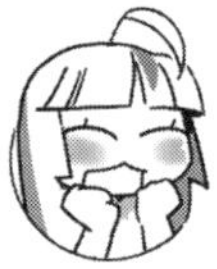

So there was a history of the universe even before the Big Bang?

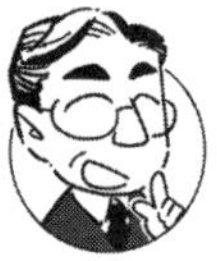

This part is still not perfectly understood. A particularly mysterious interval is the time from when the universe was born until it was 10^{-43} of a second old, which is called the *Planck epoch*. Since nothing at all existed to measure time or gravity, no physics has been created yet for describing what occurred and how the universe evolved during this period. Therefore, nothing can be said other than that the universe was born and took a certain form by the time 10^{-43} seconds had elapsed.

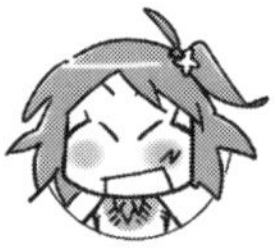

So there is even something that the Professor doesn't understand very well!

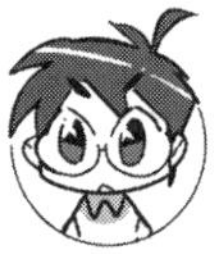

You seem so happy to hear that!

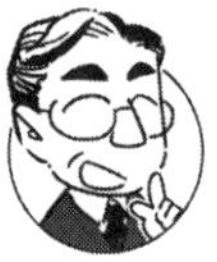

There are a great many things I don't understand. In terms of the balloon we talked about earlier, the birth of the universe corresponds to the time when the rubber that the balloon is made from was formed. However, there are various hypotheses about why that rubber suddenly expanded.

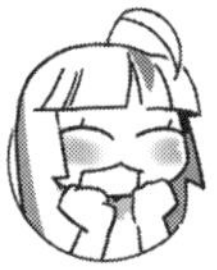

That period of expansion is called *inflation*, right?

Yes, you're right! Many astrophysicists seem to think that inflation occurred because of the energy created after the Big Bang.

 Do you also believe this, Professor?

 Although I support parts of the cosmic inflation theory, I have objections to a theory that holds that the universe was simply born from a state of "emptiness" in which nothing at all existed—and that not only space, but also time was born there.

 So time was born, too?

 Since the Planck epoch occurred immediately after the universe was born, we probably have to say that time was also born there. But I think it's meaningless to say that there was a time when there was no time. However, the universe was born into a state before which *nothing* existed. If we label this kind of activity "change" and refer to it as "time," then it seems reasonable to say that the time that was born with the birth of the universe is only the time that is specific to our universe. I think we can believe this.

 That makes sense to me. Boy, I didn't realize studying the universe would bring up these kinds of philosophical problems!

THREE PIECES OF EVIDENCE FOR THE BIG BANG THEORY

At first, most scientists considered the Big Bang theory to be outlandish. But over time, observations were made that supported the theory, and more scientists began to endorse it.

Evidence of the Big Bang, Exhibit A: Cosmic Microwave Background Radiation

In 1964, the US company Bell Laboratories made an accidental discovery while monitoring radio waves reflected from radio balloons. It determined that the background interference or "noise" in the data was being caused by a microwave signal of a specific wavelength that was coming from all directions in space. This signal is now known as *cosmic microwave background radiation (CMBR)*. Scholars who advocated the Big Bang theory hypothesized that this radiation was caused by the temperature of space (approximately 3,000 Kelvin (K), a temperature equivalent to almost 5,000°F) in the era when electrons and protons, which had been flying around freely until then, began to combine (approximately 380,000 years after the Big Bang). According to the hypothesis, atoms were created, space became transparent to radiation, and the electromagnetic waves that were emitted have now reached their current places in space due to the subsequent expansion of the universe. They exist with a frequency distribution that indicates a constant absolute temperature of 3 K. Observations of a temperature of 2.725 K support this hypothesis magnificently.

The CMBR, examined carefully by the Cosmic Background Explorer (COBE) satellite launched by NASA in 1989, has helped to advance astronomers' knowledge of the early universe. The accidental discoverers of CMBR were awarded the Nobel Prize in Physics in 1978.

Evidence of the Big Bang, Exhibit B: WMAP Satellite Measurements

The Wilkinson Microwave Anisotropy Probe (WMAP), which was launched in 2001 and operated until 2010, observed cosmic microwave background radiation temperatures across the entire sky. When its data were analyzed, it became apparent that more than 72 percent of the gravitational sources or structures in the universe consist of *dark energy* (energy held by the vacuum between the visible heavenly bodies), and matter makes up no more than the remaining approximately 28 percent. Incidentally, most of the matter is dark matter, while baryonic matter that we are familiar with (that is, "normal" matter consisting of protons, neutrons, and electrons) accounts for only approximately 4.6 percent. This result is consistent with the inflation theory, which asserts that there was a sudden expansion immediately after the universe was born.

Evidence of the Big Bang, Exhibit C: Chemical Composition of Stars

Various observations have demonstrated that the chemical composition of stars is hydrogen:helium = 3:1. The most logical explanation for why hydrogen and helium are present in such large quantities and in a specific ratio ties in nicely with the Big Bang theory. Hydrogen and helium are the lightest of all the chemical elements. The relative abundance in stars of the lightest element followed by the next-lightest element is consistent with the hypothesis that the universe began with an explosion. An explosion would result in high temperatures, which in turn would cause matter to organize in a way that would produce many small particles and fewer large ones.

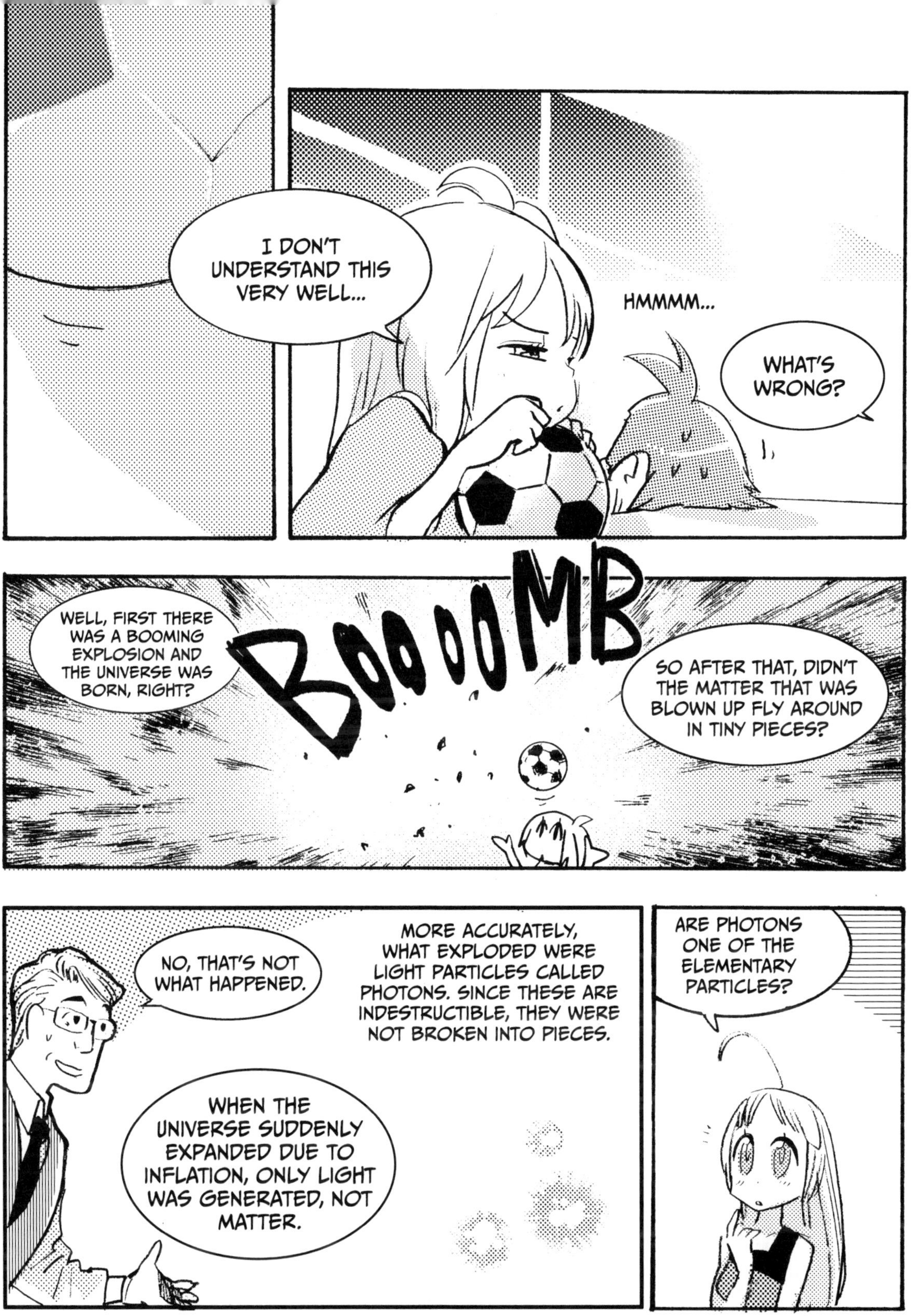
I DON'T UNDERSTAND THIS VERY WELL...
HMMMM...
WHAT'S WRONG?
WELL, FIRST THERE WAS A BOOMING EXPLOSION AND THE UNIVERSE WAS BORN, RIGHT?
BOOOOMB
SO AFTER THAT, DIDN'T THE MATTER THAT WAS BLOWN UP FLY AROUND IN TINY PIECES?
NO, THAT'S NOT WHAT HAPPENED.
MORE ACCURATELY, WHAT EXPLODED WERE LIGHT PARTICLES CALLED PHOTONS. SINCE THESE ARE INDESTRUCTIBLE, THEY WERE NOT BROKEN INTO PIECES.
WHEN THE UNIVERSE SUDDENLY EXPANDED DUE TO INFLATION, ONLY LIGHT WAS GENERATED, NOT MATTER.
ARE PHOTONS ONE OF THE ELEMENTARY PARTICLES?

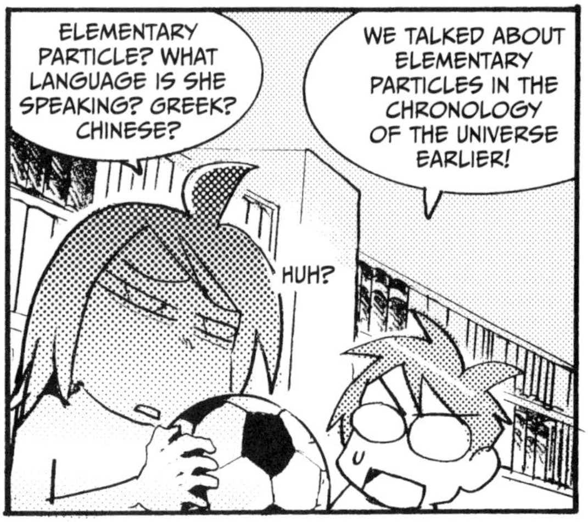
ELEMENTARY PARTICLE? WHAT LANGUAGE IS SHE SPEAKING? GREEK? CHINESE?
WE TALKED ABOUT ELEMENTARY PARTICLES IN THE CHRONOLOGY OF THE UNIVERSE EARLIER!
HUH?

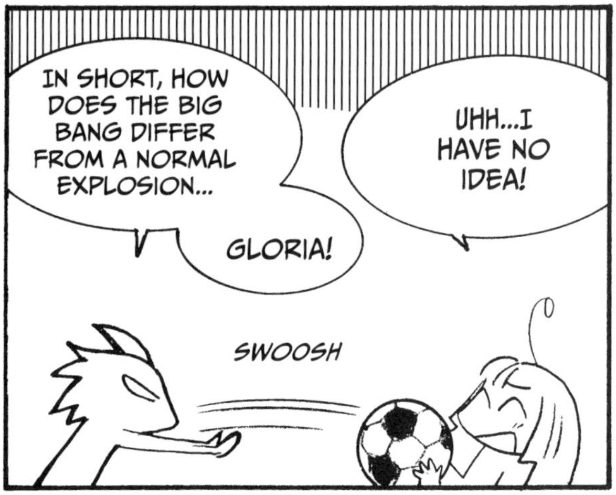
IN SHORT, HOW DOES THE BIG BANG DIFFER FROM A NORMAL EXPLOSION...
GLORIA!
UHH...I HAVE NO IDEA!
SWOOSH

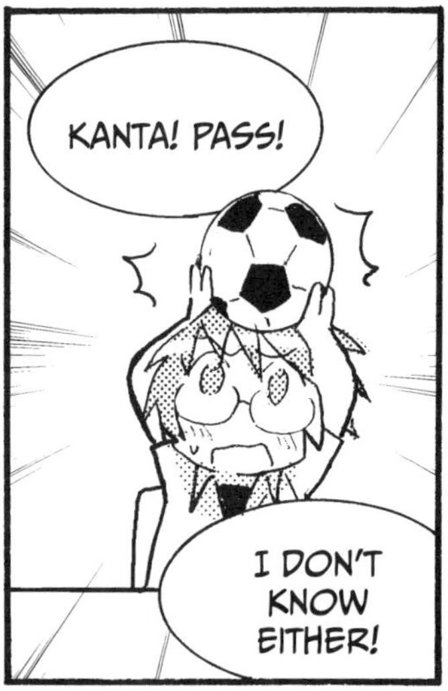
KANTA! PASS!
I DON'T KNOW EITHER!

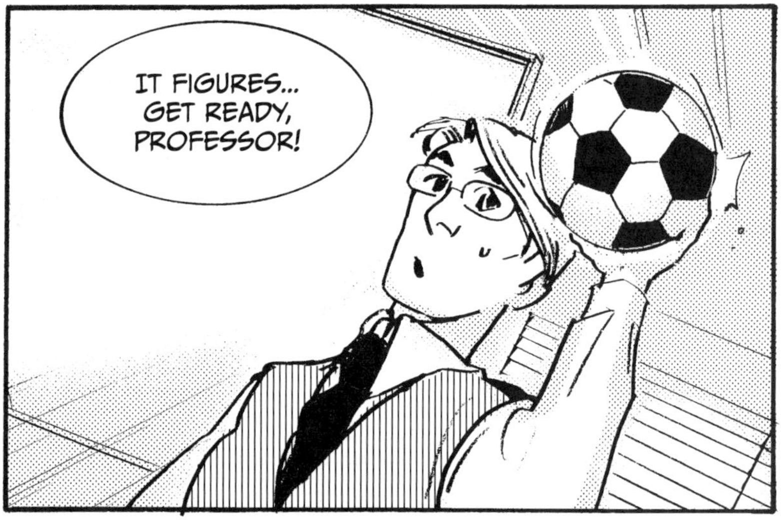
IT FIGURES... GET READY, PROFESSOR!

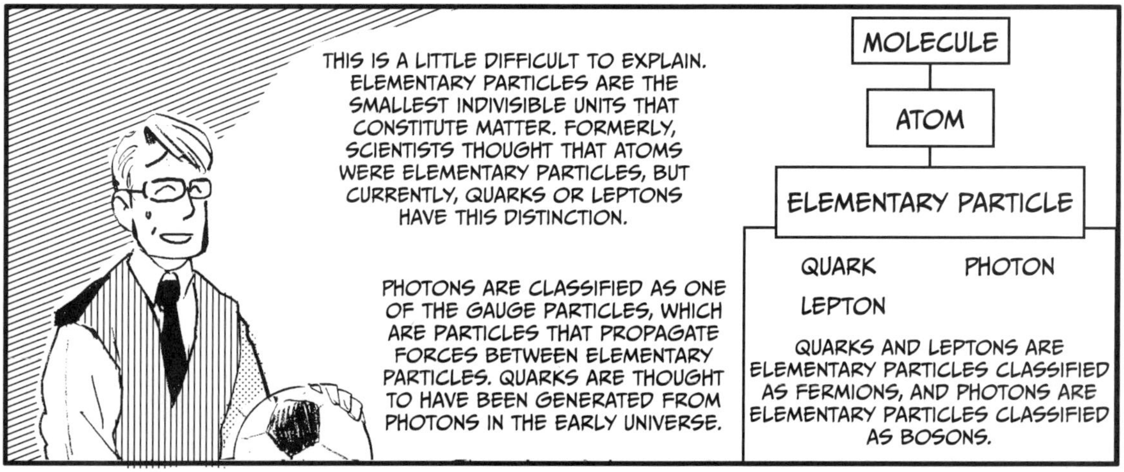
THIS IS A LITTLE DIFFICULT TO EXPLAIN. ELEMENTARY PARTICLES ARE THE SMALLEST INDIVISIBLE UNITS THAT CONSTITUTE MATTER. FORMERLY, SCIENTISTS THOUGHT THAT ATOMS WERE ELEMENTARY PARTICLES, BUT CURRENTLY, QUARKS OR LEPTONS HAVE THIS DISTINCTION.
PHOTONS ARE CLASSIFIED AS ONE OF THE GAUGE PARTICLES, WHICH ARE PARTICLES THAT PROPAGATE FORCES BETWEEN ELEMENTARY PARTICLES. QUARKS ARE THOUGHT TO HAVE BEEN GENERATED FROM PHOTONS IN THE EARLY UNIVERSE.
MOLECULE
ATOM
ELEMENTARY PARTICLE
QUARK
PHOTON
LEPTON
QUARKS AND LEPTONS ARE ELEMENTARY PARTICLES CLASSIFIED AS FERMIONS, AND PHOTONS ARE ELEMENTARY PARTICLES CLASSIFIED AS BOSONS.

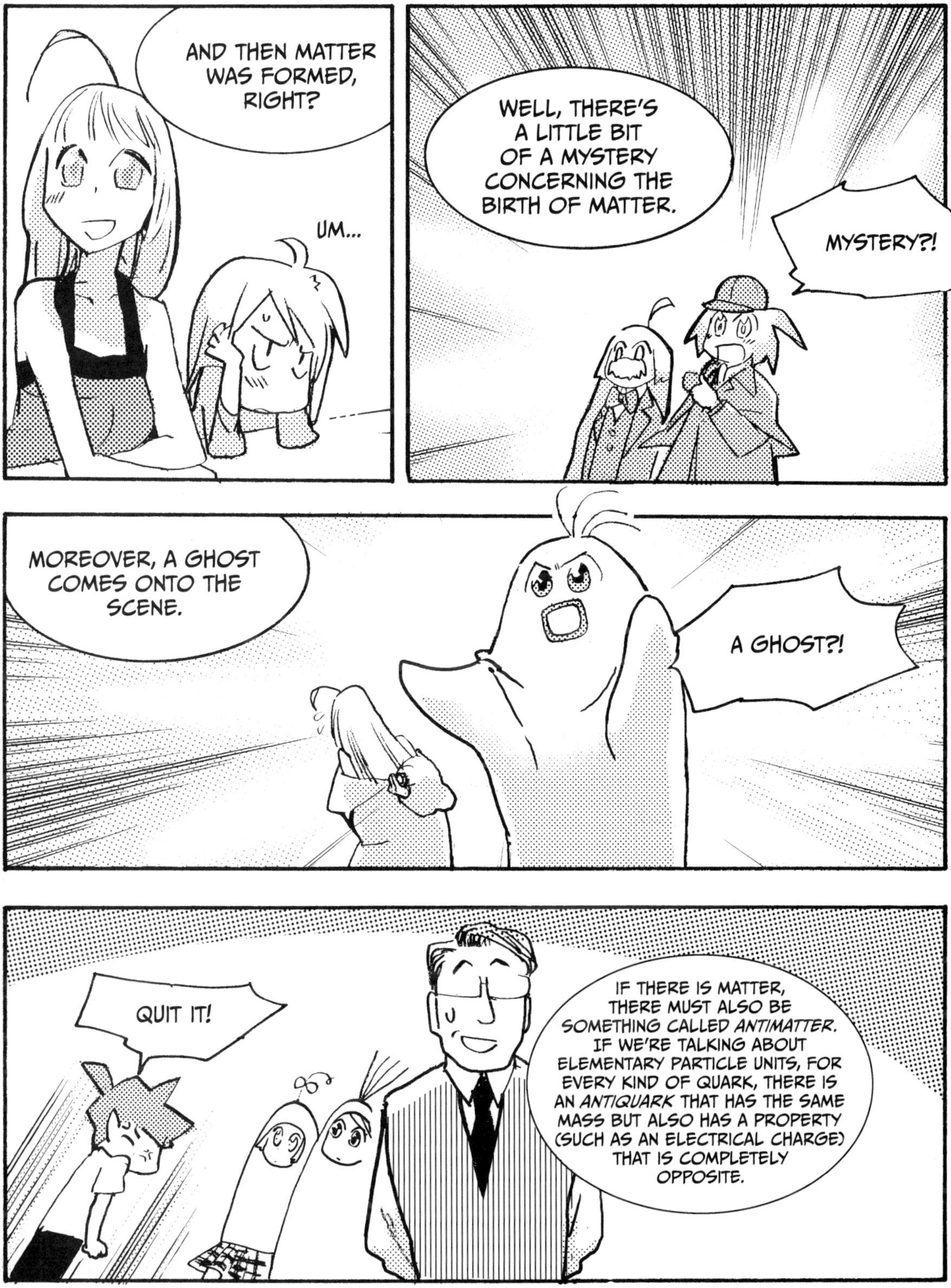
AND THEN MATTER WAS FORMED, RIGHT?
UM...
WELL, THERE'S A LITTLE BIT OF A MYSTERY CONCERNING THE BIRTH OF MATTER.
MYSTERY?!
MOREOVER, A GHOST COMES ONTO THE SCENE.
A GHOST?!
QUIT IT!
IF THERE IS MATTER, THERE MUST ALSO BE SOMETHING CALLED *ANTIMATTER*. IF WE'RE TALKING ABOUT ELEMENTARY PARTICLE UNITS, FOR EVERY KIND OF QUARK, THERE IS AN *ANTIQUARK* THAT HAS THE SAME MASS BUT ALSO HAS A PROPERTY (SUCH AS AN ELECTRICAL CHARGE) THAT IS COMPLETELY OPPOSITE.

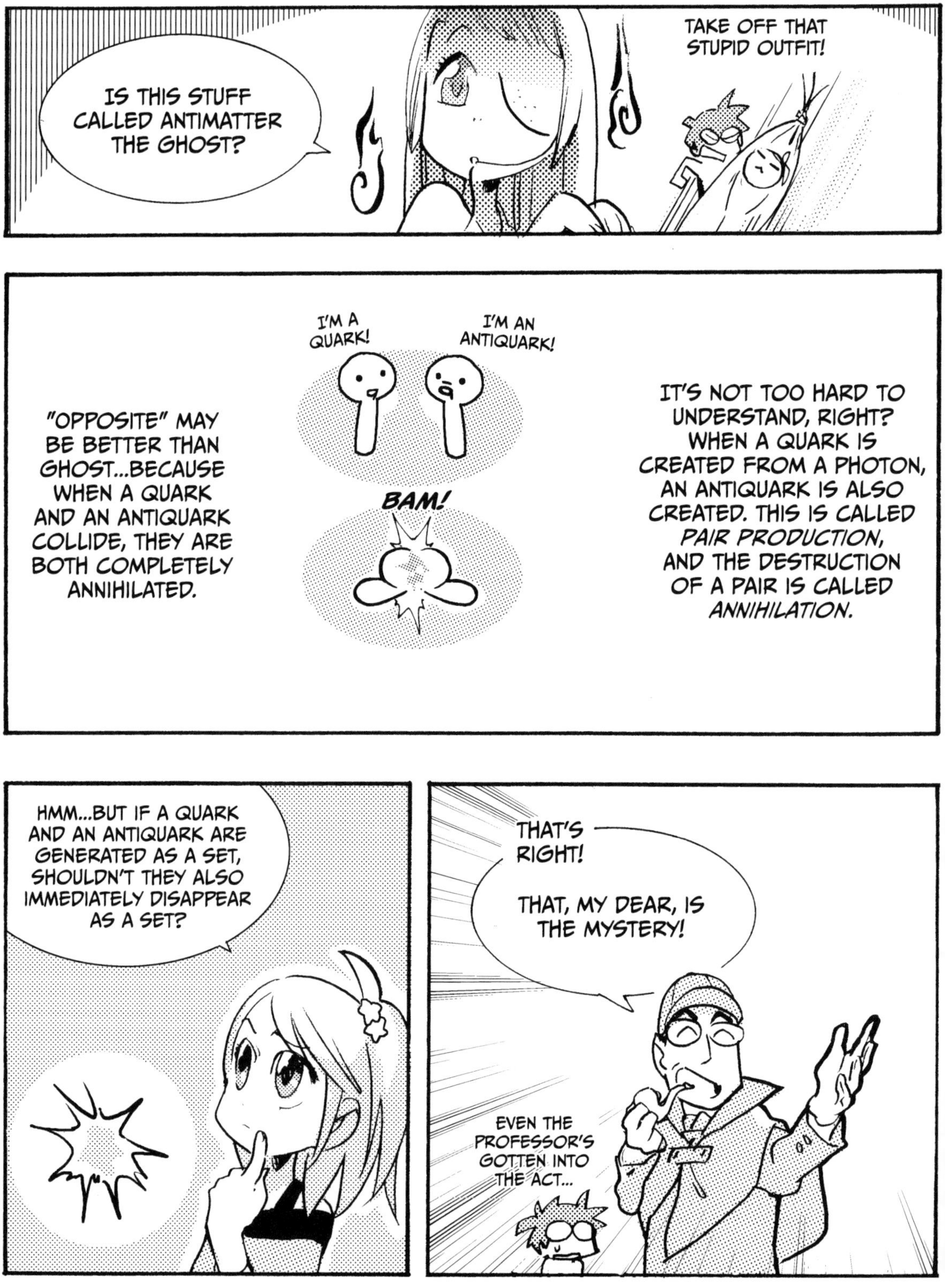
IS THIS STUFF CALLED ANTIMATTER THE GHOST?
TAKE OFF THAT STUPID OUTFIT!
"OPPOSITE" MAY BE BETTER THAN GHOST...BECAUSE WHEN A QUARK AND AN ANTIQUARK COLLIDE, THEY ARE BOTH COMPLETELY ANNIHILATED.
I'M A QUARK!
I'M AN ANTIQUARK!
BAM!
IT'S NOT TOO HARD TO UNDERSTAND, RIGHT? WHEN A QUARK IS CREATED FROM A PHOTON, AN ANTIQUARK IS ALSO CREATED. THIS IS CALLED *PAIR PRODUCTION*, AND THE DESTRUCTION OF A PAIR IS CALLED *ANNIHILATION*.
HMM...BUT IF A QUARK AND AN ANTIQUARK ARE GENERATED AS A SET, SHOULDN'T THEY ALSO IMMEDIATELY DISAPPEAR AS A SET?
THAT'S RIGHT!
THAT, MY DEAR, IS THE MYSTERY!
EVEN THE PROFESSOR'S GOTTEN INTO THE ACT...

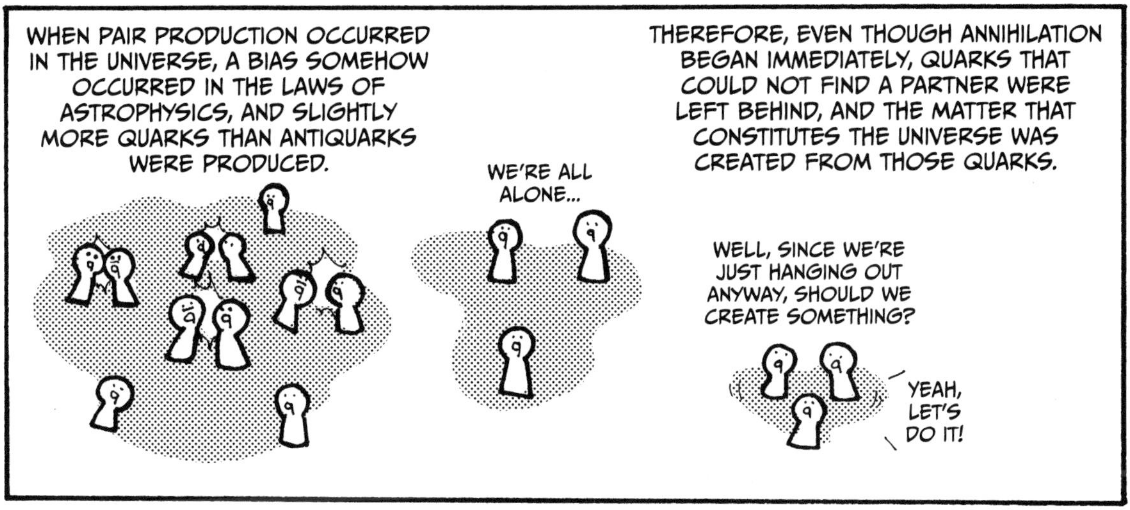
WHEN PAIR PRODUCTION OCCURRED IN THE UNIVERSE, A BIAS SOMEHOW OCCURRED IN THE LAWS OF ASTROPHYSICS, AND SLIGHTLY MORE QUARKS THAN ANTIQUARKS WERE PRODUCED.
WE'RE ALL ALONE...
THEREFORE, EVEN THOUGH ANNIHILATION BEGAN IMMEDIATELY, QUARKS THAT COULD NOT FIND A PARTNER WERE LEFT BEHIND, AND THE MATTER THAT CONSTITUTES THE UNIVERSE WAS CREATED FROM THOSE QUARKS.
WELL, SINCE WE'RE JUST HANGING OUT ANYWAY, SHOULD WE CREATE SOMETHING?
YEAH, LET'S DO IT!

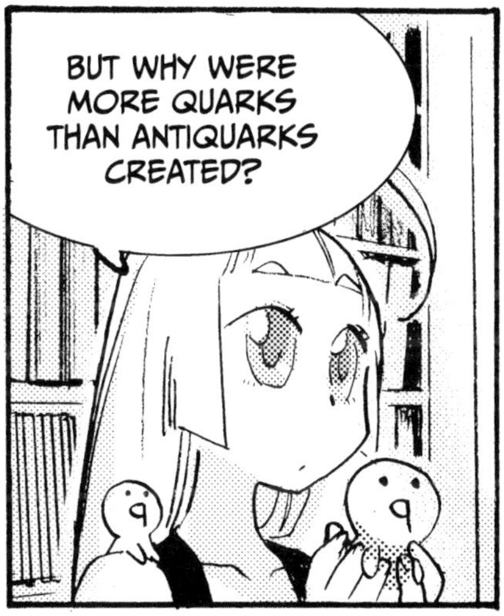
BUT WHY WERE MORE QUARKS THAN ANTIQUARKS CREATED?

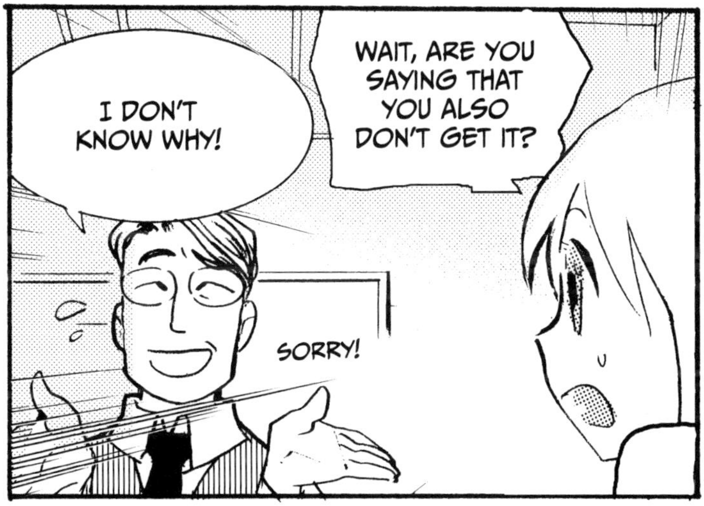
I DON'T KNOW WHY!
WAIT, ARE YOU SAYING THAT YOU ALSO DON'T GET IT?
SORRY!

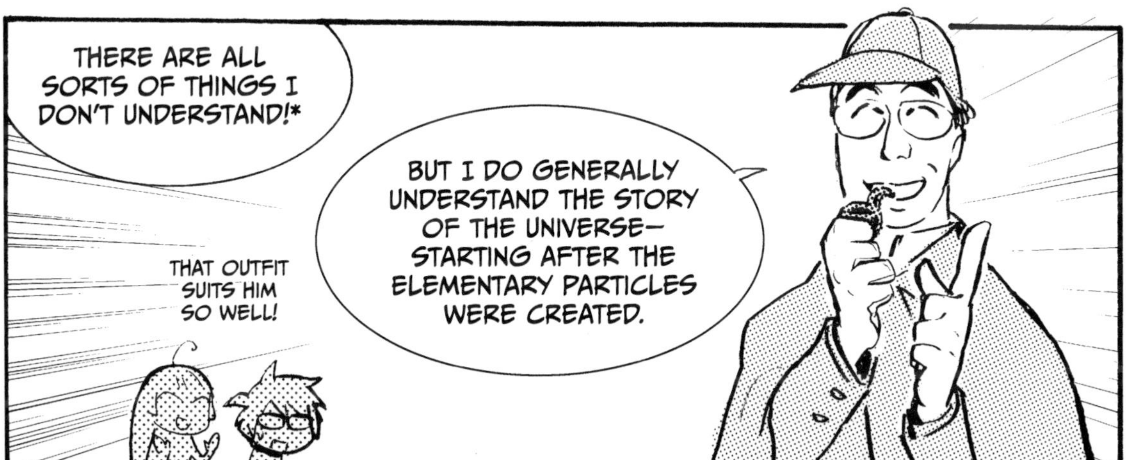
THERE ARE ALL SORTS OF THINGS I DON'T UNDERSTAND!*
BUT I DO GENERALLY UNDERSTAND THE STORY OF THE UNIVERSE— STARTING AFTER THE ELEMENTARY PARTICLES WERE CREATED.
THAT OUTFIT SUITS HIM SO WELL!

* THE RESEARCH OF THE THREE 2008 NOBEL PRIZE WINNERS IN PHYSICS, DR. YOICHIRO NAMBU, DR. MAKOTO KOBAYASHI, AND DR. TOSHIHIDE MASUKAWA, SURELY CAN PROVIDE A HINT TO SOLVING THIS PUZZLE. IF YOU'RE INTERESTED, YOU SHOULD CHECK IT OUT.

Approximately three minutes after the Big Bang occurred, space expanded exponentially at first but less rapidly later, and the temperature dropped to approximately 900 million degrees Kelvin.

 "Dropped" to 900 million degrees?! That's still incredibly hot!

 Well, since it was 150 billion degrees before that, it's actually quite a large drop!

 When the temperature dropped to this range, nuclei of hydrogen and helium, which are the simplest chemical elements, were formed. In other words, this was the birth of matter. However, the distribution was by no means homogeneous.

 That is to say, the matter was not spread uniformly throughout space. Some areas of space had more matter in them, and some had less.

 Why did that happen?

 We don't know. But when we look at various phenomena in the natural world, things almost never end up in a completely homogeneous state. Let's try a thought experiment.

 You mean an experiment we perform in our heads using our imaginations?

That's right. For example, let's assume that we scatter a lot of balls on the floor of a large room. What do you suppose will happen?

 Well, they certainly won't be spread out perfectly evenly—there will be some places where the balls are crowded together and other places where there are not many balls.

 And if the floor isn't perfectly flat, the balls will roll around and end up collecting in the low spots.

 That's a different story!

 No, no...we should also take that into consideration. I think that when matter began to form, when the universe was young, the gravitational strength in space was likely not the same everywhere—there were probably variations. Therefore, more matter might have accumulated in places where gravity was greater. If this phenomenon occurred, we cannot expect a homogeneous distribution.

 Right, because places that attract small groups will eventually attract even bigger groups!

 Exactly! When matter gathers, more matter will gather there according to the law of universal gravitation. Galaxies and clusters of galaxies are thought to have formed in this way.

Kanna's intuition is really spot on, isn't it?

Tee hee!

But something's strange...

What?

If we assume that more matter will gather where other matter has gathered before, then why didn't all the matter eventually gather in the same place?

Oh...yeah!

Kanta, what do you think?

When we're taking about this in depth, I don't understand it either. Professor...

It's quite mysterious, isn't it? Certainly, if there was a large depression in the middle of the floor, all of the balls we scattered would have gathered there.

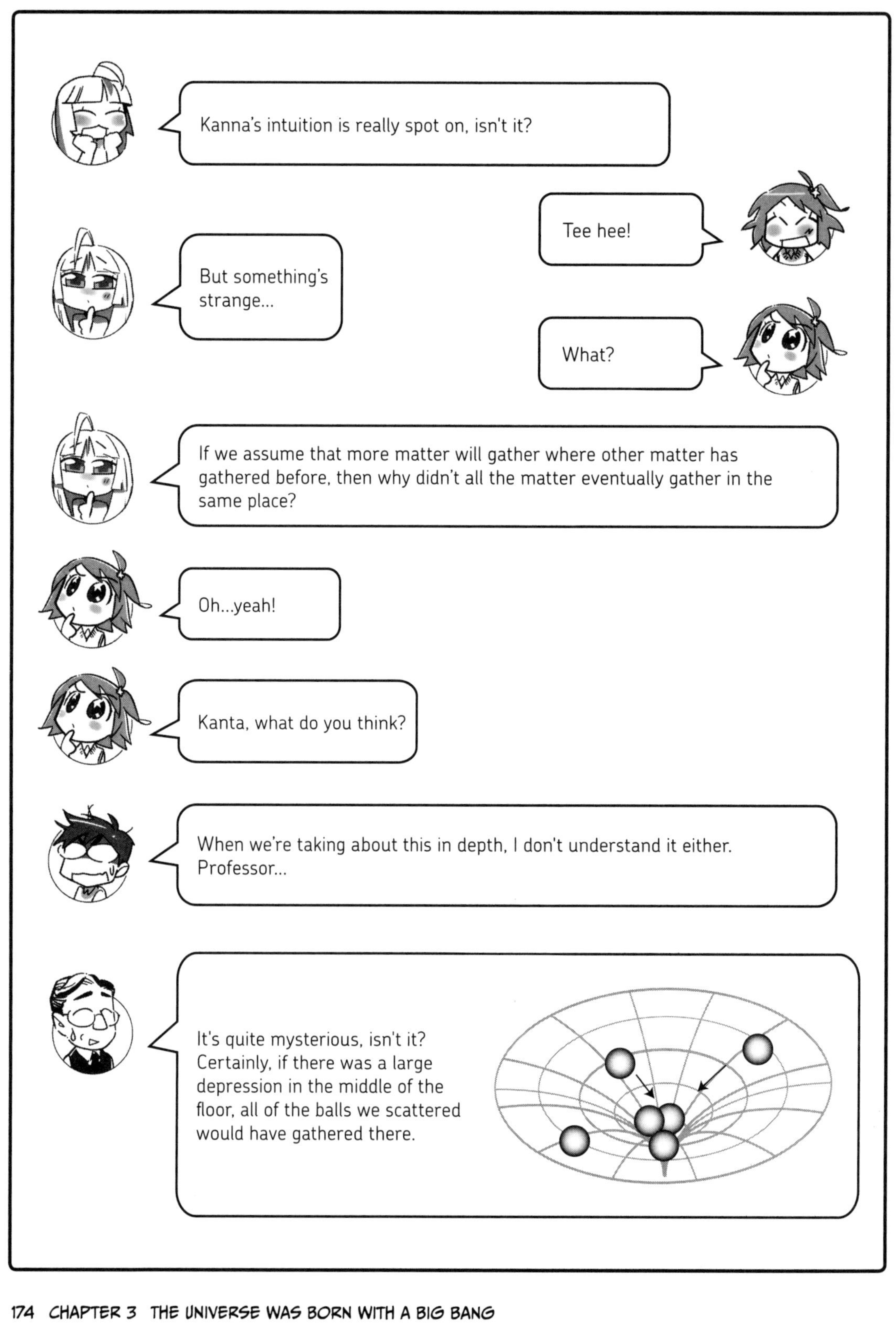

If that happened, then wouldn't the universe have only one super-large star with nothing but empty space around it?

 If all the matter in the universe had gathered in one place, the gravity of that place would be unbelievably strong, and it would have probably become an enormous black hole—so great that even light couldn't escape from it. In terms of our earlier example, so many balls would congregate in one place that their weight would cause the floor to give way, and all the balls would fall through!

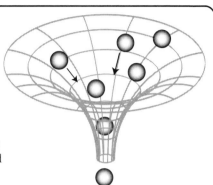

 That would sure be a terrible universe.

Totally!

 That's right. Taking the analogy one step further, we can say that the balls did not end up congregating in the same spot because the "floor" of our universe was actually dotted with countless shallow indentations, causing balls to congregate in different places. That would allow stars and black holes to coexist at a distance from each other. Then, as the universe expanded, more matter would congregate in more indentations in the "floor," thus forming large-scale structures such as galaxies, clusters of galaxies, and Great Walls.

 There are still lots of mysteries, aren't there?

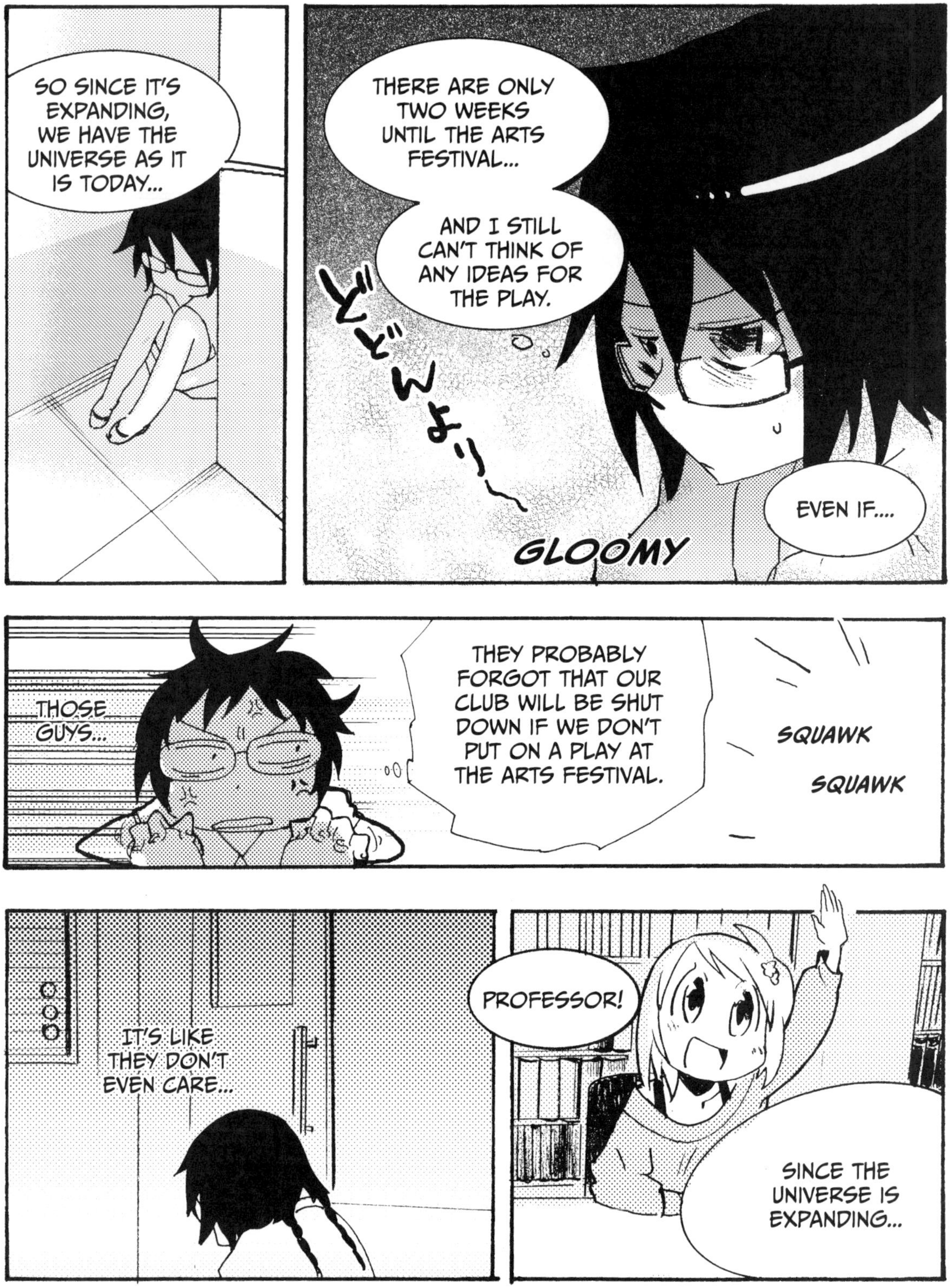
SO SINCE IT'S EXPANDING, WE HAVE THE UNIVERSE AS IT IS TODAY...
THERE ARE ONLY TWO WEEKS UNTIL THE ARTS FESTIVAL...
AND I STILL CAN'T THINK OF ANY IDEAS FOR THE PLAY.
GLOOMY
EVEN IF....
THOSE GUYS...
THEY PROBABLY FORGOT THAT OUR CLUB WILL BE SHUT DOWN IF WE DON'T PUT ON A PLAY AT THE ARTS FESTIVAL.
SQUAWK
SQUAWK
IT'S LIKE THEY DON'T EVEN CARE...
PROFESSOR!
SINCE THE UNIVERSE IS EXPANDING...

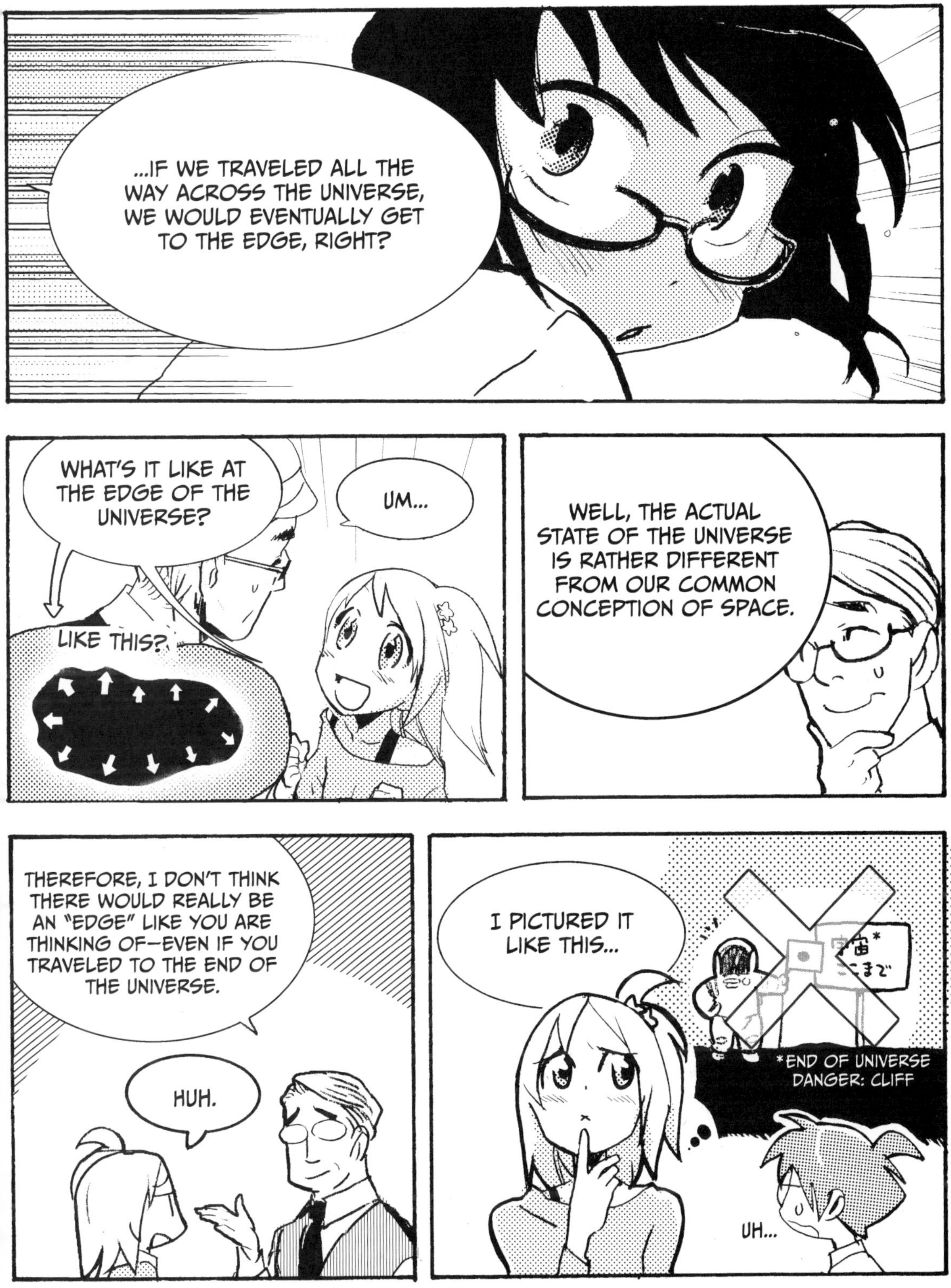
...IF WE TRAVELED ALL THE WAY ACROSS THE UNIVERSE, WE WOULD EVENTUALLY GET TO THE EDGE, RIGHT?
WHAT'S IT LIKE AT THE EDGE OF THE UNIVERSE?
UM...
LIKE THIS?
WELL, THE ACTUAL STATE OF THE UNIVERSE IS RATHER DIFFERENT FROM OUR COMMON CONCEPTION OF SPACE.
THEREFORE, I DON'T THINK THERE WOULD REALLY BE AN "EDGE" LIKE YOU ARE THINKING OF—EVEN IF YOU TRAVELED TO THE END OF THE UNIVERSE.
HUH.
I PICTURED IT LIKE THIS...
宇宙* ここまで
*END OF UNIVERSE DANGER: CLIFF
UH...

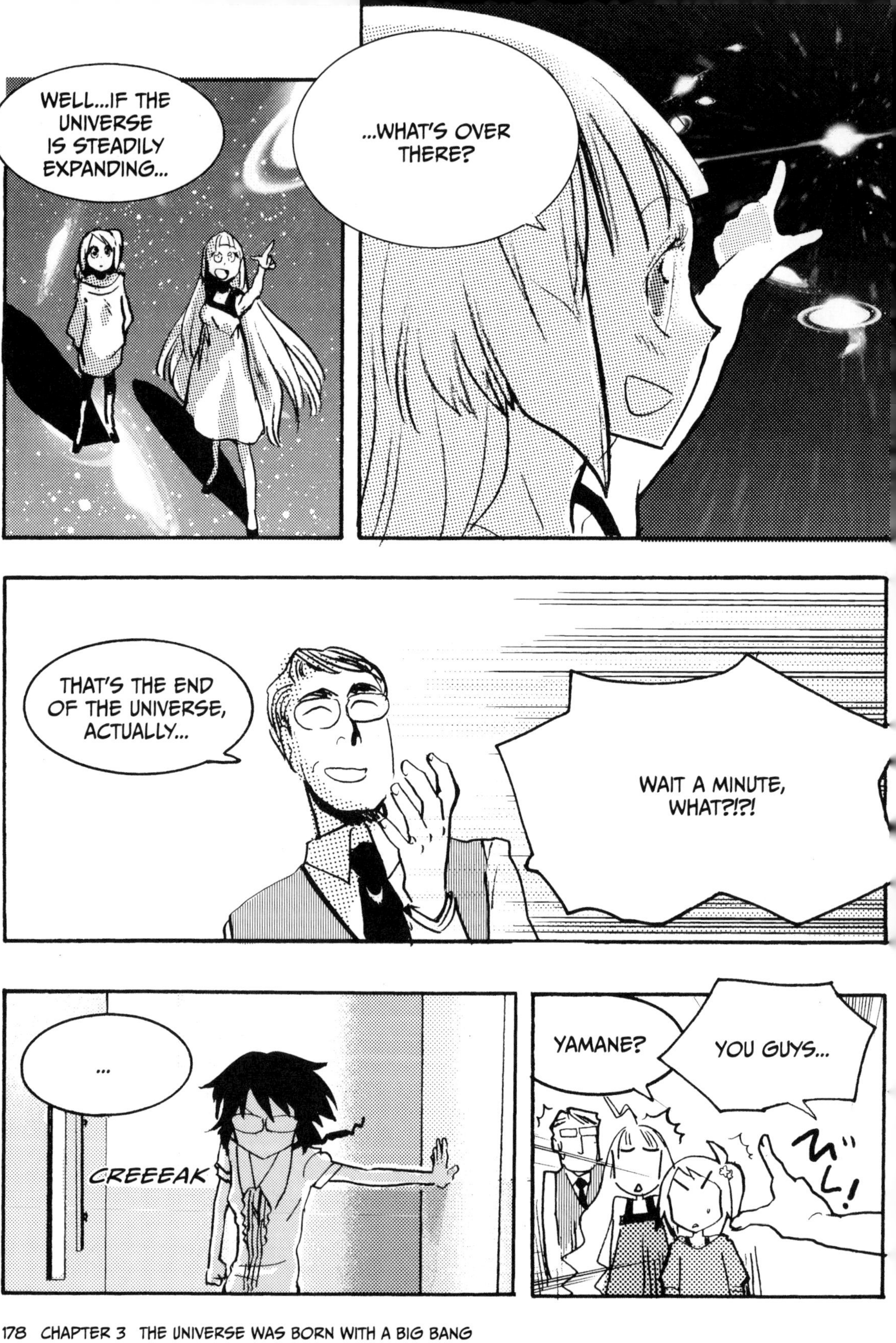
WELL...IF THE UNIVERSE IS STEADILY EXPANDING...
...WHAT'S OVER THERE?
THAT'S THE END OF THE UNIVERSE, ACTUALLY...
WAIT A MINUTE, WHAT?!?!
...
CREEEAK
YAMANE?
YOU GUYS...

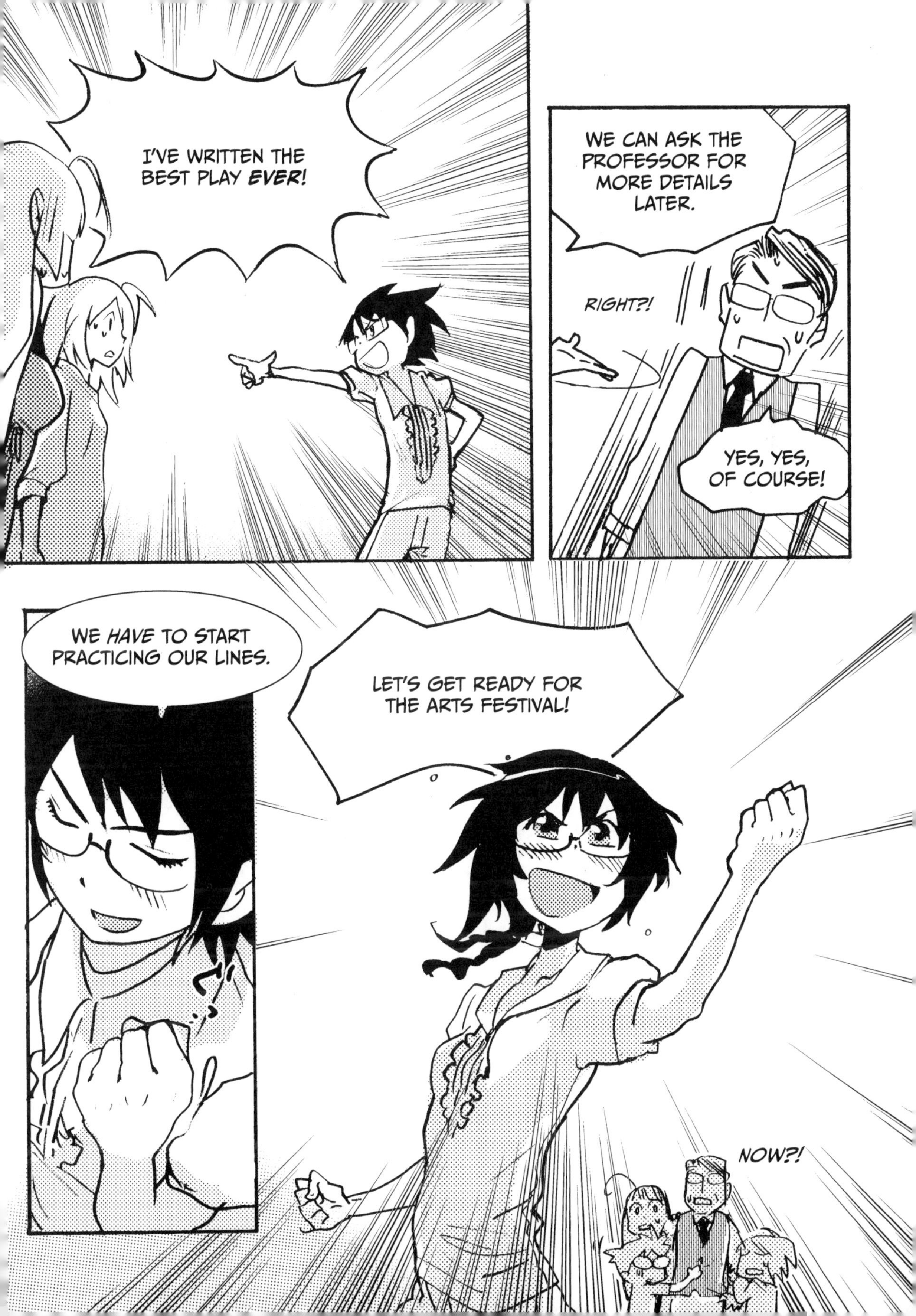
I'VE WRITTEN THE BEST PLAY *EVER*!
WE CAN ASK THE PROFESSOR FOR MORE DETAILS LATER.
RIGHT?!
YES, YES, OF COURSE!
WE *HAVE* TO START PRACTICING OUR LINES.
LET'S GET READY FOR THE ARTS FESTIVAL!
NOW?!

DO ALIENS EXIST?

In Chapter 4, we will discuss more about what it's like at the edge of the universe. But first, let's consider one common question about the universe: Do aliens exist?

We'll tell you the conclusion first. Most scientists who study the universe believe that other sentient beings like humans exist somewhere. This conclusion is based on the *Cosmological Principle*.

The Cosmological Principle is the hypothesis that the universe is homogeneous (the same kind) and isotropic (the same from any direction) on large scales. What that means is if you take a big chunk of universe, and compare it to another equally big chunk of universe, they wouldn't differ that much. This does not refer specifically to appearance—although they'll each have some areas that have a lot of clusters of galaxies, some areas with just a few galaxies, and some areas with no galaxies. It means that the same physical laws apply everywhere and don't behave differently in different places. $F = ma$ always and forever.

Since we humans, at first, were under the impression that Earth was a unique place in the universe, we believed in the geocentric theory. However, as the results of observations of the universe began to be explained more logically, a Sun-centered theory and then the heliocentric model were developed.

Following this pattern, the theory that life was created only on planet Earth began to be questioned. According to the Cosmological Principle, our Earth is by no means a special location. Therefore, there must be other planets in the universe that have environments similar to Earth's, and life must have originated there and must be evolving. So according to this argument, aliens must exist!

CALCULATING THE NUMBER OF EXTRATERRESTRIAL CIVILIZATIONS

Although the Cosmological Principle is probably correct as a framework, and space aliens may exist somewhere, there is still a question as to how common alien life is.

In 1961, the American astronomer Frank Drake (born in 1930) published an interesting equation known as the *Drake equation*, which enables us to estimate the approximate distribution of extraterrestrial civilizations in our own galaxy and determine whether we can communicate with them.

Here is the equation:

$$N = R^* \times f_p \times n_e \times f_l \times f_i \times f_c \times L$$

- N: Number of extraterrestrial civilizations in our galaxy with which communication might be possible
- R^*: Average rate at which stars are formed in our galaxy per year
- f_p: Fraction of those stars that have planets
- n_e: Average number of planets on which life can potentially exist in each star system with planets
- f_l: Fraction of the above star systems where life actually occurs
- f_i: Fraction of the above star systems where the life that occurs has evolved to intelligent life
- f_c: Fraction of those intelligent civilizations that develop interstellar communication
- L: Average length of time those civilizations perform interstellar communication

To use this equation, we must decide the values of its various parameters (variables). However, this is *very* difficult to do, since many of these factors are not known. Therefore, if we enter the numbers that Drake used in 1961, N is considerably greater than 1. In other words, he concluded that there were many highly advanced (at least having communication technology) extraterrestrial civilizations in the galaxy.

Although this equation may seem like a parlor trick, many scholars, including Carl Sagan (1934–1996), have generally approved of Drake's idea and assume there is an intriguing probability that extraterrestrials can communicate with us. (However, the various values of N obtained by the calculation range from 10 to 1,000,000.) Despite these flaws and the variations that can be generated by using differing starting parameters, aliens may be closer than we think.

EXTRATERRESTRIAL LIFE AND A WORLD-RENOWNED PHYSICIST

Since there are somewhere between 200 billion and 400 billion planetary systems like our solar system in the galaxy, it would not seem unusual for there to be planets that have environments similar to Earth's, where life has developed. However, the Italian physicist Enrico Fermi (1901–1954) directly challenged this optimistic prediction. Fermi was a Nobel Prize winner in physics who worked on developing the first atomic reactor in the world.

One day in 1950, while eating lunch with his fellow scientists, Fermi got into a discussion concerning the existence of aliens. The Drake equation was still 11 years from being published, yet astronomers in those days were already confident that the existence of extraterrestrial civilizations was highly probable, and scholars in other fields, like Fermi, were also interested in this topic.

Perhaps they had considered the possibility of extraterrestrial life from various parameters, as Drake did, but Fermi wanted to look into the idea further—and specifically, to think about *where* aliens might exist.

Although it's a simple question, it points directly to the heart of the matter.

If there are many extraterrestrial civilizations in the galaxy, even if it is difficult to encounter their spaceships, we should at least detect radio waves used in their communications. However, these traces have never been found.

Fermi was a "thought and action" person who not only had many historical achievements in theoretical physics, but also made many contributions in experimental physics. No matter how likely it seems that aliens might exist, there is no evidence of contact. This is the *Fermi paradox*.

Although many people have since tried to show the existence of extraterrestrial civilizations using methods such as the Drake equation, all have run up against the question of why there is no evidence of their existence. The Fermi paradox demands careful consideration.

The Arecibo Message was an attempt to broadcast news of our civilization to aliens. Sent in 1973 to the Messier 13 cluster, and written by Dr. Drake himself, the message includes information about elements and DNA, as well as figures of a human, our solar system, and the Arecibo radio telescope.

HAS LIFE BEEN CREATED OFTEN?

Living creatures exist everywhere on Earth—but we have come to know this only quite recently. In 1977, scientists who had been investigating deep-sea hydrothermal vents in the Pacific Ocean discovered strange creatures. One of these was the tube worm.

A *hydrothermal vent* is an opening on the ocean floor from which geothermally heated water gushes. Since the surrounding area is often teeming with poisonous substances such as hydrogen sulfide, it had previously been believed that there were no living creatures there. However, tube worms have a symbiotic relationship with chemosynthetic bacteria living inside them—the bacteria use hydrogen sulfide as an energy source, and they produce organic matter that the tube worms use as nutrition. Thus, the tube worms have propagated even in the ocean depths. Besides tube worms, many living creatures such as fish or crabs that live near hydrothermal vents comprise independent ecosystems. Such ecosystems were a major discovery.

The study of these kinds of extreme creatures continues even today, and many scholars claim that some kind of living creature exists almost everywhere on Earth, from the tops of the highest mountains to the depths of the oceans and even underground. It is even said that microbes that ate lava existed 3.5 billion years ago.

Tube worms

The fact that living creatures can survive in such an extreme environment is surely good news for people who believe in the existence of extraterrestrial life. For example, the surface of Europa (one of Jupiter's moons) is covered with ice, but since volcanic activity has been verified there, there is a good possibility that there are oceans with hydrothermal vents under the ice. If this is true, then there may be living creatures such as tube worms there.

Europa

Now, this hypothesis naturally has a rebuttal.

If living creatures can continue to propagate even in harsh environments, why didn't the moon rocks brought back by the Apollo spacecraft show any traces of living creatures? Why were no living creatures detected on Mars, which is certainly believed to contain water?

Even microbes have not been found, which means that primordial life (the basis of evolution) might not exist on either the Moon or Mars, and might never have existed there. This suggests that the probability of the existence of life on any given moon or planet might be smaller than we would think. In other words, even if a planet currently has an environment that could support living creatures, life might not necessarily exist there.

By the way, some scholars advocate the theory that life on Earth originated from organisms that were transported across the universe by meteorites. Thus, to some extent, the search for alien life is also research into the origin of life on Earth—not just astronomy.

WHICH IS THE CLOSEST STAR SYSTEM THAT COULD SUPPORT EXTRATERRESTRIAL LIFE?

Although some of the previous discussion may have been somewhat pessimistic, we will now try to find a heavenly body where extraterrestrial life may exist, taking into consideration the environmental aspects.

Inside the solar system, Ganymede (a satellite of Jupiter, just like Europa) and Titan (a satellite of Saturn) are promising candidates. This is because there is a high likelihood that ice or water exists on both of them. In that light, the possibility of life on Mars where an ice lake is reported to have been detected cannot yet be discarded.

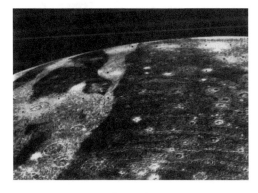
Ganymede

Outside of the solar system, there are two planetary systems that have stars similar to our Sun. Tau Ceti, which is approximately 12 light-years away, and Epsilon Eridani, which is approximately 10.5 light-years away, might have planets with environments similar to Earth's. Radio telescopes are being used to observe these stars on a continual basis, and the Search for Extra-Terrestrial Intelligence (SETI) Institute, founded by Frank Drake, the creator of the Drake equation, has projects established in Japan, the United States, and Europe. The day when space aliens are detected may not be so far away.

NASA's Kepler spacecraft was launched in 2009 with the goal of finding Earth-sized (and larger) planets in or near what astronomers call the *habitable zone*. Planets that orbit their star in the habitable zone are the proper distance away from their star to sustain life on their surface—that is, these planets are not too hot or too cold. Kepler continually observes 145,000 stars in a specific field of view for periodic changes in a star's brightness. This dimming indicates that one or more planets have moved in front of the star temporarily.

The Kepler mission announced in February 2011 that its initial data had found 1,235 planet candidates, 68 of them Earth sized. Of the total number of new candidates, 54 of them were orbiting in the habitable zone, and 5 of those were less than twice the size of Earth. From these results, mission scientists estimate that there are at least 50 billion planets in the Milky Way and at least 500 million of those planets orbit their stars in the habitable zone. Learn more at *http://kepler.nasa.gov/*.

CAN WE CONTACT AN EXTRATERRESTRIAL CIVILIZATION?

Let's consider whether we have a method of contacting an extraterrestrial civilization if we were to detect one.

If we were able to verify the existence of aliens who had some kind of advanced civilization, detection would most likely occur via radio waves. Since radio waves for broadcasts or communications clearly differ from naturally emitted electromagnetic waves, if we happened to find such radio waves, we could try to contact the civilization producing those radio waves by sending a message in its direction. However, the problem is that since even stars in the Cetus or Eridanus constellations are more than 10 light-years away from Earth, it would take more than 20 years to just exchange simple greetings like "Best wishes!" and "Nice to meet you!"

Cetus Constellation
More than 10 light-years
Best wishes!
Nice to meet you!
Earth

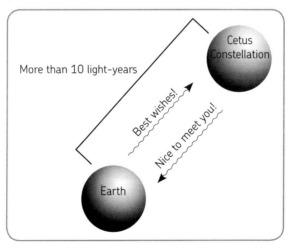

This simple exchange would take more than 20 years.

Since Alpha Centauri, which is the closest star to our solar system, is 4.22 light-years away, an exchange of messages would require a great deal of patience. This seems a little like a relationship between friends who only exchange Christmas cards—we'd only be able to correspond once every 10 years.

Mere communication is somewhat limited—isn't there some method of actually visiting in a spaceship?

Because a spaceship that could achieve a speed close to the speed of light theoretically could be developed if enough time and effort were committed, some people might volunteer to travel to a star approximately 10 light-years away (the round trip would take approximately 20 years). However, the struggle with gravity would be a problem.

Earthlings can only live in an environment with the 1G gravitational acceleration of Earth. As a result, even when traveling into space only for the elapsed time of the space shuttle flights, astronauts train every day so that their muscles do not weaken (even so, gravity seems intense to them when they return to Earth).

The spaceship that appeared in the movie *2001: A Space Odyssey* generated artificial gravity by using centrifugal force (this is the force that seems to work on a body that is rotating about a central point) as it revolved. This would probably be the best method to use, but launching such a large spaceship from Earth would be a problem. A sequential procedure would probably be needed, in which a base would first be created in space or on the surface of the Moon, the large spaceship would be assembled there, and then it would finally set off for someplace beyond our solar system.

TARDIGRADES (WATER BEARS) ARE THE TOUGHEST ASTRONAUTS

It seems that it will take a little more time for humans to pursue direct contact with extraterrestrial civilizations or to venture out into the universe. However, the most promising "astronaut" candidate to go in our place is the creature called the *tardigrade* (commonly known as the *water bear*).

A tardigrade is a tiny animal with a length of up to 1.5 mm. Although it somewhat resembles a bug, it is not really an insect at all. It ambles along on four pairs of stocky legs.

The tardigrade is noteworthy because it is a polyextremophile that can survive almost anything. These amazing creatures can live in various extreme situations that would kill anything else. A tardigrade can stay alive for close to 100 years even when it's extremely dehydrated (*anhydrobiosis* is the name of this state). In addition, water bears can withstand temperatures from −273°C to 150°C, pressures from the vacuum of space up to 75,000 times atmospheric pressure, and doses of radiation from X-rays of more than 1,000 times the lethal dose for humans.

Tardigrade

The tardigrade resuscitates its tough body and lives in a variety of environments on Earth from the tropics to the North Pole, on the highest mountains and in the ocean depths, and even in the boiling water of hot springs. There are large numbers of these commonplace yet amazing creatures. In September 2008, a Swedish and German research team did an experiment in which tardigrades were exposed for 10 days in space. The results showed that some were able to endure a vacuum, extremely low temperatures, and ultraviolet rays from the Sun.

Making a voyage into space means engaging in a challenging struggle with a harsh environment. Although creating a spaceship on which humans can safely travel is enormously difficult, if the travelers were tardigrades, the planning would be much simpler because they would be able to travel while in an anhydrobiotic state and then be resuscitated many years later at a distant star. If this were done, then in due course, animals that evolved from tardigrades might be active throughout the universe just as they live everywhere here on Earth.

A THIRD METHOD OF MEASURING THE SIZE OF THE UNIVERSE: IF YOU KNOW THE PROPERTIES OF A STAR, CAN YOU FIGURE OUT HOW FAR AWAY IT IS?

We have already mentioned that the distance to a heavenly body can be calculated via triangulation, using annual parallax and the distance between Earth and the Sun. But, if you recall, the parallax angles are so small that we can only measure those angles for stars 1,000 light-years away. Since the diameter of the Milky Way galaxy is approximately 100,000 light-years, this distance is less than 2 percent of the distance across our galaxy. There are over 100 billion galaxies in the universe! What should we do to learn about the universe beyond this distance?

One simple method is to compare the physical properties of other stars with those of our Sun.

Contraction of interstellar gas
Small stars
Solar-type stars
Giant stars
Contraction
Expansion
Expansion
White Dwarf
Red Dwarf
Red Giant
Red Giant
Contraction
Black Dwarf
Brown Dwarf
White Dwarf

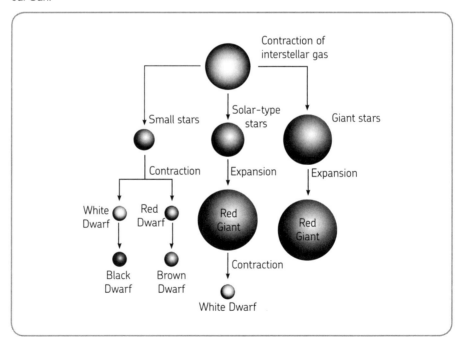

Stellar classifications

Although stars like the Sun shine by releasing energy due to a nuclear fusion reaction, the type of reaction that occurs is determined by the star's mass (that is, gravity). Therefore, if the color emitted by stars is the same, the basic brightness (absolute magnitude) of those stars is also generally the same.

The color of a star is directly related to its surface temperature. This is intuitive if you consider a flame: a hotter flame burns blue, while a cooler flame burns orange or red. So a hot star burns blue, while a cooler star burns redder. The luminosity of a star, or the amount of light a star emits per second, depends on the size of the star but also on the temperature of the star. Therefore, if two stars of the same size appear to be the same color, then the luminosity of those stars is also generally the same.

The luminosity of a star is an absolute physical value, independent of its distance from an observer. However, if we were to move away from a star, it would appear to dim—and the further we moved away from it, the more its brightness would decrease. In fact, if we doubled our distance from the star, we'd receive only one-fourth of the original amount of light we measured. Because of this relationship, we can determine how far away a star is by measuring its apparent brightness and comparing it to the star's luminosity.

The energy of the light that arrives in this case is 1.
Light source
Earth
Since the distance is doubled, the energy of the light that arrives in this case is 1/4.
Light source
Earth
Since the distance is tripled, the energy of the light that arrives in this case is 1/9.
Light source
Earth
The light energy measured by the observer is inversely proportional to the square of the distance from the observer to the light source.

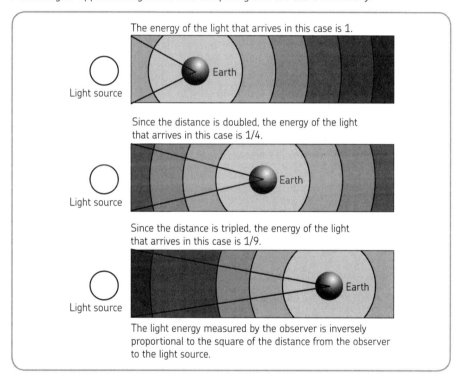

Distance measurement according to absolute magnitude

These relationships are shown together in the *Hertzsprung-Russell diagram (H-R diagram)*. This diagram, which was independently proposed by the Danish astronomer Ejnar Hertzsprung and the American astronomer Henry Norris Russell, is a distribution diagram that uses the spectral type (color = surface temperature) of a star for its horizontal axis and the star's absolute magnitude for its vertical axis.

NOTE *The absolute magnitude of a star is what its apparent magnitude would be if the star were placed 10 parsecs away from us.*

Even if a star does not exceed the measurement limit of annual parallax, if it has the same spectrum as the Sun, for example, we can determine its absolute magnitude and then estimate its distance by using its apparent magnitude (its brightness as seen from Earth).

To a certain extent, the relationship between absolute magnitude and spectrum is not that strict. If any interstellar matter or dust blocks the light along the way, the brightness of a star will not accurately represent its absolute magnitude and distance, and significant error may occur. So astronomers use certain measurements and models in their equations to correct for this.

Bright
Dark
Absolute magnitude
0
2
4
6
8
10
12
14
16
18
Red
Giants
Main
Sequence
White
Dwarfs
High
Low
Surface temperature

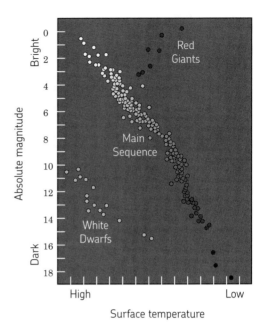

Hertzsprung-Russell diagram

Interestingly enough, the spectral classifications that Hertzsprung and Russell both used for the vertical axis were developed in the early 1900s by a woman named Annie Jump Cannon. At the time, many female astronomers were working for their male counterparts collecting observational data and processing these data. The "Harvard computers," as these workers were then called, were often paid very little yet performed much of the work that led to major discoveries by the likes of Shapley and Hubble. Cannon was one of these computers. She was the first female astronomer named as an officer in the American Astronomical Society, and she catalogued more stellar bodies than any other person to date.

STARS WITH VARYING BRIGHTNESS ARE "LIGHTHOUSES OF THE UNIVERSE"

Isn't there a more accurate method of measuring distance? The person who found the answer to this question and brought about later significant advances in astronomy is the American astronomer Harlow Shapley (1885–1972).

What Shapley noticed was the light from variable stars. There are several reasons why the light from a star will vary. In some cases, it may be the result of a supernova explosion caused by the death of a giant star, and in others, it may be coming from a seemingly variable star that is actually two stars: a bright star and dark star rotating as a pair. However, in most cases it is caused by a star whose brightness varies regularly because the surface layer is swelling and shrinking periodically. These are called *pulsating variable stars*.

The pulsations are caused, of course, by the nuclear fusion reaction. For stars called *Cepheid variable stars*, helium nuclei are fused together to form heavier carbon or oxygen nuclei, causing the entire star to shrink. Since the outer layer is unstable, the star pulsates. Cepheid variable stars with a longer period of light variation also have a greater absolute magnitude.

But it was the astronomer Henrietta Leavitt at the Harvard College Observatory, another "computer," who discovered this period-luminosity relationship. While cataloguing the magnitudes of stars, she noticed this pattern in the variable stars: the longer the variation is, the brighter the star is. She first published her findings in 1908 and confirmed them in 1912, long before Cepheid variables were used by Shapley in the Great Debate.

Shapley studied this relationship and concluded that it could be used for distance measurement if one measured the apparent magnitude and light variability period. By observing Cepheid variable stars in globular clusters in the Milky Way galaxy, he realized that the solar system was not in the center of the galaxy.

After astronomers were able to use Cepheid variable stars to measure distances to heavenly bodies, the positions of heavenly bodies at distances of approximately 10 million light-years could be accurately determined to a certain extent, and the map of the universe was significantly redrawn. Shapley and Leavitt's work certainly led to discoveries like the fact that there are many galaxies outside of the Milky Way and the redshift-based evidence that the universe is expanding.

METHODS OF MEASURING EVEN GREATER DISTANCES

By using Cepheid variable stars and subsequent advances in observational techniques, astronomers could measure the distance to heavenly bodies to approximately 100 million light-years, as long as they resigned themselves to a certain margin of error. However, even this distance still covered only approximately 1 percent of the visible universe. Since the area of the universe that we can physically observe extends out to a radius of approximately 15 billion light-years from Earth, enlarging the measurement range to that distance was one of the dreams of astronomers.

Some other measurement methods that have already been devised are introduced below.

MEASUREMENT BASED ON SUPERNOVAS

A *Type Ia supernova* (an evolved binary star system consisting of a giant or super giant and a white dwarf) has the property that its peak absolute magnitude is practically constant. Moreover, its brightness is approximately 100,000 times greater than that of a Cepheid variable star! Since it emits as much light as a galaxy, the distance to which it can be measured is extremely far. However, a supernova can only be observed at the instant it explodes, meaning the moment at which the life of that star ends.

MEASUREMENT BASED ON REDSHIFT

If we consider the fact that more distant heavenly bodies are receding from Earth at a faster rate than closer bodies, then redshift due to cosmic expansion increases in proportion to distance. Therefore, by observing the shift in the wavelengths of the spectral lines of a galaxy, the velocity of that galaxy (and its distance from Earth) can be known.

4
WHAT IS IT LIKE AT THE EDGE OF THE UNIVERSE?

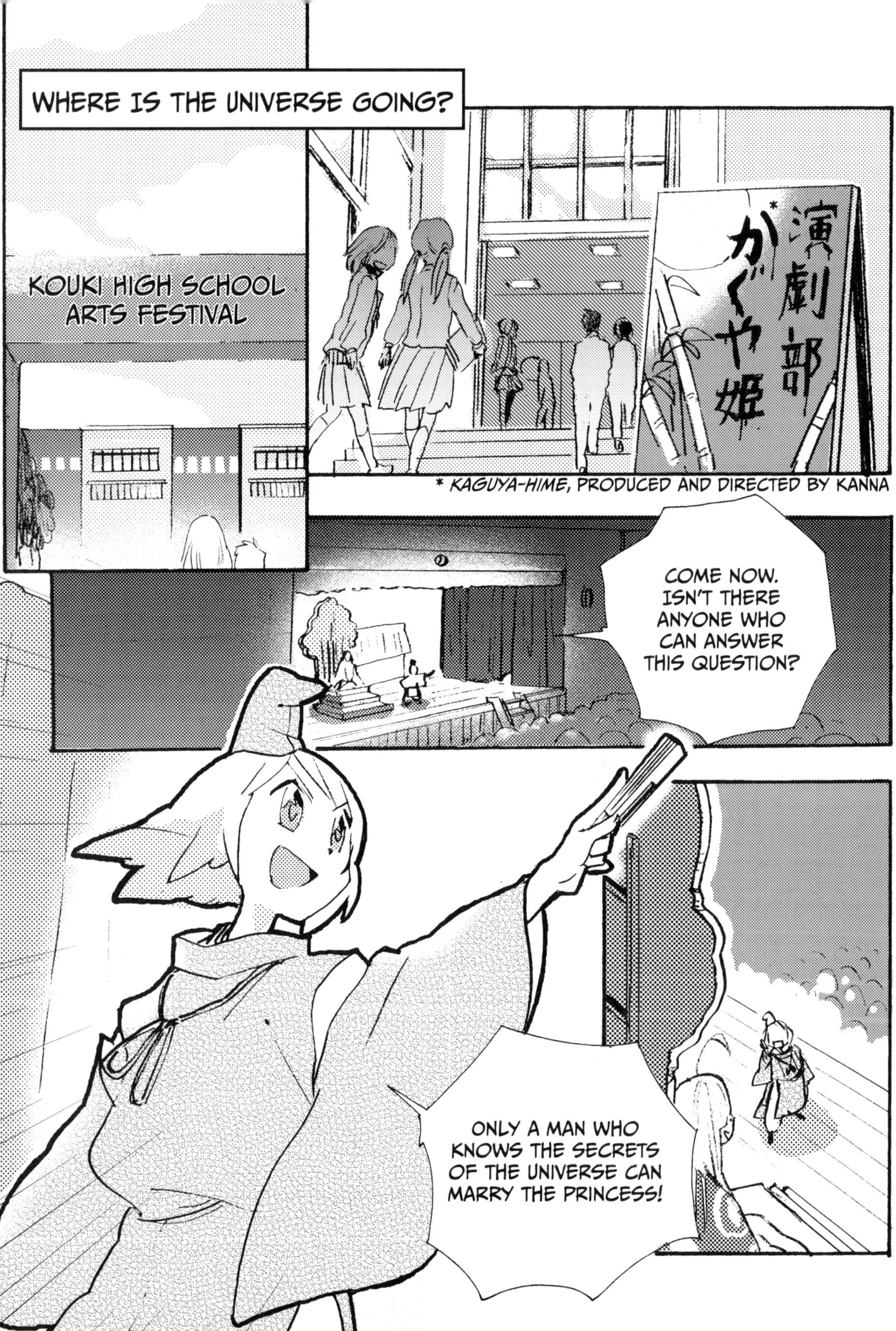
WHERE IS THE UNIVERSE GOING?
KOUKI HIGH SCHOOL ARTS FESTIVAL
演劇部
*かぐや姫
* KAGUYA-HIME, PRODUCED AND DIRECTED BY KANNA
COME NOW. ISN'T THERE ANYONE WHO CAN ANSWER THIS QUESTION?
ONLY A MAN WHO KNOWS THE SECRETS OF THE UNIVERSE CAN MARRY THE PRINCESS!

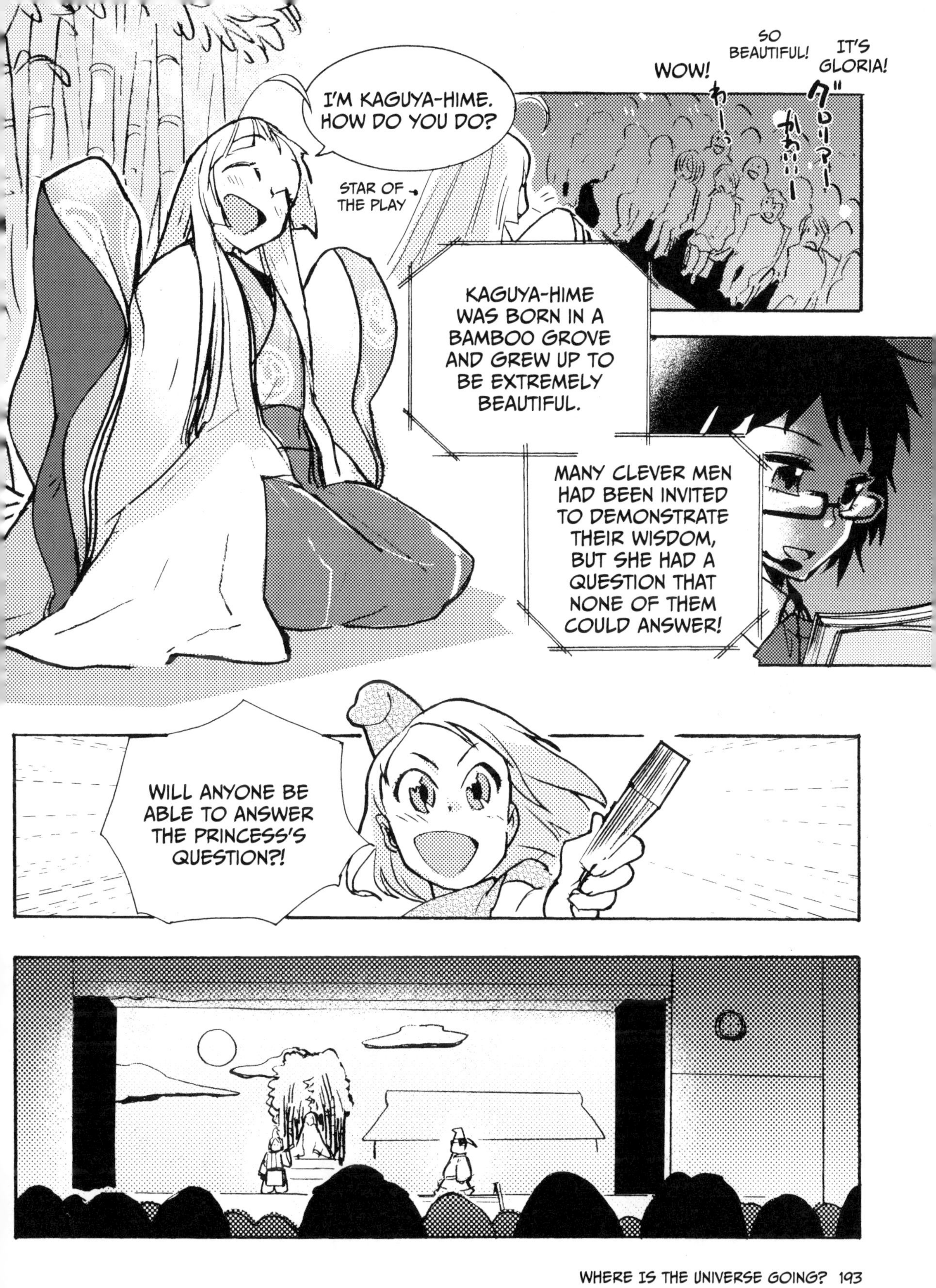
I'M KAGUYA-HIME. HOW DO YOU DO?
STAR OF THE PLAY
WOW!
SO BEAUTIFUL!
IT'S GLORIA!
KAGUYA-HIME WAS BORN IN A BAMBOO GROVE AND GREW UP TO BE EXTREMELY BEAUTIFUL.
MANY CLEVER MEN HAD BEEN INVITED TO DEMONSTRATE THEIR WISDOM, BUT SHE HAD A QUESTION THAT NONE OF THEM COULD ANSWER!
WILL ANYONE BE ABLE TO ANSWER THE PRINCESS'S QUESTION?!

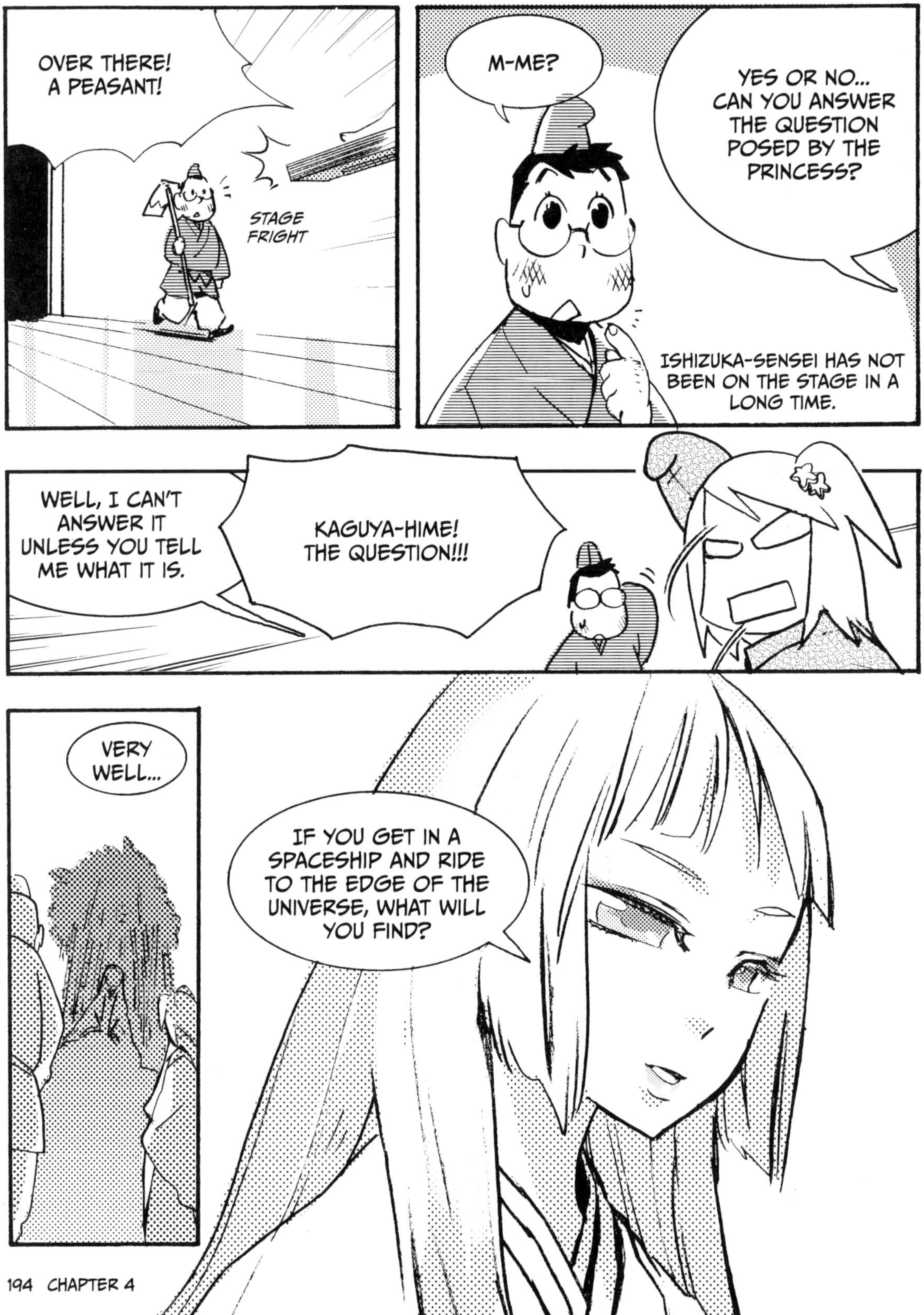

OVER THERE! A PEASANT!
STAGE FRIGHT
M-ME?
YES OR NO... CAN YOU ANSWER THE QUESTION POSED BY THE PRINCESS?
ISHIZUKA-SENSEI HAS NOT BEEN ON THE STAGE IN A LONG TIME.
WELL, I CAN'T ANSWER IT UNLESS YOU TELL ME WHAT IT IS.
KAGUYA-HIME! THE QUESTION!!!
VERY WELL...
IF YOU GET IN A SPACESHIP AND RIDE TO THE EDGE OF THE UNIVERSE, WHAT WILL YOU FIND?

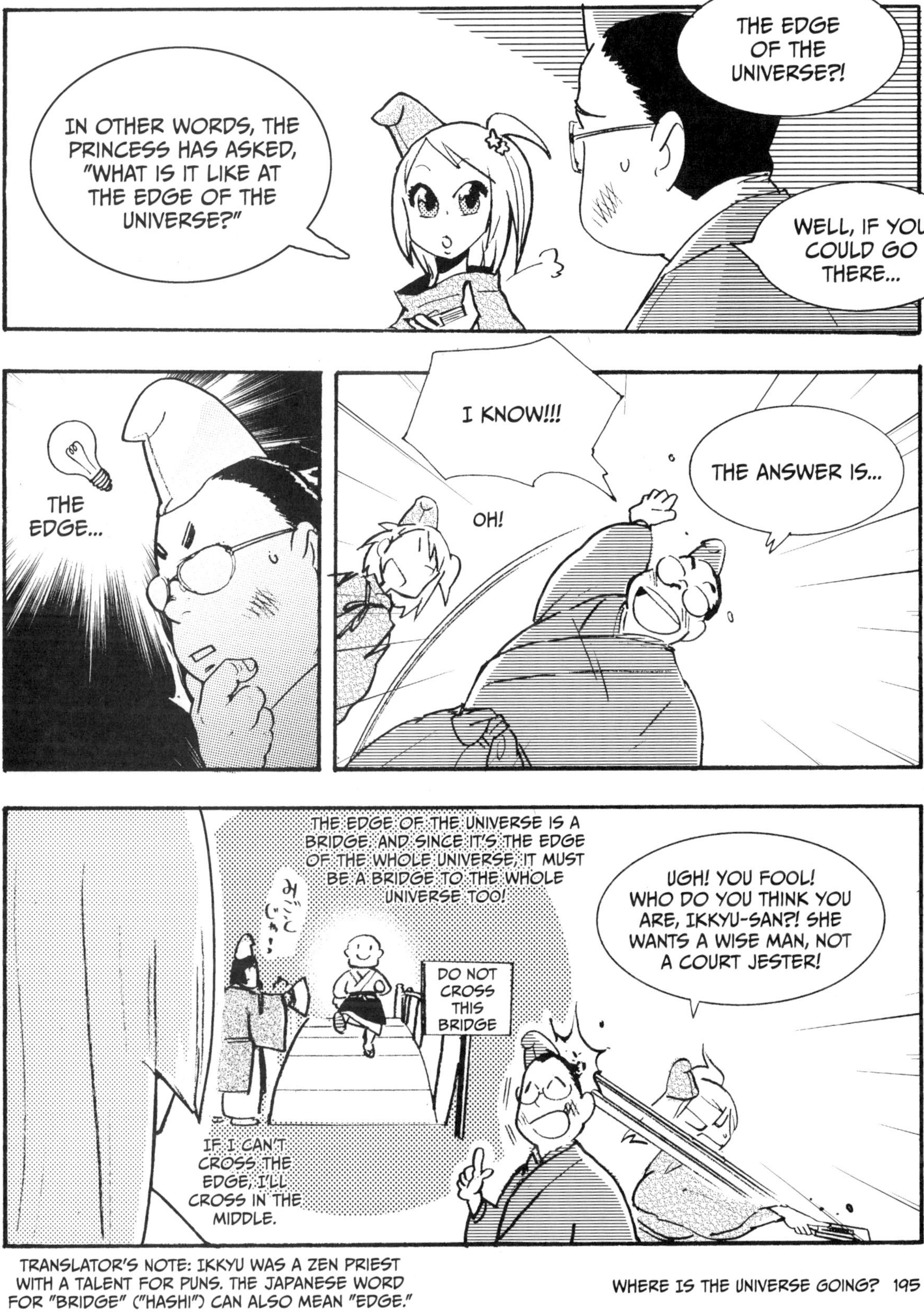
IN OTHER WORDS, THE PRINCESS HAS ASKED, "WHAT IS IT LIKE AT THE EDGE OF THE UNIVERSE?"
THE EDGE OF THE UNIVERSE?!
WELL, IF YOU COULD GO THERE...
THE EDGE...
I KNOW!!!
OH!
THE ANSWER IS...
THE EDGE OF THE UNIVERSE IS A BRIDGE. AND SINCE IT'S THE EDGE OF THE WHOLE UNIVERSE, IT MUST BE A BRIDGE TO THE WHOLE UNIVERSE TOO!
みごとじゃ!
DO NOT CROSS THIS BRIDGE
IF I CAN'T CROSS THE EDGE, I'LL CROSS IN THE MIDDLE.
UGH! YOU FOOL! WHO DO YOU THINK YOU ARE, IKKYU-SAN?! SHE WANTS A WISE MAN, NOT A COURT JESTER!

TRANSLATOR'S NOTE: IKKYU WAS A ZEN PRIEST WITH A TALENT FOR PUNS. THE JAPANESE WORD FOR "BRIDGE" ("HASHI") CAN ALSO MEAN "EDGE."

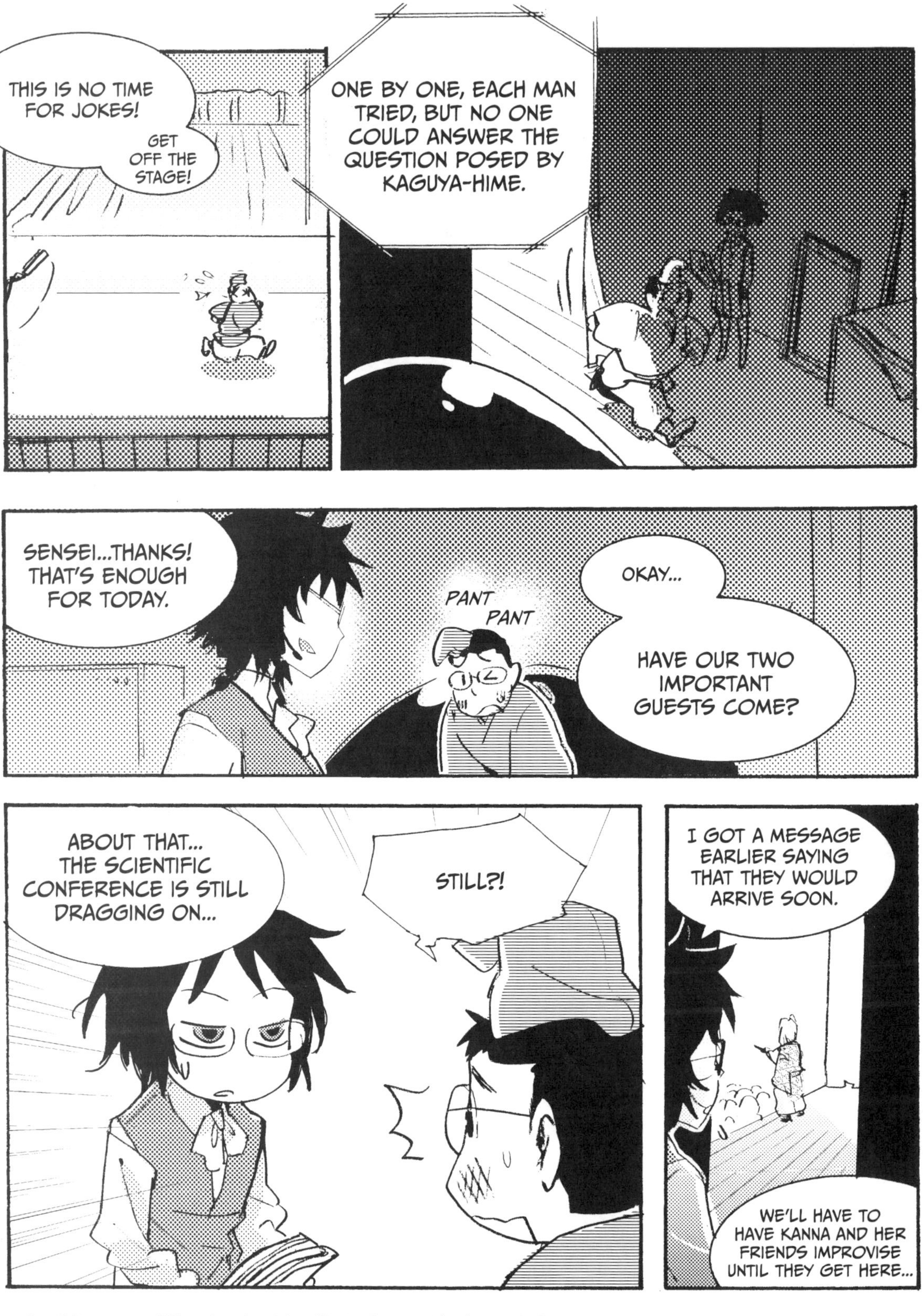
THIS IS NO TIME FOR JOKES!
GET OFF THE STAGE!
ONE BY ONE, EACH MAN TRIED, BUT NO ONE COULD ANSWER THE QUESTION POSED BY KAGUYA-HIME.
SENSEI...THANKS! THAT'S ENOUGH FOR TODAY.
PANT PANT
OKAY...
HAVE OUR TWO IMPORTANT GUESTS COME?
ABOUT THAT... THE SCIENTIFIC CONFERENCE IS STILL DRAGGING ON...
STILL?!
I GOT A MESSAGE EARLIER SAYING THAT THEY WOULD ARRIVE SOON.
WE'LL HAVE TO HAVE KANNA AND HER FRIENDS IMPROVISE UNTIL THEY GET HERE...

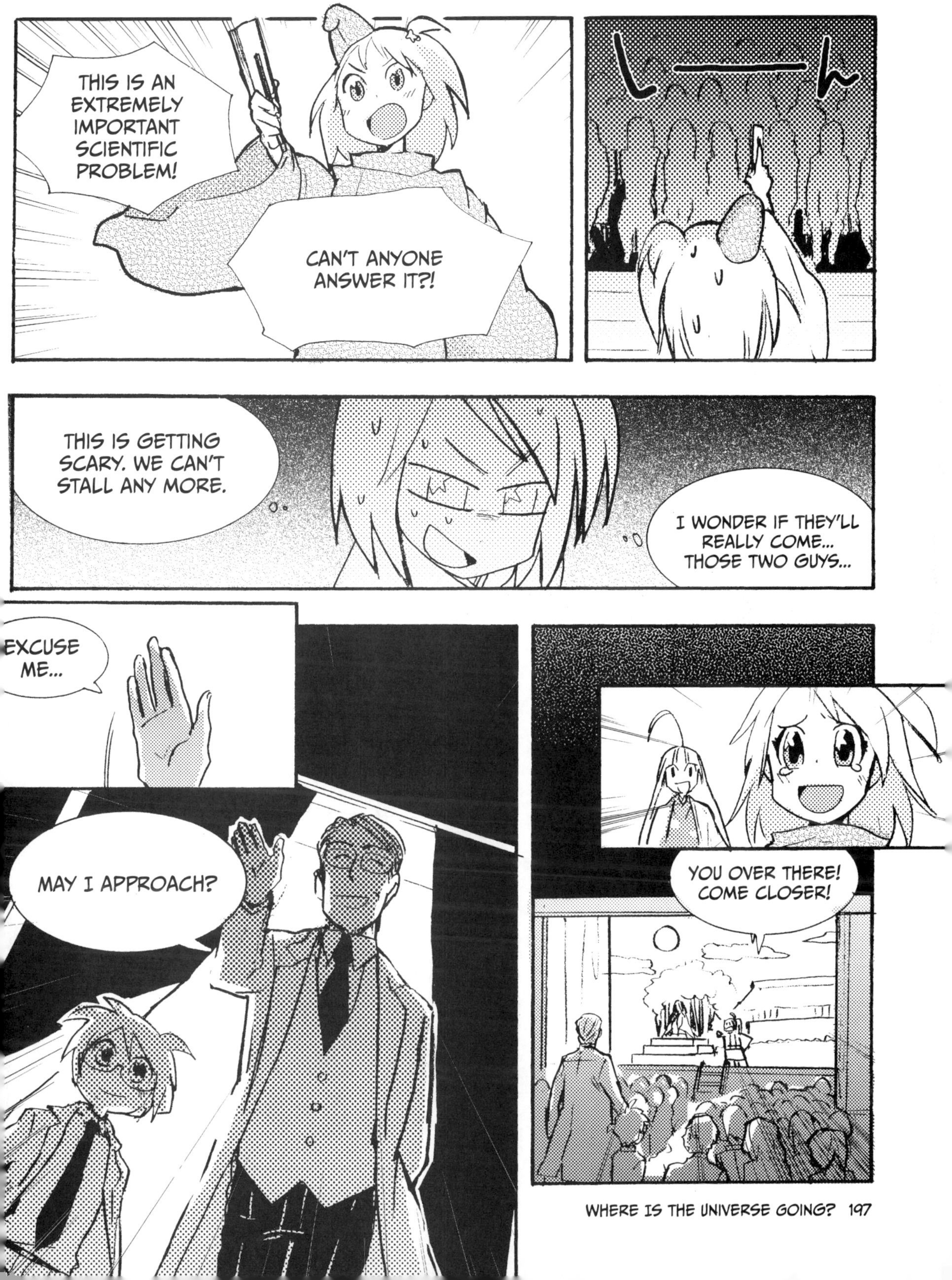
THIS IS AN EXTREMELY IMPORTANT SCIENTIFIC PROBLEM!
CAN'T ANYONE ANSWER IT?!
し———ん
THIS IS GETTING SCARY. WE CAN'T STALL ANY MORE.
I WONDER IF THEY'LL REALLY COME... THOSE TWO GUYS...
EXCUSE ME...
MAY I APPROACH?
YOU OVER THERE! COME CLOSER!

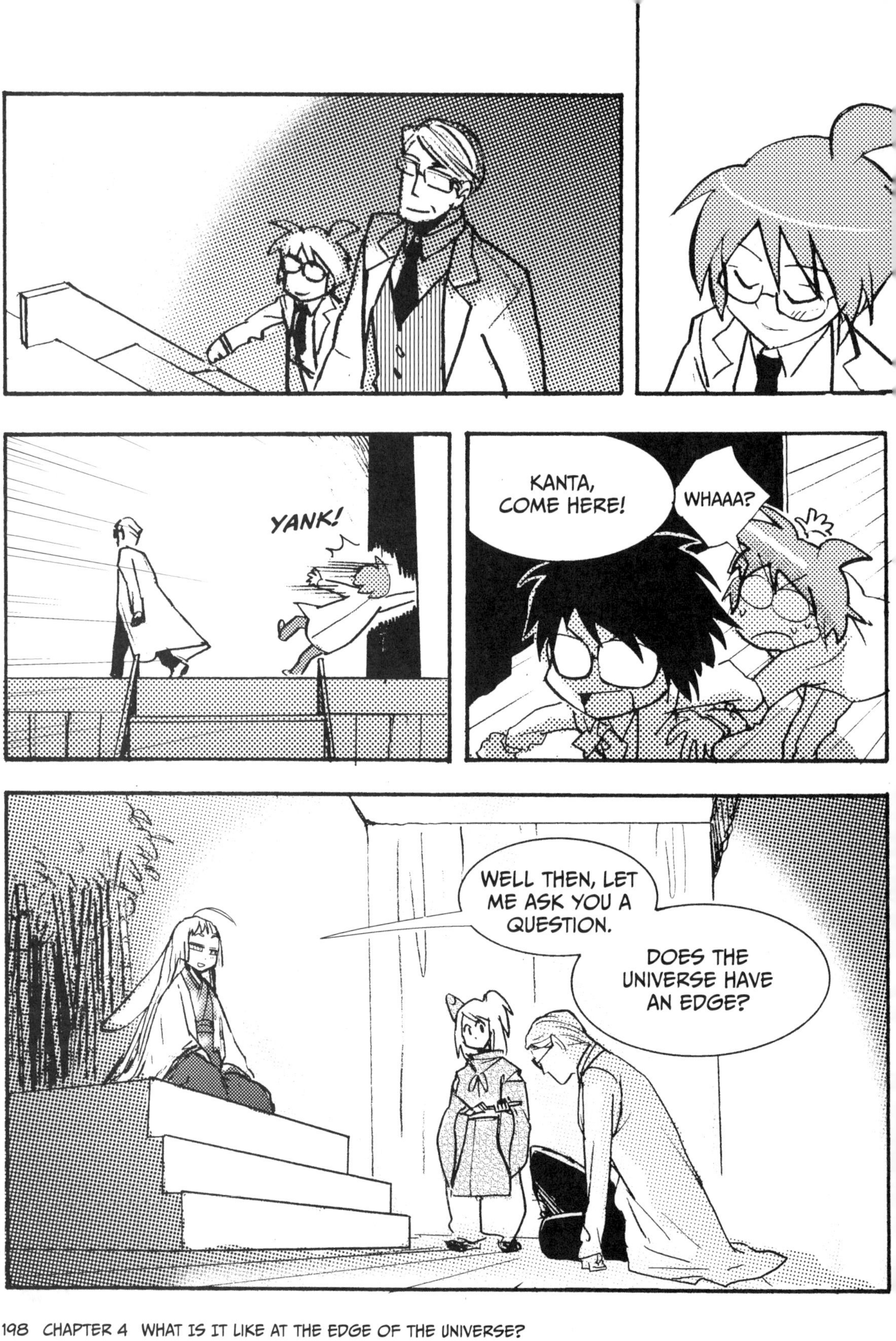
YANK!
KANTA, COME HERE!
WHAAA?
WELL THEN, LET ME ASK YOU A QUESTION.
DOES THE UNIVERSE HAVE AN EDGE?

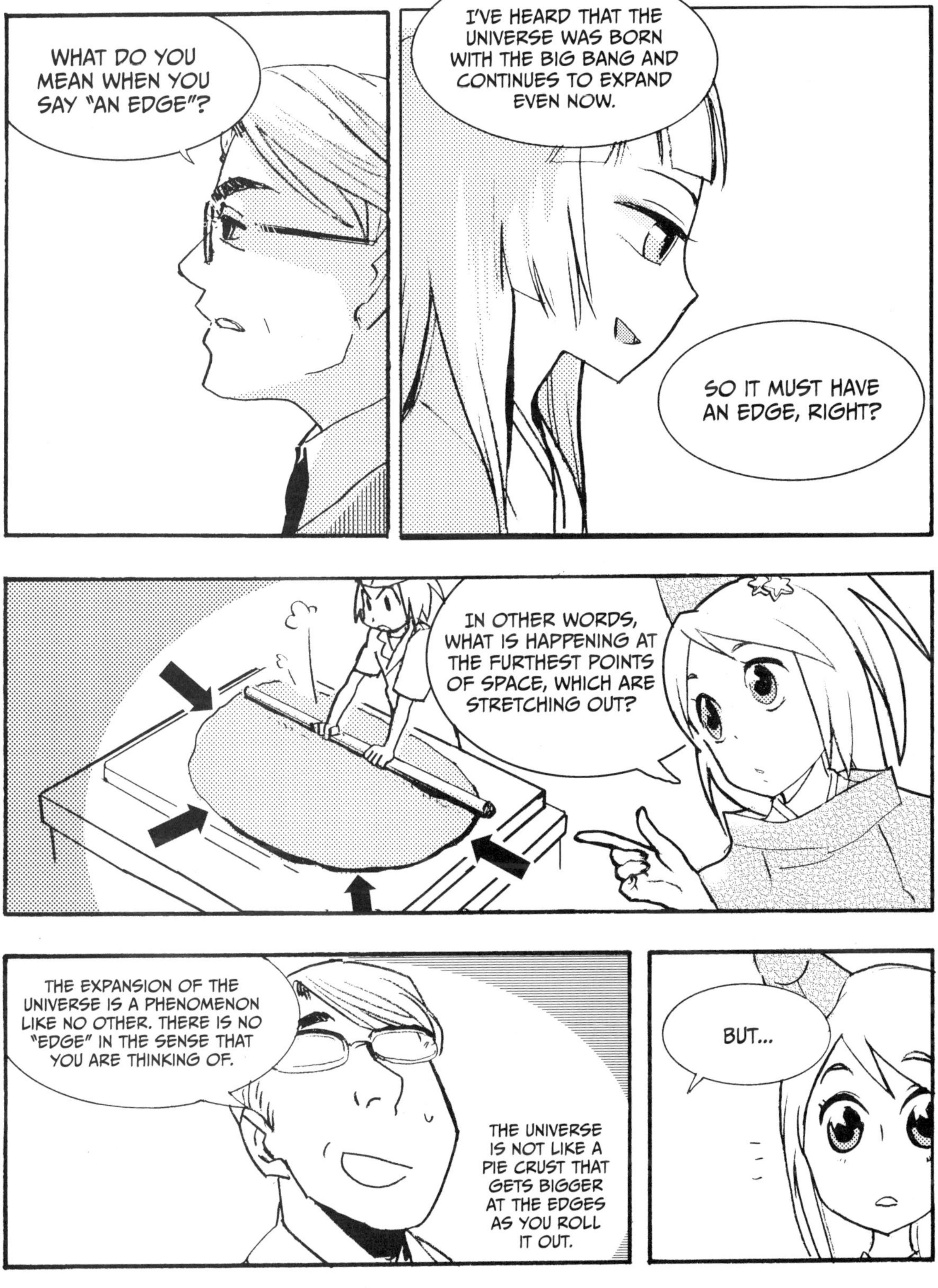
WHAT DO YOU MEAN WHEN YOU SAY "AN EDGE"?
I'VE HEARD THAT THE UNIVERSE WAS BORN WITH THE BIG BANG AND CONTINUES TO EXPAND EVEN NOW.
SO IT MUST HAVE AN EDGE, RIGHT?
IN OTHER WORDS, WHAT IS HAPPENING AT THE FURTHEST POINTS OF SPACE, WHICH ARE STRETCHING OUT?
THE EXPANSION OF THE UNIVERSE IS A PHENOMENON LIKE NO OTHER. THERE IS NO "EDGE" IN THE SENSE THAT YOU ARE THINKING OF.
THE UNIVERSE IS NOT LIKE A PIE CRUST THAT GETS BIGGER AT THE EDGES AS YOU ROLL IT OUT.
BUT...

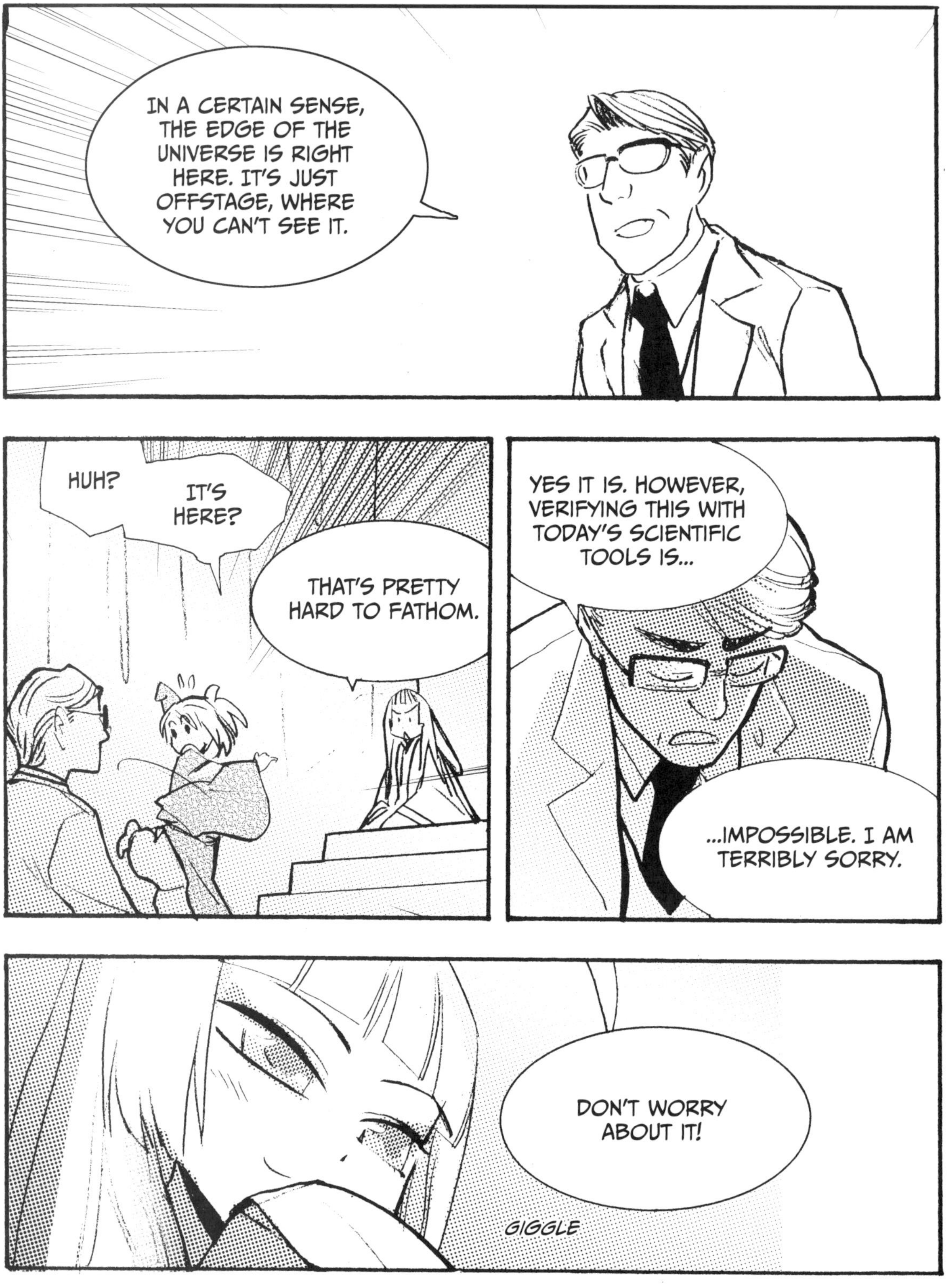
IN A CERTAIN SENSE, THE EDGE OF THE UNIVERSE IS RIGHT HERE. IT'S JUST OFFSTAGE, WHERE YOU CAN'T SEE IT.
HUH?
IT'S HERE?
THAT'S PRETTY HARD TO FATHOM.
YES IT IS. HOWEVER, VERIFYING THIS WITH TODAY'S SCIENTIFIC TOOLS IS...
...IMPOSSIBLE. I AM TERRIBLY SORRY.
DON'T WORRY ABOUT IT!
GIGGLE

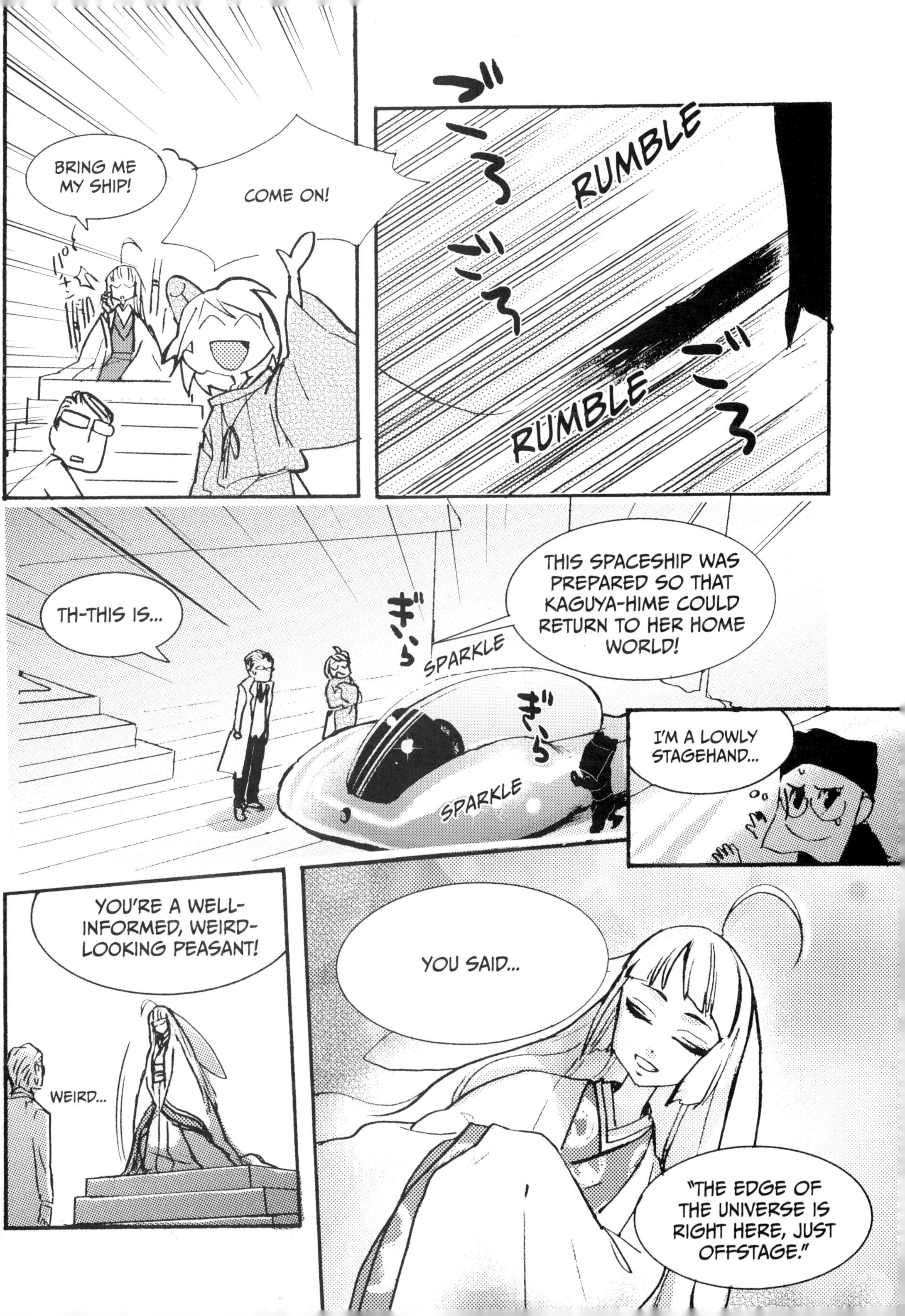
BRING ME MY SHIP!
COME ON!
ごーろ
ごろ
RUMBLE
ごーろ
RUMBLE
TH-THIS IS...
ぎら
SPARKLE
THIS SPACESHIP WAS PREPARED SO THAT KAGUYA-HIME COULD RETURN TO HER HOME WORLD!
ぎら
SPARKLE
I'M A LOWLY STAGEHAND...
YOU'RE A WELL-INFORMED, WEIRD-LOOKING PEASANT!
WEIRD...
YOU SAID...
"THE EDGE OF THE UNIVERSE IS RIGHT HERE, JUST OFFSTAGE."

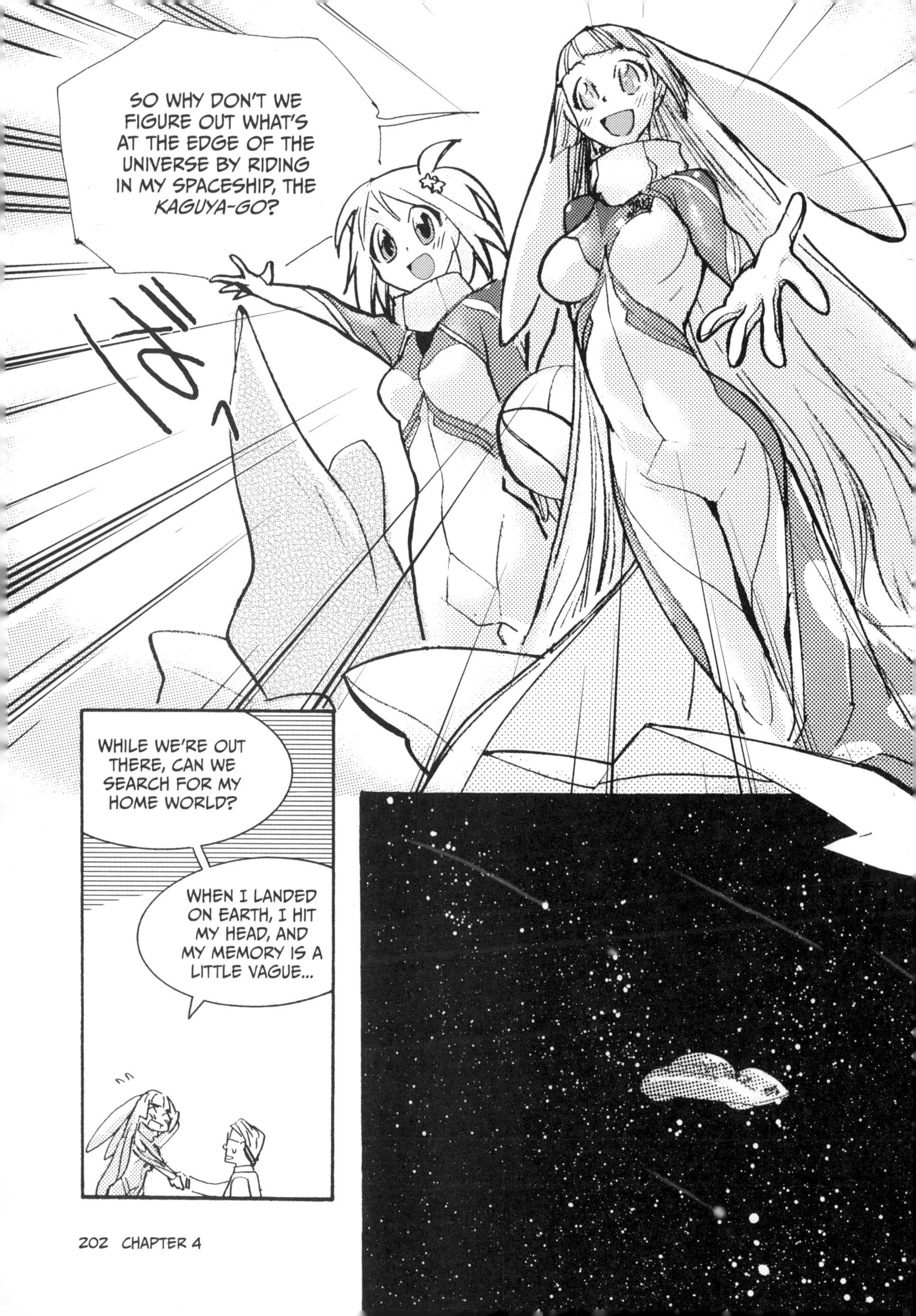
SO WHY DON'T WE FIGURE OUT WHAT'S AT THE EDGE OF THE UNIVERSE BY RIDING IN MY SPACESHIP, THE *KAGUYA-GO*?
WHILE WE'RE OUT THERE, CAN WE SEARCH FOR MY HOME WORLD?
WHEN I LANDED ON EARTH, I HIT MY HEAD, AND MY MEMORY IS A LITTLE VAGUE...

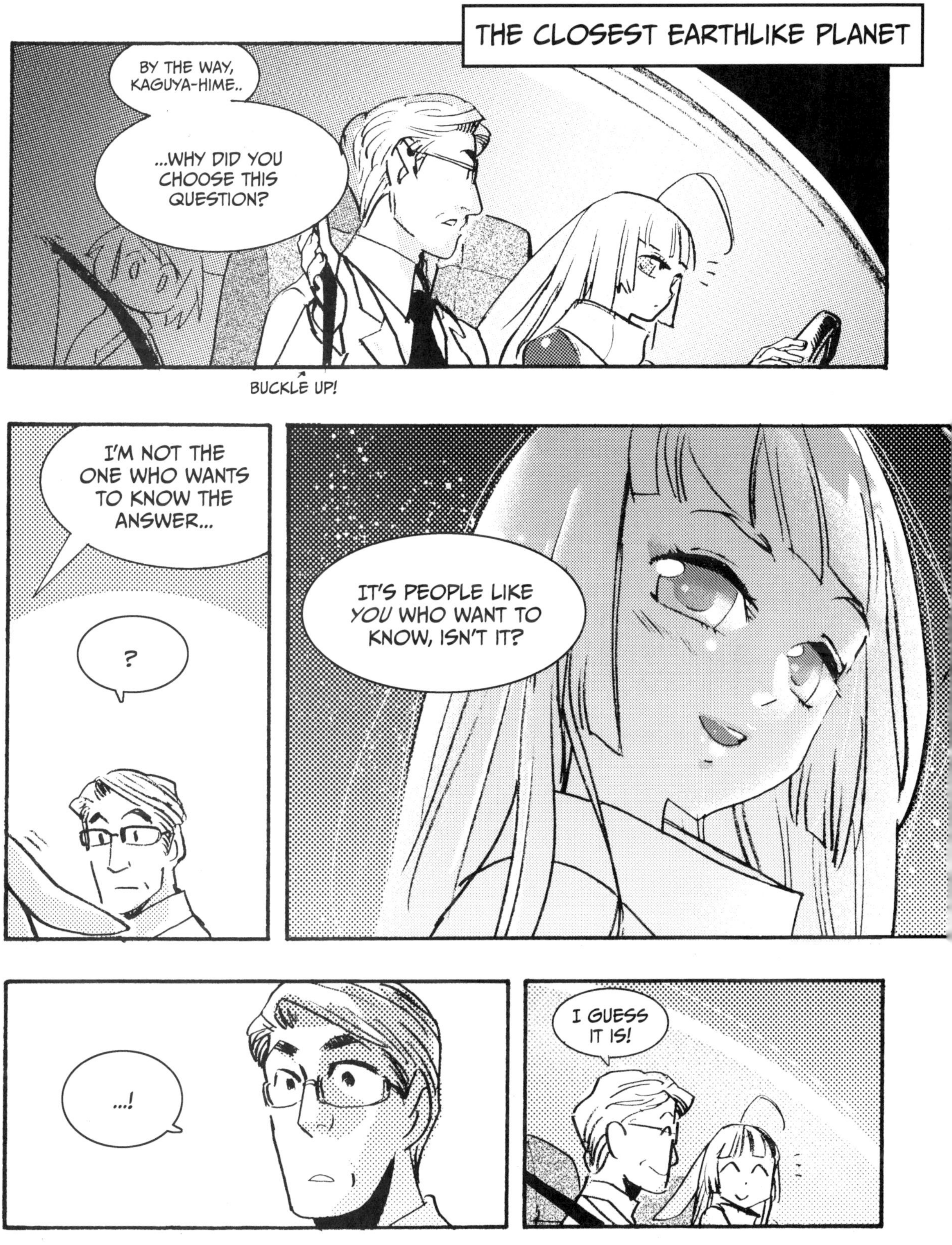
THE CLOSEST EARTHLIKE PLANET
BY THE WAY, KAGUYA-HIME..
...WHY DID YOU CHOOSE THIS QUESTION?
BUCKLE UP!
I'M NOT THE ONE WHO WANTS TO KNOW THE ANSWER...
?
IT'S PEOPLE LIKE *YOU* WHO WANT TO KNOW, ISN'T IT?
...!
I GUESS IT IS!

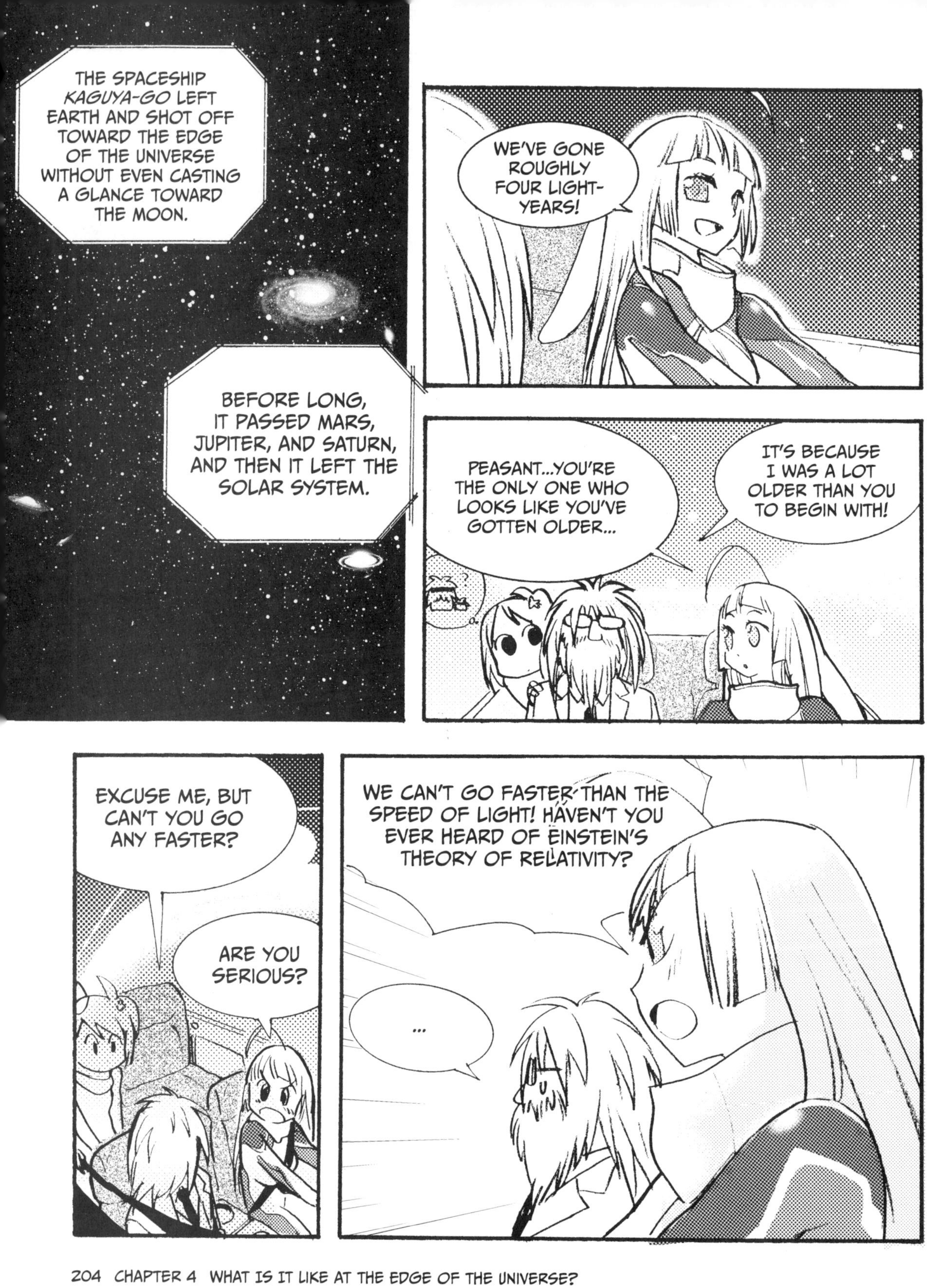
THE SPACESHIP *KAGUYA-GO* LEFT EARTH AND SHOT OFF TOWARD THE EDGE OF THE UNIVERSE WITHOUT EVEN CASTING A GLANCE TOWARD THE MOON.
BEFORE LONG, IT PASSED MARS, JUPITER, AND SATURN, AND THEN IT LEFT THE SOLAR SYSTEM.
WE'VE GONE ROUGHLY FOUR LIGHT-YEARS!
PEASANT...YOU'RE THE ONLY ONE WHO LOOKS LIKE YOU'VE GOTTEN OLDER...
IT'S BECAUSE I WAS A LOT OLDER THAN YOU TO BEGIN WITH!
EXCUSE ME, BUT CAN'T YOU GO ANY FASTER?
ARE YOU SERIOUS?
WE CAN'T GO FASTER THAN THE SPEED OF LIGHT! HAVEN'T YOU EVER HEARD OF EINSTEIN'S THEORY OF RELATIVITY?
...

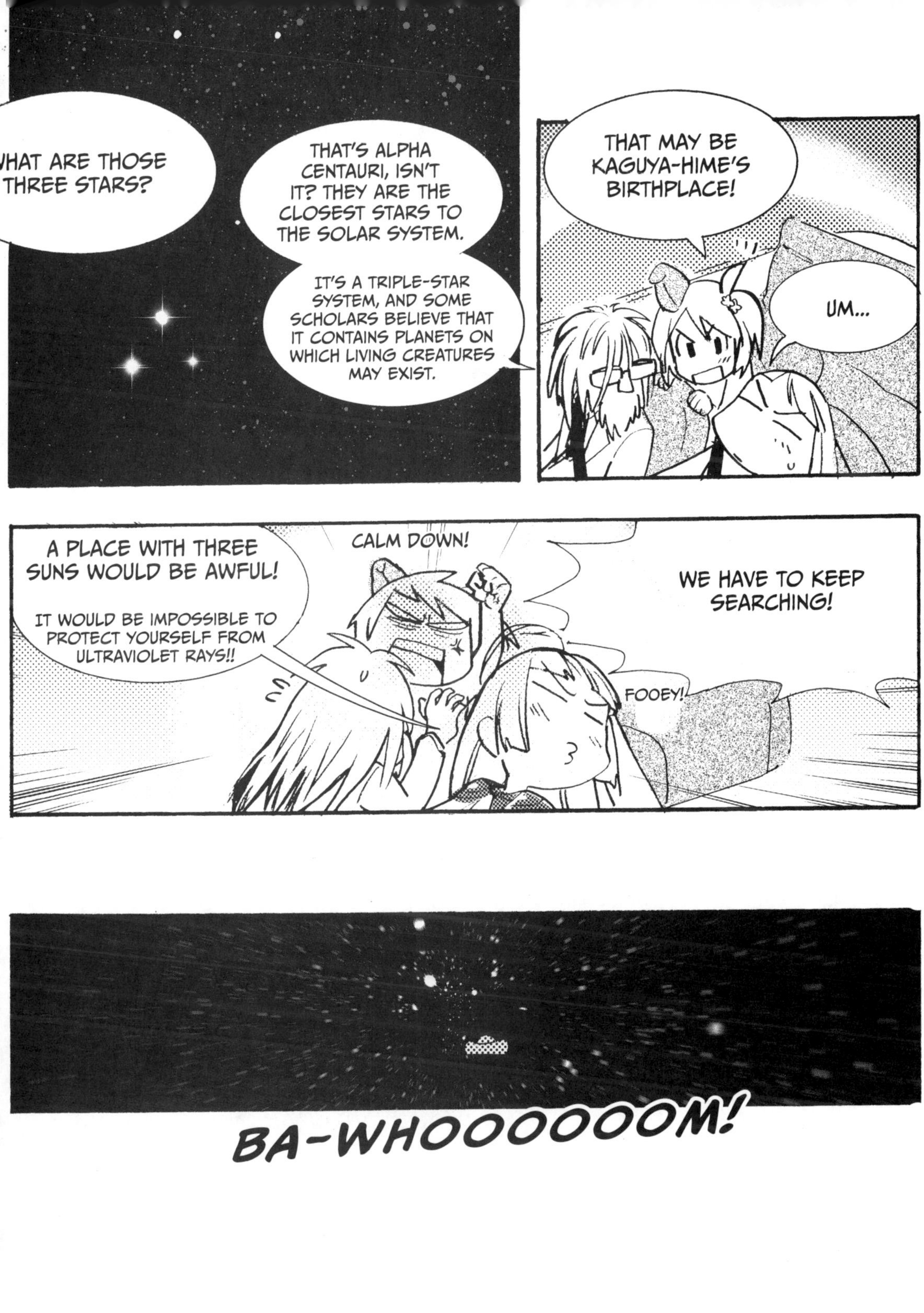
WHAT ARE THOSE THREE STARS?
THAT'S ALPHA CENTAURI, ISN'T IT? THEY ARE THE CLOSEST STARS TO THE SOLAR SYSTEM.
IT'S A TRIPLE-STAR SYSTEM, AND SOME SCHOLARS BELIEVE THAT IT CONTAINS PLANETS ON WHICH LIVING CREATURES MAY EXIST.
THAT MAY BE KAGUYA-HIME'S BIRTHPLACE!
UM...
A PLACE WITH THREE SUNS WOULD BE AWFUL!
IT WOULD BE IMPOSSIBLE TO PROTECT YOURSELF FROM ULTRAVIOLET RAYS!!
CALM DOWN!
FOOEY!
WE HAVE TO KEEP SEARCHING!
BA-WHOOOOOOOM!

The *Kaguya-go* Journey Board Game

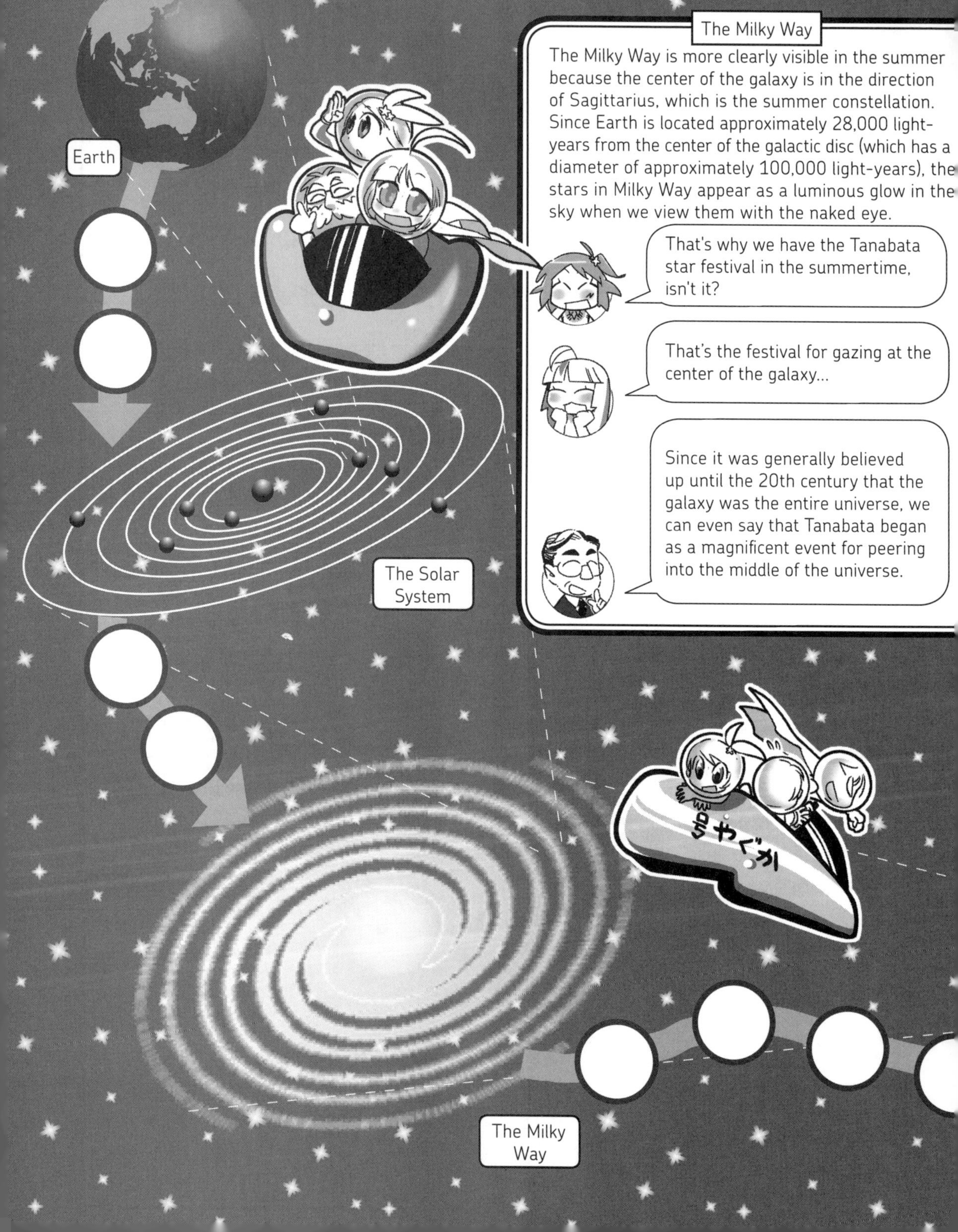

Start
Earth
The Solar System
The Milky Way

The Milky Way

The Milky Way is more clearly visible in the summer because the center of the galaxy is in the direction of Sagittarius, which is the summer constellation. Since Earth is located approximately 28,000 light-years from the center of the galactic disc (which has a diameter of approximately 100,000 light-years), the stars in Milky Way appear as a luminous glow in the sky when we view them with the naked eye.

That's why we have the Tanabata star festival in the summertime, isn't it?

That's the festival for gazing at the center of the galaxy...

Since it was generally believed up until the 20th century that the galaxy was the entire universe, we can even say that Tanabata began as a magnificent event for peering into the middle of the universe.

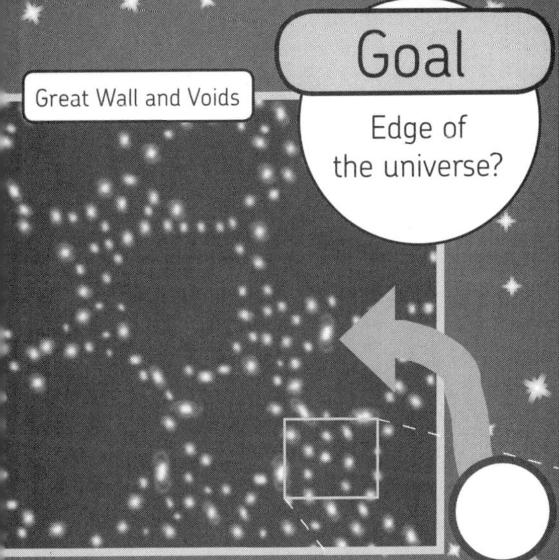

Goal

Edge of the universe?

Great Wall and Voids

Great Wall and Voids

Although galaxies form clusters and superclusters, when we look at all of space as a whole, the galaxies are arranged in a netlike pattern. In other words, many bubbles are collected together—the surfaces of those bubbles are galaxies, and the interiors of the bubbles are voids. Since the galaxies seem to create a large wall when observed from Earth, this kind of large-scale structure of the universe is called the Great Wall.

Nowadays, the mesh formed by the Great Wall and voids is said to be the largest structure of the universe.

And the same kind of structure continues, no matter how far we go, right?

Local Supercluster (Virgo Supercluster)

A supercluster is formed by an aggregation of clusters or groups of galaxies and has a diameter of more than 100 million light-years. This is truly a cluster of super-large heavenly bodies. The supercluster to which our galactic system (that is, the Local Group) belongs is called the Local Supercluster. It is also known as the Virgo Supercluster.

Since the Local Group that contains Earth is located toward the edge of the Virgo Supercluster, it is approximately 60 million light-years to the M87 galaxy in the Virgo constellation near the center of the Virgo Supercluster. The diameter of the Virgo Supercluster is said to be approximately 200 million light-years.

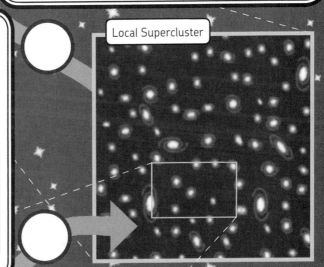

Local Supercluster

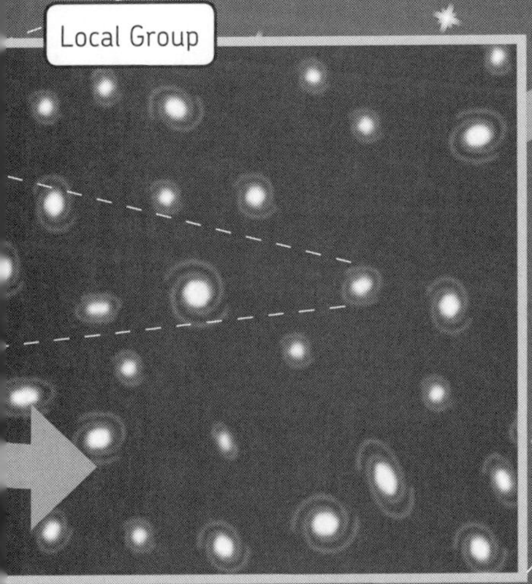

Local Group

Local Group

Galaxies create aggregations called groups or clusters of galaxies in space. The group of galaxies to which our galaxy (the Milky Way) belongs is called the Local Group. This group contains approximately 40 galaxies. The largest of these is the Andromeda galaxy; the diameter of its disc is approximately 130,000 light-years.

According to scientists' calculations, the diameter of the Local Group is 2.4 to 3.6 Mpc (megaparsecs).

A parsec is the distance to a heavenly body for which the annual parallax is 1 second. If I remember correctly, 1 pc = 3.26 light-years...therefore, the diameter is 7.8 to 11.7 **million** light-years!

ARRIVAL AT THE "EDGE" OF THE UNIVERSE

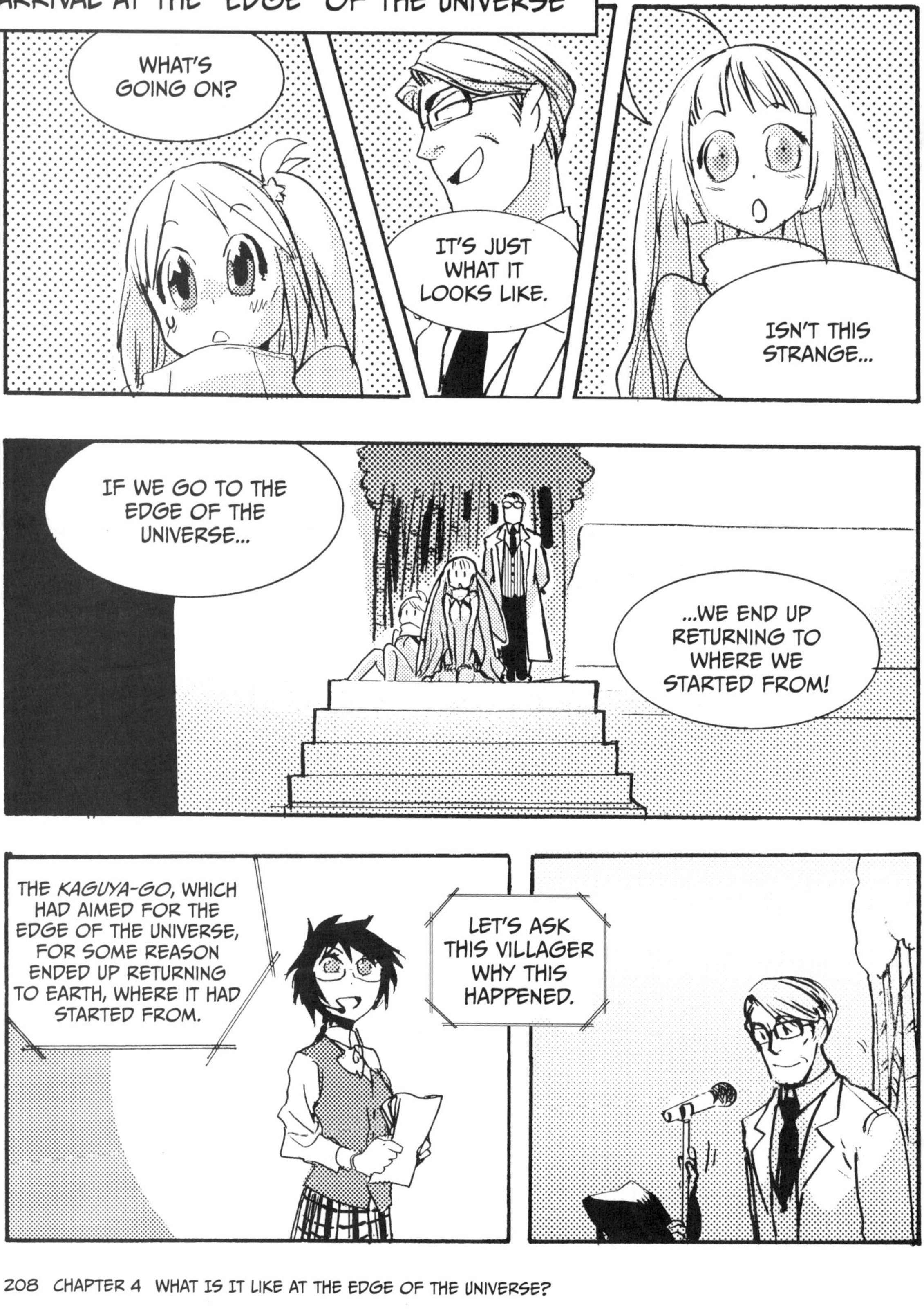
WHAT'S GOING ON?
IT'S JUST WHAT IT LOOKS LIKE.
ISN'T THIS STRANGE...
IF WE GO TO THE EDGE OF THE UNIVERSE...
...WE END UP RETURNING TO WHERE WE STARTED FROM!
THE *KAGUYA-GO*, WHICH HAD AIMED FOR THE EDGE OF THE UNIVERSE, FOR SOME REASON ENDED UP RETURNING TO EARTH, WHERE IT HAD STARTED FROM.
LET'S ASK THIS VILLAGER WHY THIS HAPPENED.

PROFESSOR SANUKI'S SOLILOQUY

Everyone has probably heard how the universe was born with the Big Bang. However, what does it mean for the universe to be born?

The universe that we recognize is three-dimensional, and it can be represented by three coordinate axes for length, width, and height. Of course, we cannot escape the bounds of it. For us, this is everything we know.

However, space with four (or more!) dimensions is called *hyperspace*, and from the perspective of hyperspace, a three-dimensional space is just a single, closed system. (By the way, the four-dimensional space I'm talking about here is space represented by four coordinate axes, not three-dimensional space plus time).

Since we cannot make an image of such a four-dimensional space, let's consider a model by looking at two dimensions from a three-dimensional perspective.

I have a balloon here, and its surface is two-dimensional. It is spatially curved and forms a three-dimensional sphere.

Spacewarp
Return to the original location
Two-dimensional rocket

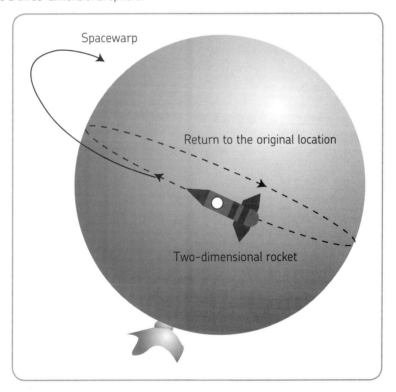

If the two-dimensional rocket aims for the edge of the balloon, it will return to its original location.

In the same way, the three-dimensional space in which we live could be four-dimensionally curved.

Since this hypothetical four-dimensional rocket simply passes beyond the edge of the three-dimensional universe, from its viewpoint (that is, if we look at our three-dimensional universe from four dimensions), the edge of the universe is everywhere. This is what I meant earlier when I said, "The edge of the universe is right here."

Incidentally, suppose that we had a spaceship with some sort of "warp drive" that allowed us to move by entering four-dimensional space and then re-entering three-dimensional space at a different location. To people observing us, our ship would appear to just vanish through a "warp" in space and then appear suddenly elsewhere.

Well, then, what is the shape of three-dimensional space?

Although I will omit the difficult explanation here, according to mathematical calculations, it is like one of the following three models.

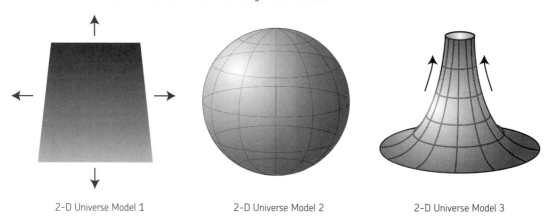

2-D Universe Model 1   2-D Universe Model 2   2-D Universe Model 3

In the first model, where the curvature of space is zero, space continues to extend no matter where you go. When illustrated in two dimensions, space is a plane that goes on forever. Although the figure appears to have an edge, since the plane actually continues in all directions, you absolutely cannot reach the edge of the universe as long as you move within three dimensions.

In the second model, where the curvature is positive, space is a spherical surface like that of a globe when it is represented as a two-dimensional model.

In the third model, where curvature is negative, space is "saddle shaped"—it curves up and it curves down.

If we consider the spherical surface with positive curvature as the model of our universe, a spaceship that travels in three dimensions while aiming for the edge of the universe will eventually return to its original location.

If we could build a spaceship that could travel faster than the speed of light, it might be able to reach hyperspace at the edge of the universe through a space warp. However, as long as we are in three-dimensional space, travel faster than the speed of light cannot occur because of the constraint of the theory of relativity. In other words, no matter how far we travel, we cannot get to the edge of the universe and, at best, will only return to our original point of departure.

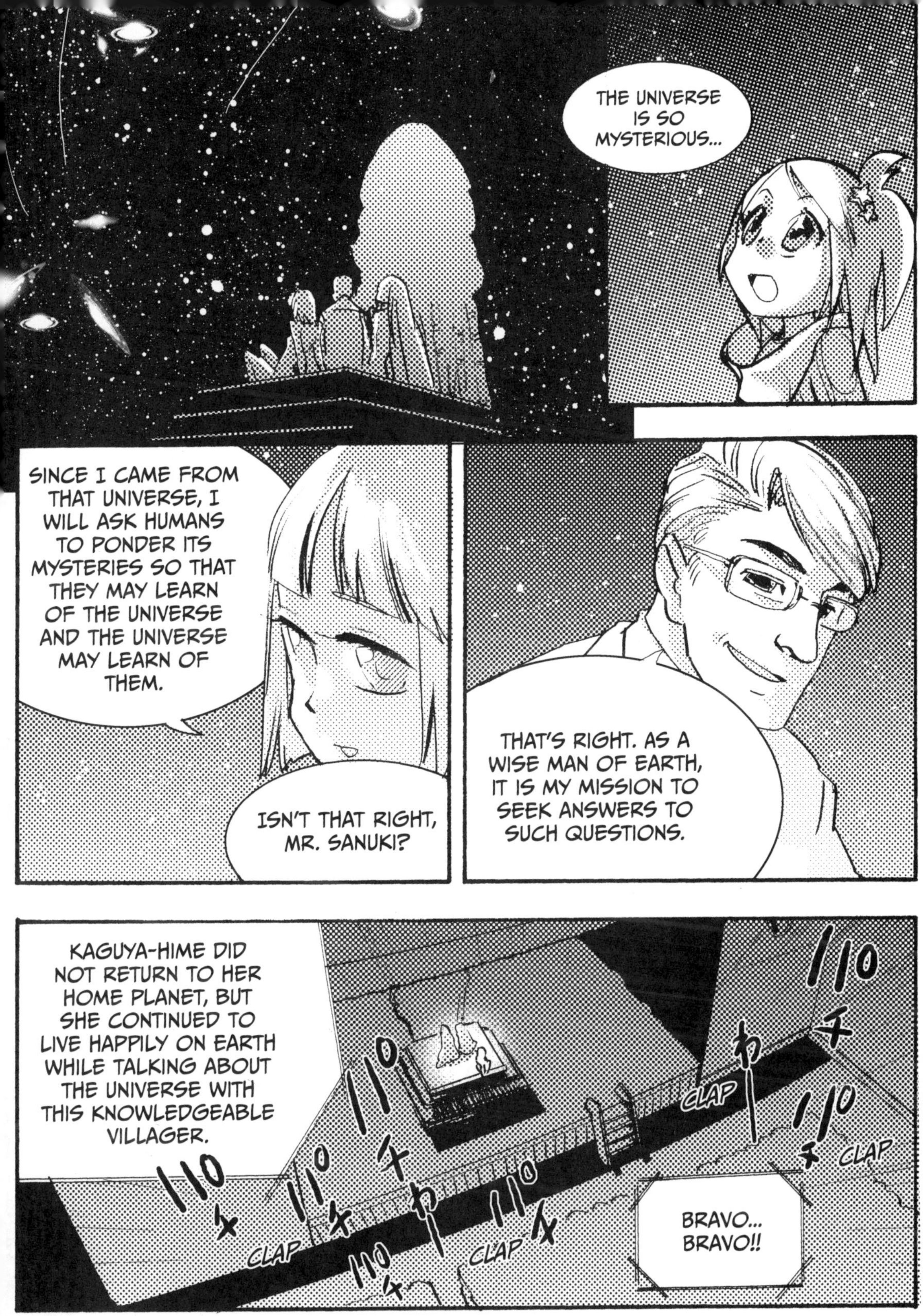
THE UNIVERSE IS SO MYSTERIOUS...
SINCE I CAME FROM THAT UNIVERSE, I WILL ASK HUMANS TO PONDER ITS MYSTERIES SO THAT THEY MAY LEARN OF THE UNIVERSE AND THE UNIVERSE MAY LEARN OF THEM.
ISN'T THAT RIGHT, MR. SANUKI?
THAT'S RIGHT. AS A WISE MAN OF EARTH, IT IS MY MISSION TO SEEK ANSWERS TO SUCH QUESTIONS.
KAGUYA-HIME DID NOT RETURN TO HER HOME PLANET, BUT SHE CONTINUED TO LIVE HAPPILY ON EARTH WHILE TALKING ABOUT THE UNIVERSE WITH THIS KNOWLEDGEABLE VILLAGER.
CLAP
CLAP
CLAP
CLAP
BRAVO... BRAVO!!

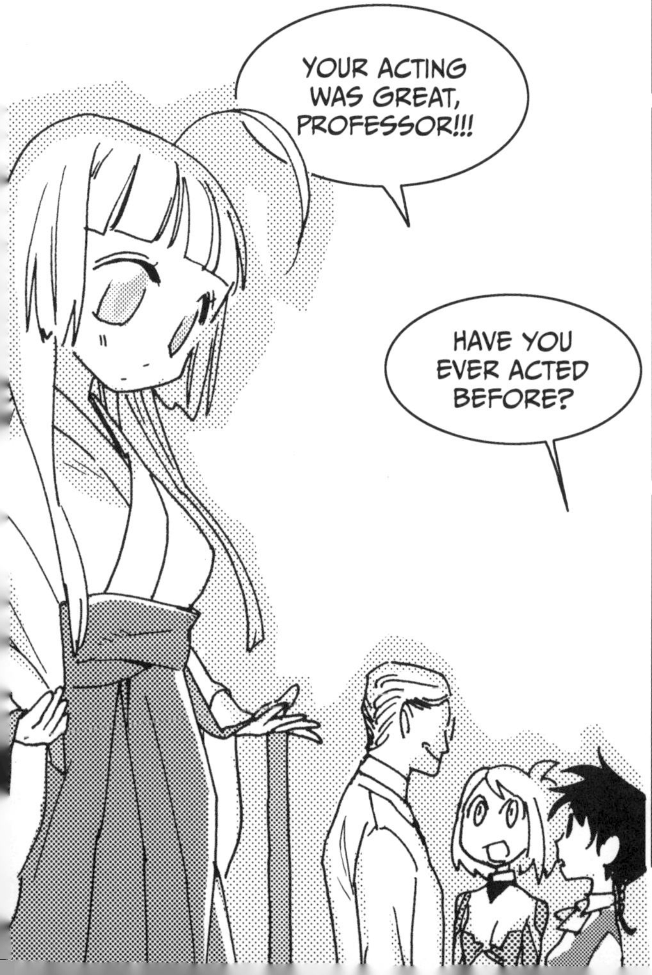
YOUR ACTING WAS GREAT, PROFESSOR!!!
HAVE YOU EVER ACTED BEFORE?

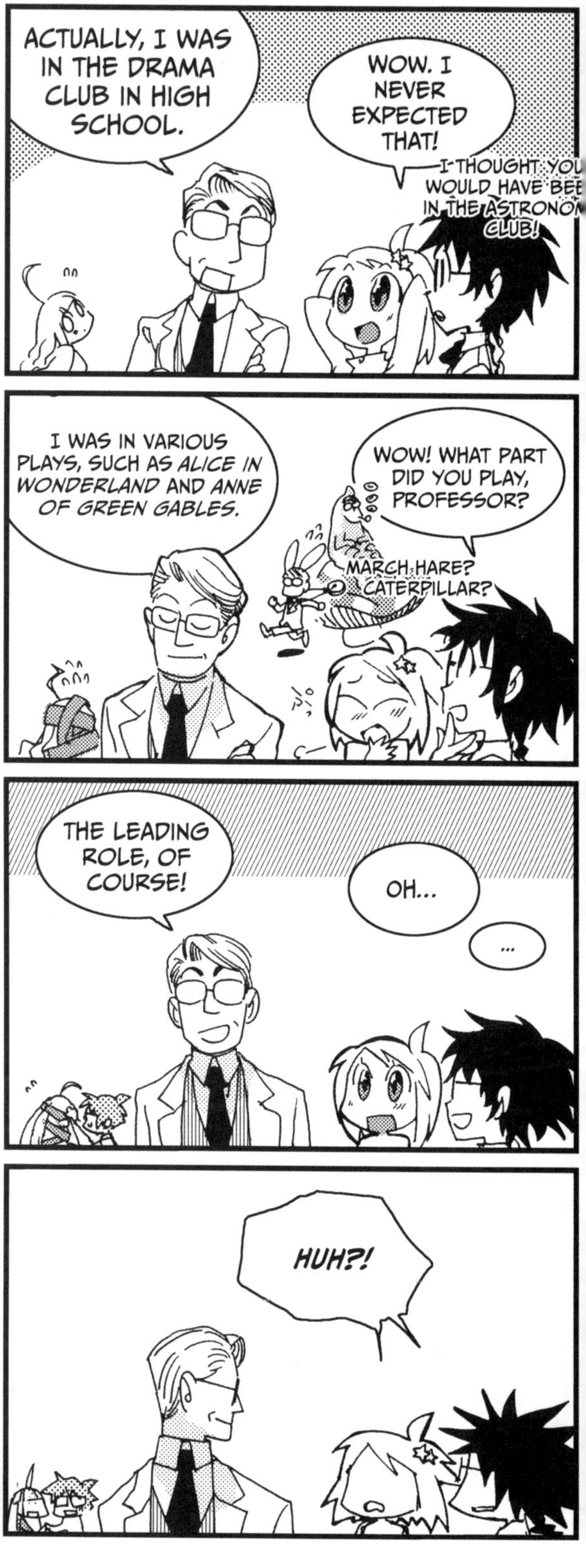
ACTUALLY, I WAS IN THE DRAMA CLUB IN HIGH SCHOOL.
WOW. I NEVER EXPECTED THAT!
I THOUGHT YOU WOULD HAVE BEE IN THE ASTRONO CLUB!
I WAS IN VARIOUS PLAYS, SUCH AS *ALICE IN WONDERLAND* AND *ANNE OF GREEN GABLES*.
WOW! WHAT PART DID YOU PLAY, PROFESSOR?
MARCH HARE?
CATERPILLAR?
THE LEADING ROLE, OF COURSE!
OH...
...
HUH?!

5
OUR EVER-EXPANDING UNIVERSE

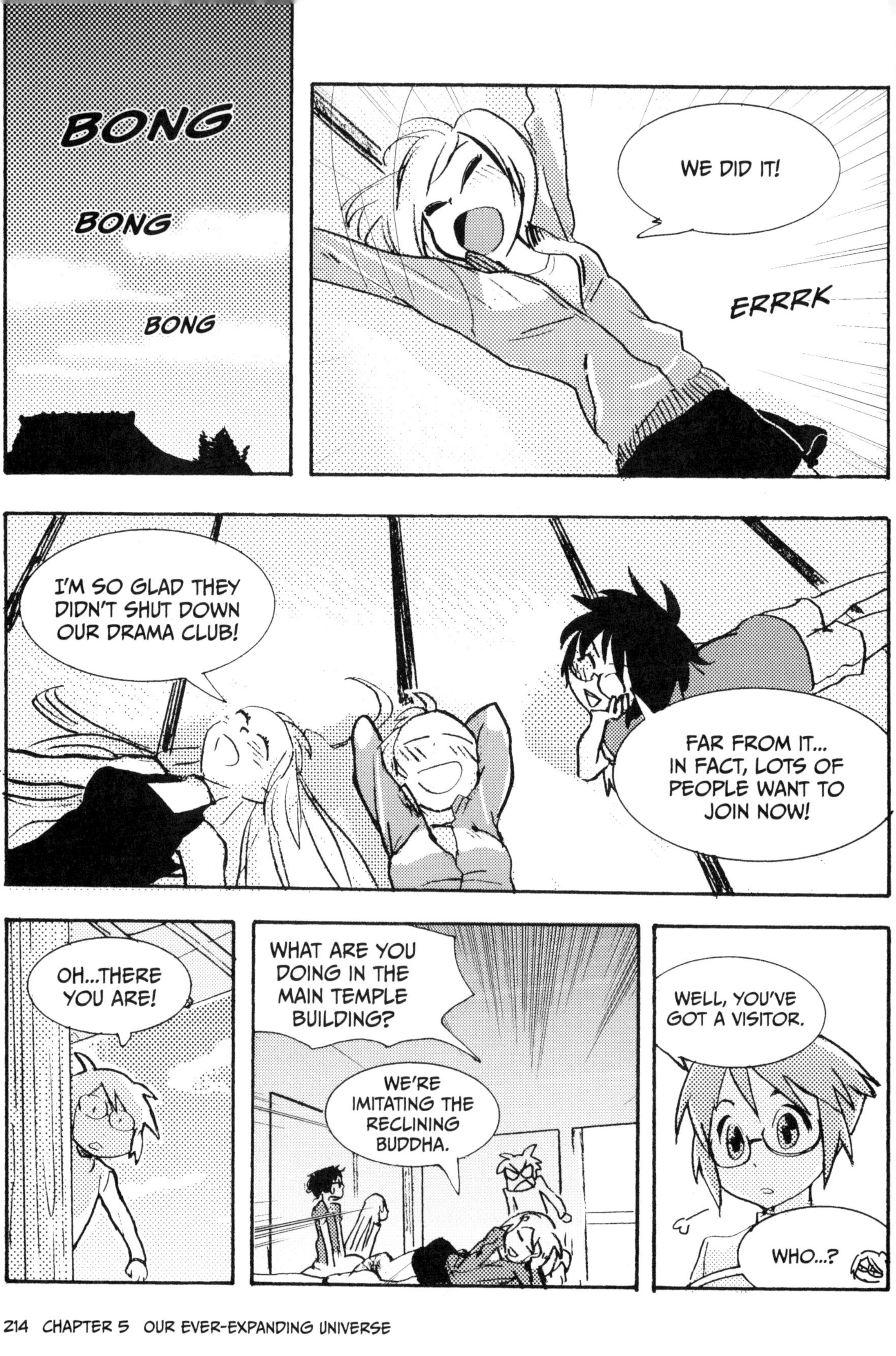
BONG
BONG
BONG
WE DID IT!
ERRRK
I'M SO GLAD THEY DIDN'T SHUT DOWN OUR DRAMA CLUB!
FAR FROM IT... IN FACT, LOTS OF PEOPLE WANT TO JOIN NOW!
OH...THERE YOU ARE!
WHAT ARE YOU DOING IN THE MAIN TEMPLE BUILDING?
WE'RE IMITATING THE RECLINING BUDDHA.
WELL, YOU'VE GOT A VISITOR.
WHO...?

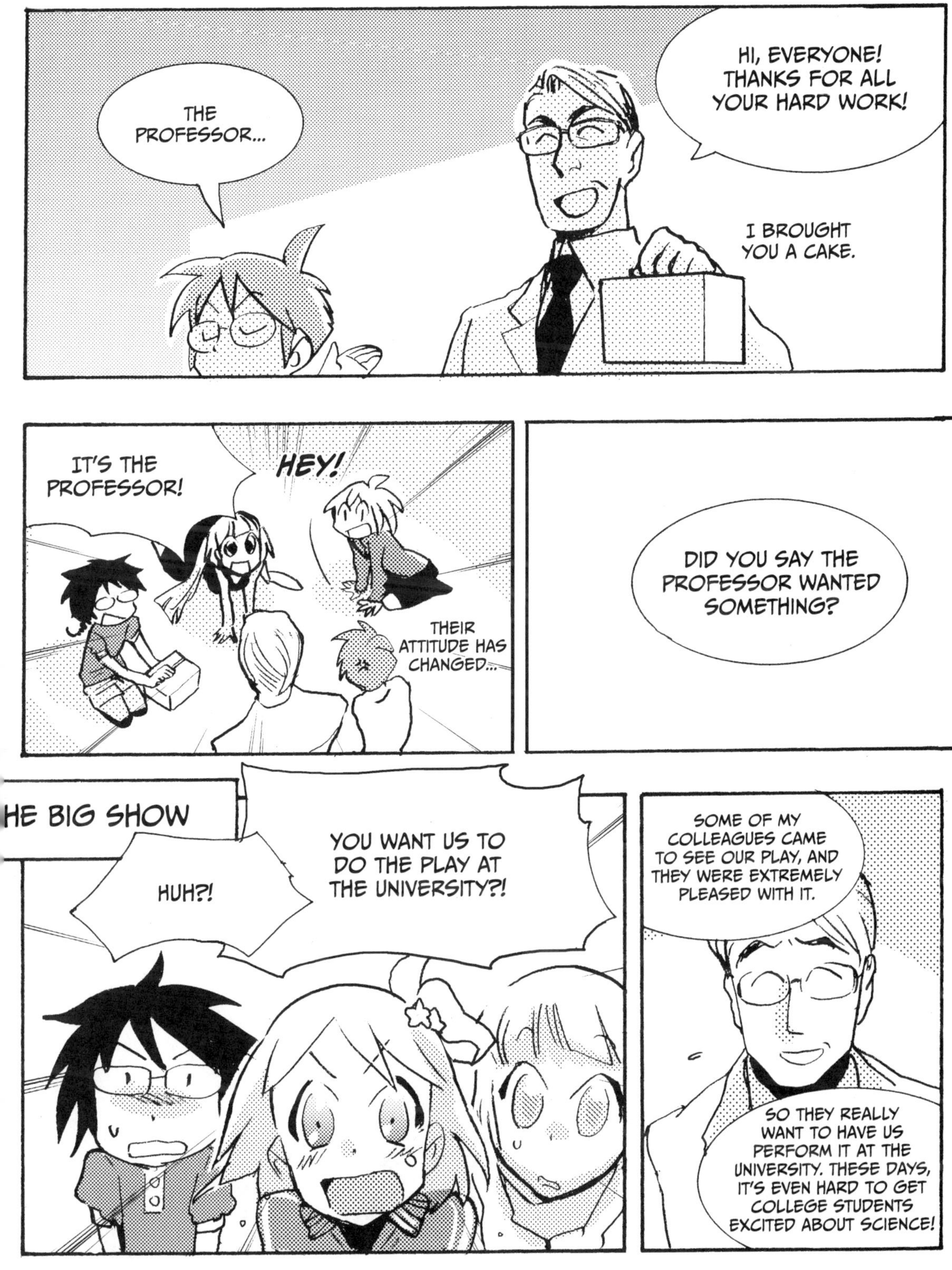
THE PROFESSOR...
HI, EVERYONE! THANKS FOR ALL YOUR HARD WORK!
I BROUGHT YOU A CAKE.
IT'S THE PROFESSOR!
HEY!
THEIR ATTITUDE HAS CHANGED...
DID YOU SAY THE PROFESSOR WANTED SOMETHING?
HE BIG SHOW
HUH?!
YOU WANT US TO DO THE PLAY AT THE UNIVERSITY?!
SOME OF MY COLLEAGUES CAME TO SEE OUR PLAY, AND THEY WERE EXTREMELY PLEASED WITH IT.
SO THEY REALLY WANT TO HAVE US PERFORM IT AT THE UNIVERSITY. THESE DAYS, IT'S EVEN HARD TO GET COLLEGE STUDENTS EXCITED ABOUT SCIENCE!

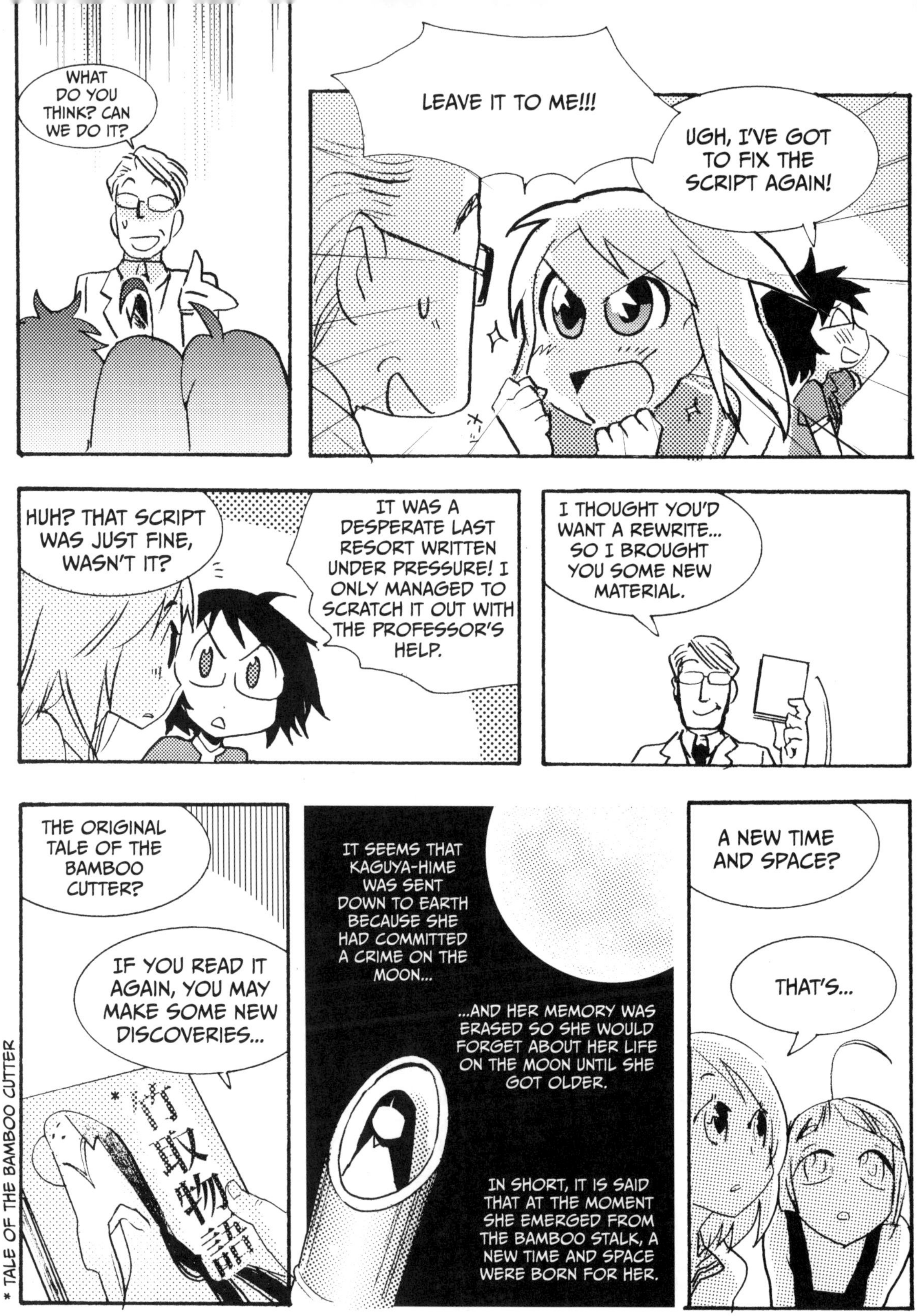
WHAT DO YOU THINK? CAN WE DO IT?
LEAVE IT TO ME!!!
UGH, I'VE GOT TO FIX THE SCRIPT AGAIN!
HUH? THAT SCRIPT WAS JUST FINE, WASN'T IT?
IT WAS A DESPERATE LAST RESORT WRITTEN UNDER PRESSURE! I ONLY MANAGED TO SCRATCH IT OUT WITH THE PROFESSOR'S HELP.
I THOUGHT YOU'D WANT A REWRITE... SO I BROUGHT YOU SOME NEW MATERIAL.
THE ORIGINAL TALE OF THE BAMBOO CUTTER?
IF YOU READ IT AGAIN, YOU MAY MAKE SOME NEW DISCOVERIES...
* TALE OF THE BAMBOO CUTTER
IT SEEMS THAT KAGUYA-HIME WAS SENT DOWN TO EARTH BECAUSE SHE HAD COMMITTED A CRIME ON THE MOON...
...AND HER MEMORY WAS ERASED SO SHE WOULD FORGET ABOUT HER LIFE ON THE MOON UNTIL SHE GOT OLDER.
IN SHORT, IT IS SAID THAT AT THE MOMENT SHE EMERGED FROM THE BAMBOO STALK, A NEW TIME AND SPACE WERE BORN FOR HER.
A NEW TIME AND SPACE?
THAT'S...

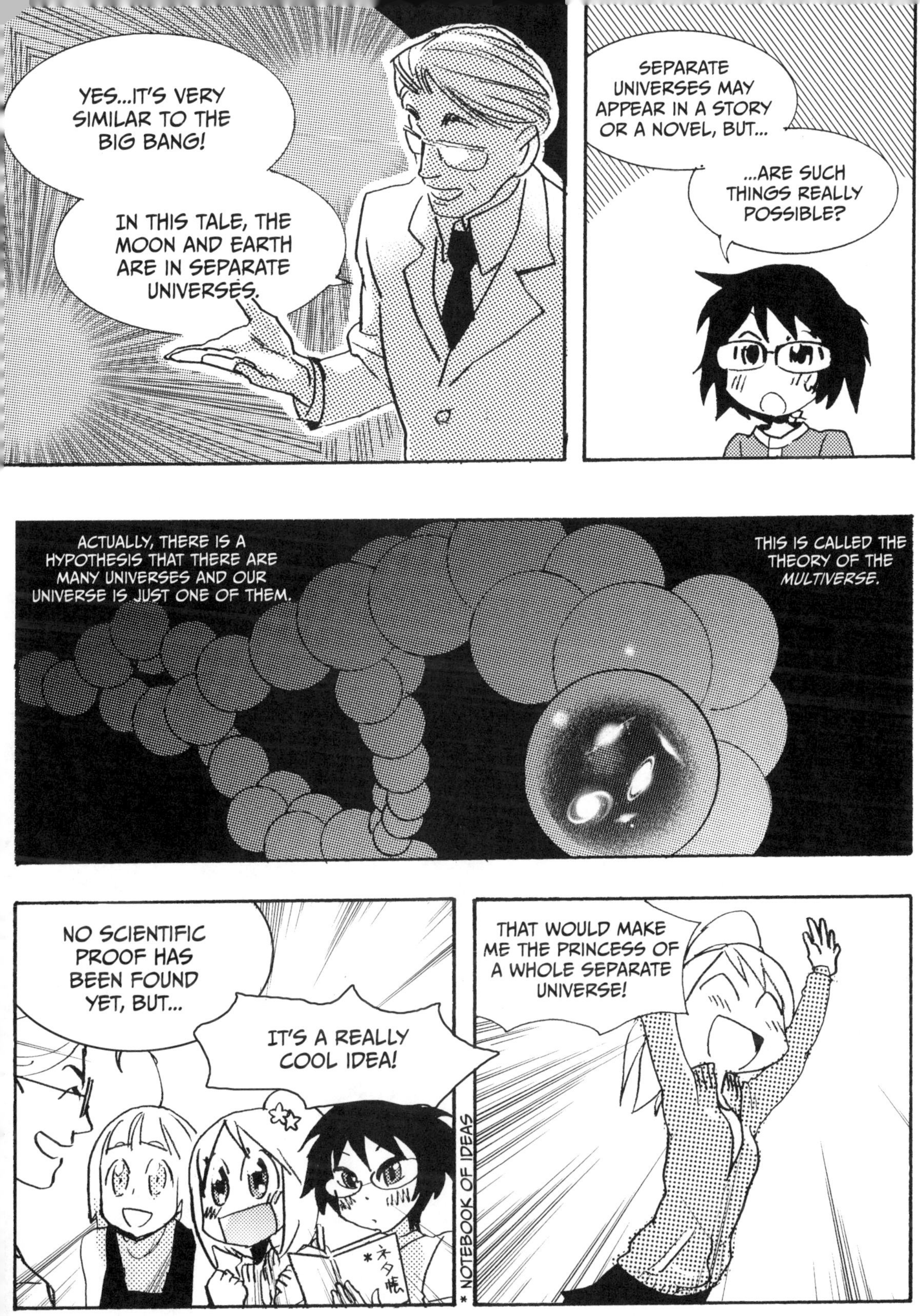
YES...IT'S VERY SIMILAR TO THE BIG BANG!
IN THIS TALE, THE MOON AND EARTH ARE IN SEPARATE UNIVERSES.
SEPARATE UNIVERSES MAY APPEAR IN A STORY OR A NOVEL, BUT...
...ARE SUCH THINGS REALLY POSSIBLE?
ACTUALLY, THERE IS A HYPOTHESIS THAT THERE ARE MANY UNIVERSES AND OUR UNIVERSE IS JUST ONE OF THEM.
THIS IS CALLED THE THEORY OF THE *MULTIVERSE*.
NO SCIENTIFIC PROOF HAS BEEN FOUND YET, BUT...
IT'S A REALLY COOL IDEA!
*ネタ帳
* NOTEBOOK OF IDEAS
THAT WOULD MAKE ME THE PRINCESS OF A WHOLE SEPARATE UNIVERSE!

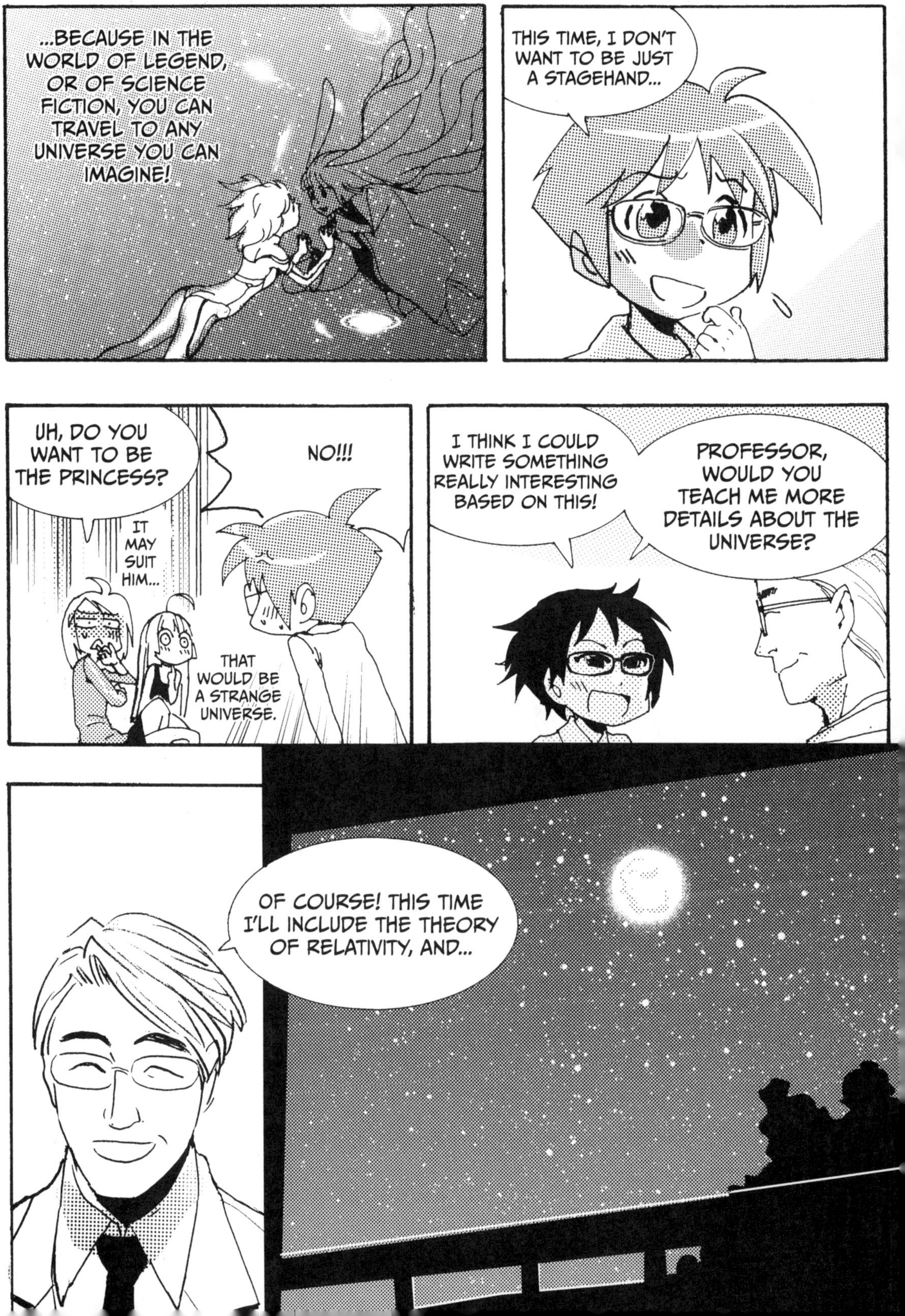
...BECAUSE IN THE WORLD OF LEGEND, OR OF SCIENCE FICTION, YOU CAN TRAVEL TO ANY UNIVERSE YOU CAN IMAGINE!
THIS TIME, I DON'T WANT TO BE JUST A STAGEHAND...
UH, DO YOU WANT TO BE THE PRINCESS?
NO!!!
IT MAY SUIT HIM...
THAT WOULD BE A STRANGE UNIVERSE.
I THINK I COULD WRITE SOMETHING REALLY INTERESTING BASED ON THIS!
PROFESSOR, WOULD YOU TEACH ME MORE DETAILS ABOUT THE UNIVERSE?
OF COURSE! THIS TIME I'LL INCLUDE THE THEORY OF RELATIVITY, AND...

THE MULTIVERSE CONTAINS NUMEROUS UNIVERSES

The theory of the *multiverse* hypothesizes that there are multiple universes outside of our own. Some people hypothesize that *hyperspace* is a receptacle for the universe (that is, space itself), and they believe that many universes are floating within hyperspace. There are many different theories about how these so-called *parallel universes* were formed and how they are related to our own universe. As the professor said, though, there is no scientific proof for any multiverse theory, so the kinds of relationships that may exist between universes are unknown. But many scientists nevertheless entertain the idea that the multiverse may form a structure larger than the observable universe.

The *cosmological principle* posits that if viewed from a sufficiently large scale, the properties of the universe are the same for all observers. This means that there is no special place in the universe, that the universe will have the same general appearance from any location, and that the same laws of physics will apply at any location. If we extend our interpretation of this principle, it would seem logical to assume that there could be countless other universes; the idea that our universe is unique is then illogical. In other words, if there were a super-cosmological principle, the idea of the multiverse would not be at all far-fetched. But philosophically, isn't it more than a little strange to believe that other universes must exist? Needless to say, this conjecture has its share of critics as well.

THE EDGE, BIRTH, AND END OF THE UNIVERSE...

The degree that space bends is called *curvature*. When we say "space," we mean everything we consider to be in our universe: planets, stars, gas, comets, and even energy bends. In the last chapter we talked about the potential shapes of our universe—let's explore that idea a bit more, revisiting the idea of *positive curvature*.

If the universe has positive curvature, a spaceship that aims for the edge of the universe and continues proceeding "forward" will eventually return to its original location. Although we can say that this will happen because space is curved, understanding what that really means can be tricky. So let's try to explore this idea.

WHY MIGHT SPACE BE CURVED?

Curvature in three-dimensional space can be difficult to comprehend. Let's begin by considering two-dimensional space instead. Two-dimensional space is like a world that exists entirely on an infinite sheet of paper, as in Figure 5-1. The position of every object in this world can be represented by using two coordinate axes.

Since that graph does not give us the feeling of "space," let's try to view our two-dimensional model in three dimensions, as in Figure 5-2.

From this perspective, we can see the two-dimensional plane is a flat world, like a board. If we assume that this world has two-dimensional inhabitants, it won't matter to them whether or not their world is bent in three dimensions. If the graph paper is curved, folded, or crumpled up into a ball, it will make no difference to them, since the world that is indicated by the *x*- and *y*-coordinates will remain the same. The inhabitants will not notice the bending of that space—or at least they won't notice without traveling very long distances.

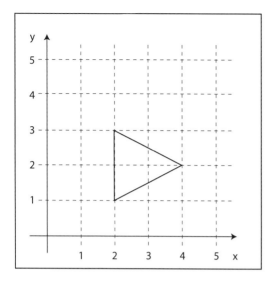

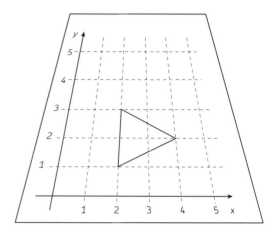

Figure 5-1: A simple plane, expanding in all directions—imagine that this model does not have an edge

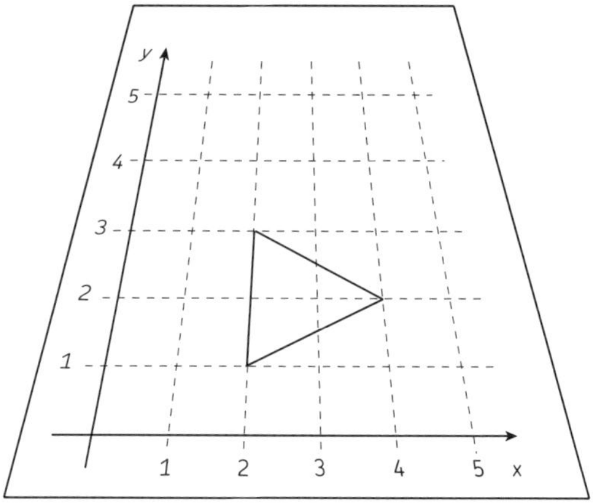

Figure 5-2: A 2-D plane viewed from a 3-D perspective

WILL YOU RETURN TO THE SAME LOCATION IN A PLANE, A CYLINDER, AND A SPHERE?

When the curvature of space is zero (meaning it's flat, like a sheet of paper), it can be drawn with straight lines. But the greater the curvature of space, the more it is bent, meaning it must be drawn using more sharply curving lines.

When a perfectly flat world like that represented on the graph paper in Figures 5-1 and 5-2 is viewed from our perspective in three dimensions, we can see a two-dimensional space with zero curvature. But even graph paper is seldom perfectly flat—it would be difficult to maintain a curvature of exactly zero.

So let's assume that the curvature is no longer zero in the x direction. In Figure 5-3, the graph paper is bent horizontally. What will happen in this case?

If two-dimensional space extends infinitely but is curved, and if we assume that the curvature is constant, then it will curl around and eventually arrive back where it started, taking on a cylindrical shape as in Figure 5-3. This cylindrical shape will be created when the x-coordinate's positive and negative directions meet.

The inhabitants of this two-dimensional world will have no idea that it is a cylinder. But if they walk in a straight line, looking for the address $x = \infty$, they will eventually experience the oddity of x becoming negative.

Moreover, a two-dimensional space that bends only in the direction of the x-axis is a rather special condition. If the curvature in the direction of the y-axis is also positive, then the two-dimensional shape that is derived is a sphere, as in Figure 5-4. Even if the curvature in the direction of the x- or y-axis is not necessarily constant, if two-dimensional space continues curving in a fixed direction, it will ultimately intersect in both the x and y directions to form a closed shape like a sphere.

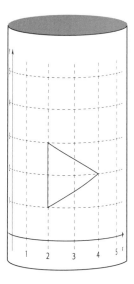
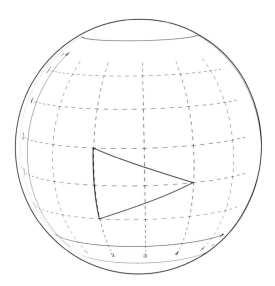

Figure 5-3: A cylinder results from a positive curvature in the x direction.

Figure 5-4: A sphere results from positive curvature in the x and y directions.

We can extend this idea to three-dimensional space and predict similar behavior. If the three x-, y-, and z-coordinate axes that we can set are perfectly straight when viewed from the fourth dimension, we can keep traveling forever and ever in the universe. But if these axes are bent just a little, as in our cylinder or our sphere, we will eventually return to the place where we started.

NEGATIVE CURVATURE

But, as we've said before, there can be three types of curvature: zero, positive, and negative. What does it mean for the value representing curvature (that is, the degree to which a curved line or curved surface bends) to be *negative*? First, recall the three diagrams of 2D universe models that appeared in the lecture given by Professor Sanuki on page 210. They were a spherical surface (positive curvature), a plane (zero curvature), and a shape that looked like a saddle (negative curvature).

Just as our flat (zero-curvature) two-dimensional plane wasn't a rectangle with defined edges, we say a universe with negative curvature is "sort of" like a horse's saddle because it doesn't have a definite edge but instead continues to spread out infinitely both vertically and horizontally.

Let's draw triangles on these three models to show the effects that differently curved types of space have on geometry. On the "plane" in model 2, the sum of the triangle's interior angles is 180°, as is normal in basic geometry.

But what happens on the sphere in model 1? Here, the sum of the interior angles is greater than 180°. And on the saddle-shaped surface in model 3, the sum of the interior angles is less than 180°.

1. Curvature is positive.
2. Curvature is zero.
3. Curvature is negative.
Sum of the interior angles is greater than 180°.
Sum of the interior angles is equal to 180°.
Sum of the interior angles is less than 180°.

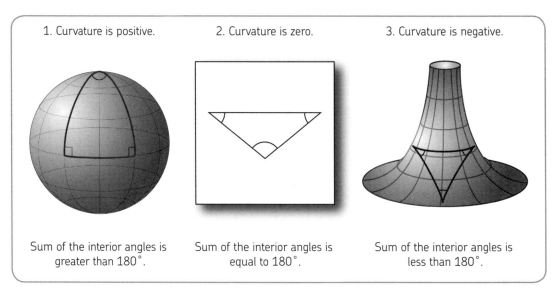

Figure 5-5: Each representation of the universe's curvature has different implications.

Consider a triangle where the apex is the North Pole of Earth and the base is on the equator (like the triangle in model 1 in Figure 5-5). In this case, the angles created by the base (equator) and the sides connecting the apex and base (that is, the meridians) are right angles (90°). Therefore, just the sum of the two interior angles created by the base is 180°, and when the interior angle of the vertex is added, the sum has to exceed 180°. You can intuitively see that the opposite is true of a triangle drawn on a plane with negative curvature, as shown in model 3.

FRIEDMANN'S DYNAMIC UNIVERSE

The three-dimensional universe that we live in could also take any of three types of shapes when viewed from the fourth dimension, with positive, zero, or negative curvature. The famous Friedmann models of the universe were created from this kind of analysis.

The Russian astrophysicist Alexander Friedmann (1888–1925) hypothesized a dynamic universe; that is, a universe that is continuously subjected to forces that cause it to expand or contract. He considered what would happen if the curvature of this dynamic space was positive, zero, or negative. The results of these three different curvatures are modeled in three dimensions in Figure 5-6. Each letter *S* affixed to the surfaces of the models represents a galaxy.

Figure 5-7 shows Friedmann's predictions about what happens in these three models over time. The *y*-axis represents the average distance between galaxies in the universe, while the *x*-axis represents time elapsed. A scale factor of 1 on the *y*-axis indicates that the distance between galaxies exists as it is now, while 2 indicates that the distance between galaxies has doubled.

Astronomers don't typically refer to the specific curvature of space but rather to the overall geometry of space. A universe with positive curvature, like a sphere, is called a *closed universe*. If you traveled in a straight line in a closed universe, your journey would be a closed loop; you'd eventually come back to your original location. As shown in Figure 5-7,

1. Curvature is positive.
Galaxy
Expansion
2. Curvature is zero.
Galaxy
Expansion
3. Curvature is negative.
Galaxy
Expansion

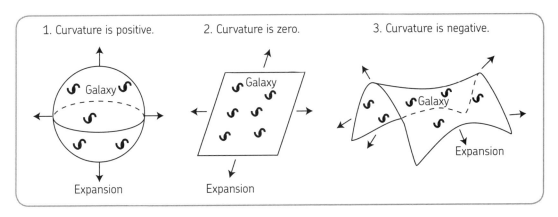

Figure 5-6: Friedmann's models of the universe—the S *shapes in each model represent galaxies.*

1. Curvature is negative
(an open universe).
2. Curvature is zero
(a flat universe).
3. Curvature is positive
(a closed universe).
Scale factor
1
Present
Time

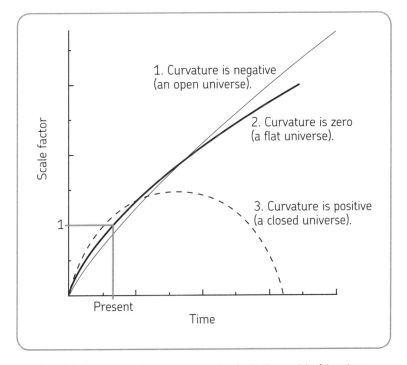

Figure 5-7: Friedmann predicted a change over time for the three models of the universe.

a closed universe will eventually collapse in on itself. A universe with a negative curvature is called an *open universe*, and a universe with zero curvature is called a *flat universe*. Figure 5-7 also demonstrates Friedmann's prediction that while a universe with zero or negative curvature would slow down its rate of expansion over time, it would continue to expand forever.

In summary, there are three ways in which space can curve: positively, not at all (zero curvature), or negatively. Those three types of curvatures give us three types of universes to consider: closed, flat, or open, respectively. For now, that's all you need to keep in mind.

BUT IS THE UNIVERSE DYNAMIC?

You learned that Hubble deduced that the universe was expanding after discovering the redshift of heavenly bodies, confirming Friedmann's theory of expansion only after his death. The person who created the motivation for Friedmann's theory was Albert Einstein (1879–1955). But Einstein believed that the universe was static and unchanging, not expanding. As a result, he made a huge blunder.

The general theory of relativity published by Einstein in 1915 stated that gravitational attraction is a physical phenomenon caused by the distortion of surrounding space produced because matter has mass. Therefore, according to Einstein's theory, gravity is considered to be an effect on space itself rather than the mutual attraction of matter postulated by Newtonian physics.

From a Newtonian perspective, objects with mass mutually attract each other (see Figure 5-8). From a perspective informed by Einstein, objects with mass actually cause indentations in space (represented by the plane in Figure 5-9).

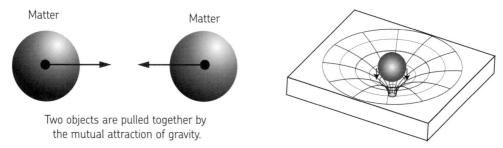

Two objects are pulled together by the mutual attraction of gravity.

Figure 5-8: A Newtonian representation of gravity

Figure 5-9: Einstein proposed that space itself became warped as a result of gravity.

However, even this way of thinking about gravity could not solve the problem of why the universe has its current shape, because if all gravity is caused by matter, then the universe should be contracting over time (even if it was initially static).

Newton, in contrast, assumed that the universe extends infinitely and is not contracting, since many heavenly bodies are exerting attraction from positions far away from each other. Many people doubted whether the universe actually was maintained in such a delicate balance. Simple calculations showed that this "balance" was not very stable and that if there were a location where matter (in this case, stars) was even a little more concentrated than in the surrounding area, matter would aggregate toward that point and increasingly collide.

Therefore, Einstein assumed that there was a mutual attraction between matter and a repulsive force causing a mutual repulsion (see Figure 5-10). He also assumed that space is static because the attractive force of gravity and the repulsive force balance each other out. This was his conclusion around 1915.

But then, cosmologists began to realize that Einstein's static universe, like the universe considered by Newton, would be in an extremely unstable equilibrium. The universe would become dynamic if a slight concentration in the density of matter were to occur anywhere, and then the universe would very quickly begin to contract or expand. This then led to the following cosmology.

Repulsive
Force
Gravity
Gravity
Gravity
Repulsive
Force
Repulsive
Force
Gravity
Gravity
Gravity

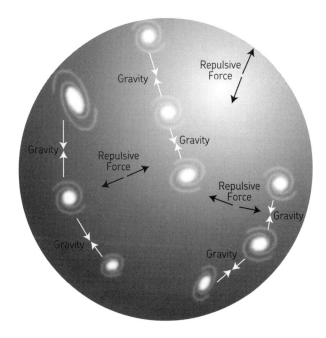

Figure 5-10: A conception of the static universe by Einstein, with gravity and a mysterious repulsive force in balance

MORE LIKE EINSTEIN THAN EINSTEIN

At first, the equations (called *gravitational field equations*) that Einstein introduced in the general theory of relativity had no factor for a repulsive force. However, when he followed his own line of thought, he realized that the gravitational forces would cause the universe to collapse. Therefore, in order to maintain the static and unchanging universe he believed in, Einstein decided that there was nothing he could do but add a constant. That constant, which is now known as the *cosmological constant*, represented the effect of the repulsive force needed to stabilize the universe and prevent it from collapsing or expanding.

Since the repulsive force was simply a hypothetical concept created in Einstein's head, there is no reason it can't be eliminated. If you believe that the universe is dynamically expanding or contracting, the cosmological constant is unnecessary.

Friedmann then derived three solutions shown in his models of the universe (Figure 5-7). If the total mass of matter in the universe is small, gravity will lose out to the force of expansion, and the universe will steadily grow larger. If the total mass of all matter in the universe is large, the universe will contract. If, by chance, the mass of matter is at the critical borderline between these two cases, then expansion will continue but the rate of expansion will eventually decrease.

Though Einstein initially denied the possibility that the universe was expanding, he eventually accepted that he was incorrect and called adding the cosmological constant to his equations the greatest blunder of his life.

DARK MATTER

Einstein's theory of general relativity says that mass curves the space around it. Therefore, a large mass can cause light traveling by it to alter its course, thus "bending" the light. Astronomers proved this in 1919 when, looking at stars near the Sun during a total solar eclipse, they observed these stars "out of position" due to their light being bent by the gravity of the Sun. Einstein went on to say that it was possible that light could be bent around an astronomical object to the extent that an observer would see multiple images of the same light source. These "gravitational lenses" bend light, just as a glass lens would. Therefore, photons of light that were originally not heading to Earth get bent toward us, thus giving us multiple images of the object that produced the photons. Astronomer Fritz Zwicky theorized that this could happen with clusters of galaxies, an idea that was confirmed in 1979, after his death.

Using gravitational lenses, it is possible to calculate the mass of a cluster of galaxies and how that mass is distributed around the cluster, because the amount of mass and where it is distributed determines how the light is bent and thus what image we perceive. After performing this calculation, astronomers were amazed to discover that there was much more mass creating the gravitational fields than just the mass from the gases and dust that were in the galaxies. Furthermore, the majority of this extra mass wasn't even where the galaxies were; instead it was between the galaxies where no heavenly bodies were evident. This mysterious matter, which neither emits nor absorbs light, is invisible to us and is therefore called *dark matter*.

This resolved another problem astronomers had in the late 1960s. They looked at how rapidly the stars in spiral galaxies were orbiting around the centers of those galaxies. Standard physical models using the mass from the visible matter in a galaxy predicted that stars further from the center of the galaxy would orbit around the center of the galaxy more slowly than those closer to the center. However, calculations showed that in fact, stars move at an almost uniform rate around the galaxy no matter how far from the center of the galaxy they are. This implied that there was more of this unseen matter around the outer edges of galaxies.

Some particle physicists believe that dark matter is a type of particle that isn't made up of the same type of subatomic particles that makes up "normal" matter. We call this normal matter *baryonic* matter and it is the matter consisting of the protons and neutrons that we, Earth, the Sun, and stars are made of. Calculations have shown that matter of all types constitutes 28 percent of the universe, but when astronomers added up the combined mass of all the baryonic matter, they discovered that it only makes up 4.6 percent of the universe. That means that dark matter makes up the other 23 percent. Dark matter is theorized to be a nonbaryonic particle called a *WIMP* or *Weakly Interacting Massive Particle*. Attempts are being made by scientists to detect WIMPs and understand the true nature of this particle. Until then, dark matter remains one of the biggest and most important mysteries in astronomy.

EINSTEIN'S BLUNDER LIVES ON

Einstein was embarrassed that he couldn't refute the theory of cosmic expansion, but he did remove the cosmological constant from his own gravitational field equations.

Until the 1990s, the standard model of the universe held that the universe was created in a Big Bang and then expanded rapidly. It was thought that the universe would continue to expand until gravity slowed the expansion and then reversed it, shrinking the universe until it collapsed into a supermassive black hole. This idea was called the *Big Crunch*. However, during the 1980s and 1990s, observations of cosmic background radiation and Type Ia supernovas in very distant galaxies (which also means the galaxies were formed relatively shortly after the Big Bang) showed cosmologists that not only was the universe expanding, but that the expansion was accelerating! This ran counter to the widely accepted belief that the expansion was slowing down. Some kind of energy source was required to account for this increasingly rapid expansion. As a result, it made sense to reincorporate the cosmological constant into the relativistic equations of gravity.

If these observations are correct, then the biggest blunder of Einstein's life may actually have been that he ended up refuting his own cosmological constant. If Einstein had declared that the cosmological constant was absolutely necessary, he might be even more renowned as a genius.

THE MYSTERY OF DARK ENERGY

So what's causing the universe to expand faster and faster? It could be that Einstein's idea of the cosmological constant is correct, meaning that some constant repulsive force is accelerating the expansion of the universe. One explanation for this is that the very vacuum of empty space has some energy that drives the acceleration. Astronomers call this unknown energy *dark energy*. That energy may well be the cosmological constant (denoted with the Greek letter lambda, Λ).

By observing distant supernovas, astronomers have been able to look back in time and discover that the universe is acting as predicted in the models in Figure 5-7. In each model, the gravity exerted by galaxies slows the universe's expansion early in its life. This happens because the galaxies are close together, so the gravitational pull they exert on each other is very strong. This makes it difficult for them to move away from each other, thus slowing the universe's expansion. The gravitational pull among galaxies is so strong that even though dark energy existed in the early universe and was trying to drive the galaxies apart, it couldn't completely overcome the force of their gravity. But eventually, the galaxies moved far enough apart that their gravitational pull was weakened to a point at which the effect of dark energy began to exceed the gravitational pull. At this point, the dark energy began to drive the galaxies even farther apart, thus expanding the universe at an accelerated rate.

WHAT WILL ULTIMATELY BECOME OF THE UNIVERSE?

The universe seems to be changing dynamically, after all. So what will ultimately happen to the universe after a great deal of time passes? First, we have to know what kind of universe we live in, because the type of universe we live in will tell us its eventual fate.

We will consider the *Friedmann-Lemaitre-Robertson-Walker (FLRW) model of the universe*, in which the theory of the Belgian astrophysicist Georges Lemaitre (1894–1966), who was also one of the advocates of the theory of cosmic expansion, is added to Friedmann's three models of the universe that we looked at earlier. However, a little background knowledge is required first.

The story is simple—the fate of the universe depends on the curvature of space, and that curvature has a one-to-one correspondence with the average density ρ_m of matter that currently exists in the universe (the Greek letter rho, ρ, is used as the symbol for density). The average density of matter in the universe that would be needed to halt the expansion of the universe at some point in the future is called the *critical density*, or ρ_c. To determine the curvature of space, researchers often use the equation $\Omega_M = \rho_m / \rho_c$, which is the ratio of the average density to the critical density, rather than just ρ_m itself. The symbol used to represent this *mass density ratio* is the Greek letter omega, Ω. Thus, the equation $\Omega_M = \rho_m / \rho_c$ states that if ρ_m is greater than the critical density ρ_c, $\Omega_M > 1$ and the curvature of space is positive; if the density averages are the same, then $\Omega_M = 1$ and the curvature is zero; and if ρ is less than ρ_c, $\Omega_M < 1$, the curvature is negative. If we follow this convention, then the curvature of space, represented by the letter k, is positive if $\Omega_M > 1$, zero if $\Omega_M = 1$, and negative if $\Omega_M < 1$.

Using these conventions, we can get one of several outcomes. We can use the information we currently have to get an idea of what the universe actually looks like, that is, which of the three models that we saw in Figure 5-7 accurately represents our universe.

- If $\Omega_M > 1$, then $k = +1$ (positive curvature), and we have a closed universe that is positively curved like the surface of a sphere (see model a in Figure 5-11). Because $\Omega_M > 1$, the universe has more than the amount of matter needed to stop the expansion of the universe; it actually has enough to reverse it! The attractive force of gravity will eventually collapse the universe into the Big Crunch.
- If $\Omega_M = 1$, then $k = 0$ (zero curvature), and we have a flat universe similar to the so-called static universe considered by Einstein. This universe begins with the Big Bang and expands forever, slowing down until it reaches a fixed size when time reaches infinity (see model b in Figure 5-11).
- If $\Omega_M < 1$, then $k = -1$ (negative curvature), and we have an open universe that is negatively curved like a saddle (see model c in Figure 5-11). This universe will expand forever with its rate of expansion barely slowing down. Eventually, the universe will expand so much that its temperature will drop to the point that it is too cold to sustain life, and then it will drop even lower to near absolute zero (0 degrees Kelvin).
- If $\Omega_M = 0$, then that means there is no matter in the universe, that it is completely empty and void. This model (model d in Figure 5-11) would cause the universe to expand at a constant rate because there would be no gravity to slow it down. Of course, this model doesn't fit our universe because we exist in it and it obviously isn't devoid of all matter.

However, none of the first four models in Figure 5-11 (models a through d) match our observations of our universe. This is because they don't take into account dark energy, the mysterious energy that is accelerating the expansion of our universe.

The FLRW equations can be simplified to:

$$\Omega = \Omega_M + \Omega_\Lambda$$

Ω_M is, again, the ratio of the average density of all matter (which includes the normal baryonic matter that makes up stars and galaxies and also nonbaryonic matter like dark matter) to the critical density needed to halt the expansion of the Friedmann universe. Ω_Λ is the ratio of the average energy density to the critical density. Ω_Λ contains the cosmological constant Λ, meaning the dark energy that drives the acceleration of the expansion of the universe.

Ω_M and Ω_Λ added together give Ω, which is the universe's *density constant*. This is the true constant that determines the curvature of space. Models a through d in Figure 5-11 plot the effects of different values of Ω_M, but because these models don't include the effect of dark energy, Ω_Λ becomes zero and Ω_M is equal to Ω. But, because dark energy is affecting the expansion of the universe, we must take that into account when plotting the model.

Model e in Figure 5-11 shows the effects of dark energy. We have an open universe that was born with the Big Bang, and after decelerating due to gravity as in the Friedmann

(e)
Ω = 1
Λ > 0
(d)
Ω = 0
(c)
Ω < 1
(b)
Ω = 1
Now
(a)
Ω > 1
Average distance between galaxies
Time

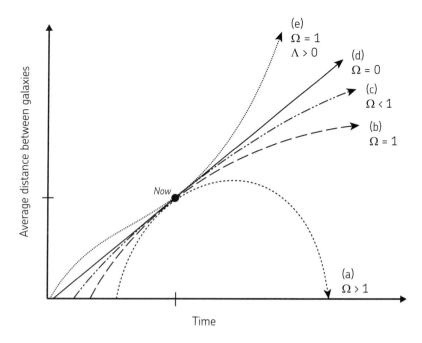

Figure 5-11: Changes in the Friedmann-Lemaitre-Robertson-Walker universe models over time

universes, its expansion began to accelerate due to the effects of the cosmological constant. It continues to expand perpetually (see Figure 5-12 for a full timeline).

Astronomers now know that for our universe, $\Omega = 1$. This means that our universe is flat. But how do we know that, and why doesn't the model for our universe look like model *b*, the flat universe, which shows a slowly decreasing rate of expansion?

WMAP AND OUR FLAT UNIVERSE

The majority of the data that we use to verify the curvature of the universe comes from images and other information taken from the *Wilkinson Microwave Anisotropy Probe*, or *WMAP*. WMAP measured differences in the sky's temperature, as seen in Figure 5-13, via *cosmic microwave background radiation (CMBR)* (refer to Chapter 3). For almost 400,000 years after the Big Bang, the entire universe was an opaque, hot, and dense fog of photons and baryons. Eventually, this fog cooled enough to start forming atoms, thus making the fog more transparent (meaning visible light could penetrate it). The CMBR that we observe is made of those photons from the early universe, although they have redshifted from visible wavelengths to microwave wavelengths. Agreeing with the cosmological principle (which states that no place in the universe is special and that the universe appears to be the same from every direction), the CMBR shows the homogenous nature of the universe in that the temperature of the universe is observed to be 2.725 Kelvin from all directions, with only a 0.003 Kelvin difference between the hottest and coldest parts of the sky. WMAP was able to show us those almost indiscernible differences in temperature. By understanding these differences, scientists can determine all sorts of information, including how it is that our universe has very little curvature.

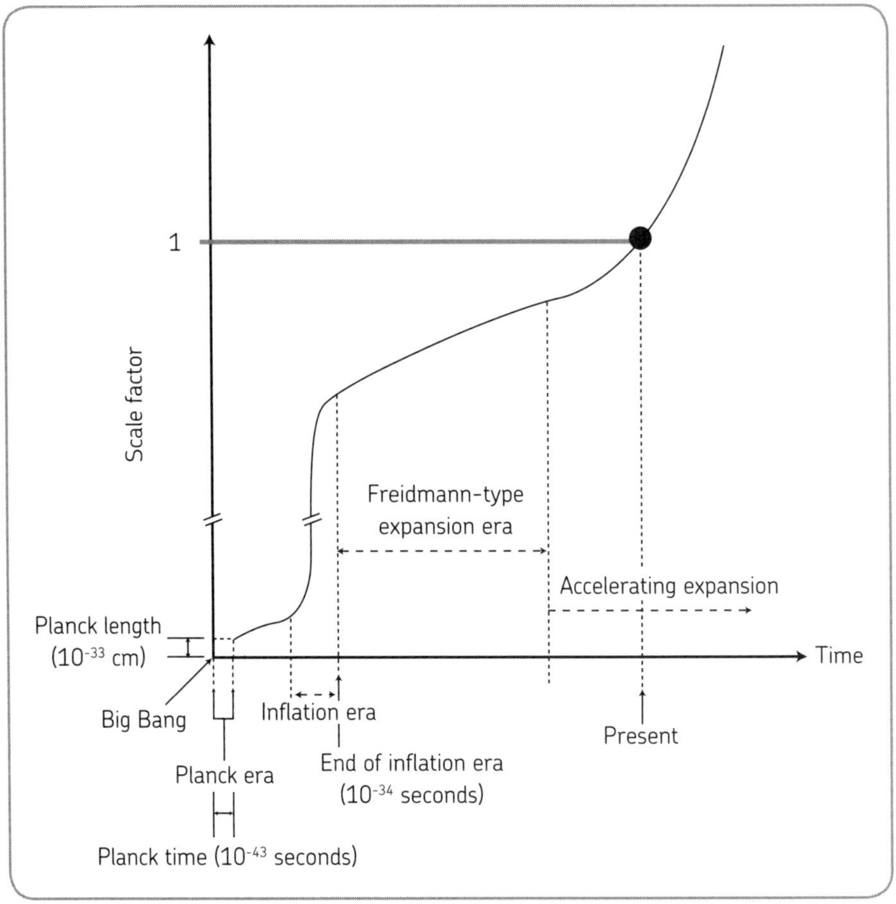
Scale factor
1
Freidmann-type
expansion era
Accelerating expansion
Planck length
(10-33 cm)
Time
Big Bang
Inflation era
Present
Planck era
End of inflation era
(10-34 seconds)
Planck time (10-43 seconds)

Figure 5-12: A timeline of an essentially flat universe

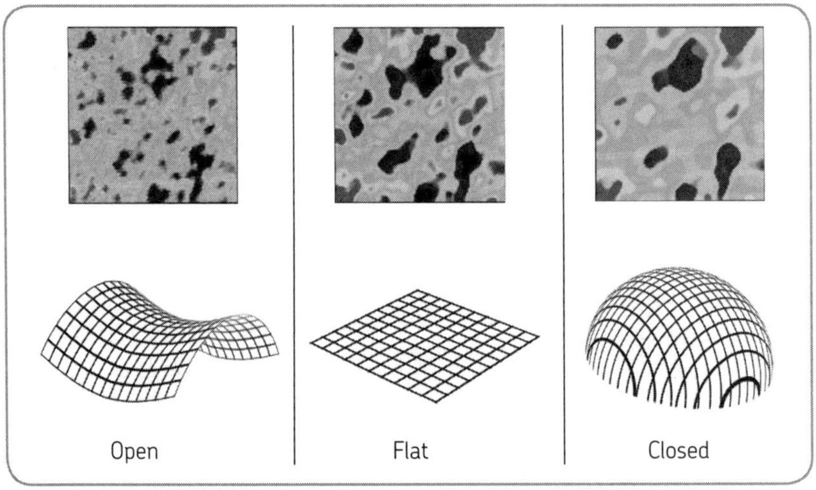
Open
Flat
Closed

Figure 5-13: The varying shades in WMAP's readings represent the varying temperatures across the universe—WMAP's actual measurements are consistent with a flat universe.

Before WMAP and the precise measurement of cosmic background radiation, astronomers were limited by the *horizon problem*. Observing a distant heavenly body with a telescope will provide us with the latest information about the universe as soon as possible, even if that light was emitted, for example, several tens of thousands of years or several hundreds of millions of years ago (depending on its distance in light-years from Earth). The speed of light creates an effective horizon when observing the universe, just like the horizon we encounter when trying to see objects on the surface of Earth.

The CMBR is uniform across the sky; as far as cosmic background radiation is concerned, one part of the sky looks exactly like another. The only way that two regions of space that are now at a great distance from each other could have the same radiological conditions is if they were once close enough to each other for a sufficiently long period of time to exchange energy (like light, heat, etc.) so that they could equalize. However, according to the theory of relativity, energy cannot be conveyed from one location to another faster than the speed of light. This is problematic because it is impossible for two different points of the sky that are separated by an angle of 1 degree or more to have ever been connected, since light from either of those two points could not have reached the other. Thus, there is no way that the energy in those two areas of the sky could have equalized. The regions are said to be beyond each other's light horizon. So how could the CMBR in those regions (and all other regions of the sky, for that matter) be homogenous?

To correct for this problem, the Big Bang theory was modified. Immediately after the Big Bang, the universe was a dense bundle of plasma made of photons and baryons that was going through all sorts of random quantum fluctuations. Then the universe expanded very quickly in a process called *inflation*. Initially, the diameter of the universe was 10^{12} times smaller than a proton, but then it suddenly expanded to roughly 1 meter across. This Inflation theory corrects the horizon problem present in the original Big Bang theory by saying that before inflation, the universe was small enough that it was all connected and able to transfer energy among all its parts to equalize the physical properties of each region. During inflation, the sudden expansion of space essentially froze those properties at all locations and smoothed out the largest temperature differences.

The temperature differences we see in the CMBR are due to the random quantum fluctuations of density that were happening in the hot dense plasma fog, which was then enlarged during inflation to an astronomical size. As the universe cooled, those denser areas began to attract matter gravitationally. The addition of that matter increased the density of those areas further, which in turn increased their gravitational pull so that they attracted more and more matter. Gas gathered and formed stars, stars gathered to make galaxies, galaxies gathered to create clusters of galaxies, clusters of galaxies gathered to make superclusters, superclusters gathered to form walls and filaments, and so on. Thus, the random quantum fluctuations that happened when the universe was miniscule are responsible for making the largest structures in the universe.

But how can we use the CMBR to determine the curvature of the universe? Some theories hold that areas of major temperature difference would typically measure 1 degree across when observed from Earth. These areas were created by sound (pressure) waves that moved through the hot gas in the early universe at a known speed (the speed of sound) and for a known time (approximately 400,000 years). The distance these waves moved is distance = speed × time, so we know how big these areas should be. Because the light from these areas is coming from the edge of our horizon limit, we should be able to observe the effects of the curvature of space on this light.

We have already seen that different curvatures of space cause the geometry of a triangle to change (see Figure 5-5). On a flat surface, parallel lines remain parallel, meaning that the edges of a triangle are straight and the angles of the triangle add up to 180 degrees. On a closed (spherical) surface, parallel lines converge so that a triangle's lines bend outward and the angles of the triangle add up to something greater than 180 degrees. On an open (saddle-shaped) surface, parallel lines bend apart so that a triangle's lines bend inward and the angles of the triangle add up to less than 180 degrees.

The lines of the triangle formed by our observations while looking at an area of temperature fluctuation will conform to the curvature of space and thus determine how the area we're observing appears to us. If the universe is flat ($\Omega = 1$), the image will appear normal, and the area will measure 1 degree across. A closed universe ($\Omega > 1$) would bend the light coming from the area outward, magnifying the image and making the area look 1.5 degrees wide. An open universe ($\Omega < 1$) would bend the light inward, demagnifying the image and making the area look 0.5 degrees wide. The WMAP calculations showed that our universe is indeed flat.

Because we now know that our $\Omega = 1$, we can determine how much dark energy makes up the universe. Early calculations showed that $\Omega_M = 0.30$, meaning that that the universe is 30 percent matter. This includes the 5 percent that is normal, baryonic matter and the 25 percent made up of dark matter. As for the rest of the makeup of the universe, Ω_Λ gives us how much dark energy makes up the universe. Since $\Omega = \Omega_M + \Omega_\Lambda$ and we know that $\Omega = 1$ and $\Omega_M = 0.30$, simple math allowed astronomers to predict that $\Omega_\Lambda = 0.70$, meaning that 70 percent of the universe is dark energy. This was later proven by WMAP when it determined that $\Omega_\Lambda = 0.72$.

Since only 5 percent of the total mass-energy density of the universe is the baryonic matter with which we are familiar, 95 percent of the universe consists of things we don't yet understand.

NOTE *Scientists project that if the rate of expansion continues to accelerate, we will eventually enter what's called a* de Sitter universe, *one in which everything has been stretched out so far that there won't be matter in the sense that we know it. No more planets, stars, or even spread-out particles. The only thing left will be the cosmological constant.*

THE TRUE AGE OF THE UNIVERSE

Now we seem to have a better idea of what our universe really looks like and how it seems to be changing over time: it is essentially flat and expanding outward. To get a deeper understanding of what the universe looks like, though, we should also know how old it is.

We've talked a lot about the Big Bang, but when did it happen? The estimated age of the universe (that is, the time since the Big Bang) is 13.7 billion years. This number has been determined from theoretical calculations and has been verified numerically from observations that WMAP provides. Other observations also help to confirm the age of the universe. It makes sense that the universe has to be at least as old as the oldest thing in it, right? Well, in 2009, the Swift Gamma-Ray Burst Mission observed a gamma-ray burst that was 13 billion years old! This gamma-ray burst happened when a star that was roughly 200 times more massive than our Sun used up all its nuclear fuel and collapsed into a black hole. When it collapsed, it created a *hypernova*, an explosion over 100 times more energetic than a normal supernova. This explosion happened only 600 million years after the Big Bang, which told astronomers that the star only lived for about 500 million years. Compare that to the 10 billion years that our Sun will live.

By using known facts to make conjectures about things that cannot be known absolutely, we can continue to formulate theories and develop them little by little as new observations, experimental results, and ideas come to light. This is what science is all about.

There are still many things that we don't know about our universe. This is not limited to how we describe our universe physically. We can physically describe the universe in the same way we could physically describe a person, but humans don't consist of just their physical descriptions. Many small details make each person an individual and different from the next person. We cannot absolutely understand what goes on in another person's mind. Yet we happily live our lives, making friends or falling in love with people we can never completely understand. Likewise, we will never lose our fascination with the universe—from the familiar romance of our own Moon to the infinite possibilities of the multiverse. We will continue to observe, explore, and hypothesize as we seek a more perfect understanding.

INDEX

A

age of universe, 232–233
aliens. *See* extraterrestrial life
Alpha Centauri (binary star system), 185, 205
ancient beliefs about universe, 18–19
Andromeda galaxy
    collision with Milky Way galaxy, 109
    discovery as galaxy, 144–145
    distance from Earth, 144
    early observation of, 120
annihilation, 170
annual parallax, 126, 186–188
antimatter, 169–171
antiquarks, 169–171
aperture of telescopes, 121
Apollo space missions, 13, 67
Arecibo Message, 182
Aristarchus, 40–41, 45–47, 125
Aristotle, 39
atoms, 168
average density of matter, 227–228

B

Babylonian beliefs about universe, 19
Bamboo-Cutter, The Tale of the, 10–11
bar galaxy, 106
barred spiral galaxies, 106
baryonic matter, 166, 226
beliefs about universe, ancient, 18–19
Bell Laboratories, 166
Big Bang, 155–167
    and antimatter, 169–171
    appearance of elementary particles, 167–168
    birth and distribution of matter after, 172–175
    chronology after, 163
    evidence supporting, 166
    inflation after, 231
    overview, 155–159
    Planck epoch, 164–165
    temperature after, 172
    when occurred, 161, 232

Big Crunch, 226
black holes, 108–109
board game, Kaguya-go Journey, 206–207
bosons, 168

C

Cannon, Annie Jump, 188
capture hypothesis, 93
celestial sphere, 19
Cepheid variable stars, 188–189
CfA2 Great Wall, 141
China, development of astronomy in, 19
closed universe, 222–223, 228, 230, 232
clusters of galaxies, 111
    formation of, 173, 175
    Local Group, 140–141, 207
    mass of, 226
    number of galaxies in, 140
    superclusters, 111, 141, 207
CMBR (cosmic microwave background radiation), 166, 229, 231
COBE (Cosmic Background Explorer) satellite, 166
Copernicus, Nicolaus, 39, 53, 69, 71
corner cube
    mirrors, 67
    prism, 67–68
Cosmic Background Explorer (COBE) satellite, 166
cosmic background radiation, 227
cosmic inflation theory, 164–165
cosmic microwave background radiation (CMBR), 166, 229, 231
cosmological constant, 225–227, 232
cosmological principle, 180, 219, 229
critical density, 227–228
Curtis, Heber Doust, 144
curvature of space
    degree of, 229–232
    effect of returning to same place, 220–221
    and fate of universe, 227–229
    general explanation, 219–220

    negative curvature, 221–222
    positive curvature, 219–221
cylindrical-shaped space, 220–221

D

dark energy, 109, 166, 227–228
dark matter, 109, 166, 226
de Sitter universe, 232
Democritus, 116, 120
density constant, 228
density of matter, average, 227–228
disc-shaped galactic model, 117–119
distances. *See* measuring distances
Doppler effect, 147
Drake, Frank, 180–181, 184
Drake equation, 180–181
dynamic universe, 222–227

E

Earth
    appearance of rotation of Sun and Moon around, 34–39
    in diagram of solar system, 64
    diameter of, 60
    distance from Moon
        baseball field comparison, 61–62
        as basis for figuring distance to Sun, 125
        determining using corner cube mirrors, 67–69
        figuring by triangulation, 41–45
    distance from Sun, 41–45, 64
    facts about, 91
    formation of, 163
    and formation of Moon, 92–93
    orbit of, 77
    radius of, 66
    size of
        calculation of, 20–21
        compared to Sun/Moon, 45–49
    as spherical, discovery of, 20–22
    tides on, 94
eccentricity, 74–75

eclipses, lunar, 46
"edge" of universe, 177–178, 199–200, 209–210
Egyptian beliefs about universe, ancient, 18
Einstein, Albert, 224–227
elementary particles, 167–168
elliptical orbits, 71–72
    and Kepler's First Law, 73–75
    and Kepler's Second Law, 75–76
energy, dark, 109, 166, 227–228
Epsilon Eridani (planetary system), 184
Eratosthenes, 20–21, 68
Europa (satellite of Jupiter), 86, 182–183
expansion of universe
    acceleration of, 227
    cone illustration of, 156–158
    ongoing, 226–227
    redshift as proof of, 146–152
    velocity of, 162
extraterrestrial life, 180–186
    closest star system that could support, 183–184
    contacting, 184–185
    and Cosmological Principle, 180
    Fermi paradox regarding, 182
    number of possible civilizations, 180–181
    and variety of life on Earth, 182–183

F

Fermi, Enrico, 181–182
Fermi paradox, 182
fermions, 168
flat universe, 220, 223, 228–232
FLRW (Friedmann-Lemaitre-Robertson-Walker) model of universe, 227–229
four-dimensional space, 209–210
Friedmann, Alexander, 222, 225
Friedmann models of universe, 222–223
Friedmann-Lemaitre-Robertson-Walker (FLRW) model of universe, 227–229

G

galaxies. *See also* Andromeda galaxy; clusters of galaxies; Milky Way galaxy
    defined, 140

formation of, 137, 163, 173, 175
groups of, 111, 140–141, 173, 175, 207
shape of, 137
soccer game example, 133–139
Galaxy IOK-1, 123
Galilean telescope, 121
Galileo Galilei
    blunders of, 72
    discoveries of, 56–57, 72–73, 116, 120
    name of, 72
    overview, 54
game, Kaguya-go Journey, 206–207
gamma rays, 96, 232–233
Ganymede (satellite of Jupiter), 93, 183
gauge particles, 168
Geller, Margaret, 141
general theory of relativity, 224–227
geocentric model, 38–39
    disproved by Galileo's discoveries, 57
    vs. heliocentric model, 55
    planetary orbits according to, 70
    and positions of Moon and Sun, 42–46
    Ptolemy's contributions to, 51–54
    reasons for past popularity of, 69
    and size of Sun, 46–49
    and Tychonic system, 70–71
giant impact hypothesis, 93
gravitational forces
    according to general theory of relativity, 224–225
    bending of light by, 226
    and dark energy, 227
    and dark matter, 226
    in early universe, 173
    law of universal gravitation, 173
    Newtonian representation of, 224
    repulsive force, 224–225, 227
    and spaceship travel, 185
gravitational lenses, 226
Great Andromeda Nebula, 120, 144–145. *See also* Andromeda galaxy
Great Dark Spot, Neptune's, 89
Great Debate, 144
Great Red Spot, on Jupiter, 86
Great Wall, 141, 163, 207
Grecian explanation of universe, ancient, 20

groups of galaxies, 111
    difference from clusters, 140
    distribution of, 173
    and superclusters, 207

H

habitable zone, 184
Hale Telescope, Mount Palomar Observatory, 122
harvest moon festival, 11–12
heliocentric model, 39–40
    and Copernicus, 53
    diagram of, 54
    and Galileo, 57, 72–73
    vs. geocentric model, 55
    initial missing elements of, 71
    significance of, 73
    support of by invention of telescope, 57
    support of by Kepler's Laws, 58, 72
helium, 97, 151–152, 172
Herschel, Frederick William, 118–119
Hertzsprung, Ejnar, 187–188
Hertzsprung-Russell (H-R) diagram, 187–188
Hipparchus, 68
Hipparcos High Precision Parallax Collecting Satellite, 126
Hooke, Robert, 86
Hooker Telescope, Mount Wilson Observatory, 122
horizon
    curvature of, 21–22
    distance to, 66
    and measurement of cosmic background radiation, 231
H-R (Hertzsprung-Russell) diagram, 187–188
Hubble, Edwin
    brief history of, 142–143
    discovery of expansion of universe, 146–149
Hubble's constant, 162
Hubble's law, 122
Hubble Space Telescope, 122–123, 142
Huchra, John, 141
hydrogen, 97, 151–152, 172
hydrostatic equilibrium, 96–97
hydrothermal vent, 182
hypernova, 232–233
hyperspace, 209–210, 219

I

Indian beliefs about universe, ancient, 18
inflation, 164–165, 230–231
intermediate-sized black holes (ISBH), 109
IOK-1 (galaxy), 123
island universes, 120, 137, 140. *See also* galaxies
ISBH (intermediate-sized black holes), 109

J

Japan
    belief of Earth and Moon as spheres, 22
    diagram of solar system to fit map of, 64
    Kaguya satellite, 13
Jupiter
    in diagram of solar system, 64
    facts about, 86
    orbit of, 77
    satellites of, 57, 86, 93, 182–183

K

Kaguya satellite, 13
Kaguya-go Journey board game, 206–207
Kaguya-hime, story of, 10–11
Kant, Immanuel, 119–120, 140
KBOs (Kuiper Belt objects), 127
Kepler, Johannes, 58, 72, 121
Keplerian telescope, 121
Kepler's Laws, 58, 72
    First Law, 72–75
    Second Law, 72, 75–76
    Third Law, 72, 77
Kepler spacecraft, 184
Kobayashi, Makoto, 171
Kuiper Belt objects (KBOs), 127

L

Large and Small Magellanic Clouds, 120, 140
Leavitt, Henrietta, 189
Lemaitre, Georges, 227
lenses of telescopes, 121
leptons, 168
light, bending of, 226
Local Group, 140
Local Supercluster, 207
lunar eclipse, 46

M

Magellanic Clouds, Large and Small, 120, 140
magma, 91
Marius, Simon, 120
Mars
    atmosphere of, 81
    in diagram of solar system, 64
    facts about, 85
    no life found on, 183
    orbit of, 77
    surface temperature of, 81
mass density ratio, 228
Masukawa, Toshihide, 171
matter
    antimatter, 169–171
    average density of, 227–228
    baryonic, 166, 226
    birth and distribution of, 172–175
    dark, 109, 166, 226
    elementary particles, 167–168
measuring distances
    based on redshift, 189
    based on supernovas, 189
    by comparing properties of stars with Sun, 186–188
    using period-luminosity relationship, 188–189
Mercury
    in diagram of solar system, 64
    facts about, 83
    orbit of, 77
Milky Way galaxy
    center of, 106–109
    collision with Andromeda galaxy, 109
    composition of, 109
    discovery that is not whole universe, 144–145
    early beliefs about, 116
    formation of, 108, 163
    number of stars in, 106
    shape of, 108, 117–119
    size of, 104–105
    structure of, 116–119
    when most visible, 206
    why appears "milky," 99–101, 116–117
mirrors, corner cube, 67–68
mirrors in telescopes, 121–122
Mitsubishi Electric Corporation, 122
molecules, 168

Moon (Earth's)
    appearance of rotation around Earth, 34–39
    distance from Earth
        baseball field comparison, 61–62
        as basis for figuring distance to Sun, 125
        determining using corner cube mirrors, 67–69
        figuring by triangulation, 41–45
    and Earth's tides, 94
    eclipse of, 46
    facts about, 92
    formation of, 92–93
    harvest moon festival, 11–12
    Japanese satellite, 13
    mirrors on, 67–68
    no life found on, 183
    radius of orbit of, 62
    size of, 45–49, 62, 94
    and story of Kaguya-hime, 10–11
    US landing on, 13
    waxing and waning of, 40–41
motion, retrograde, 52
Mount Palomar Observatory Hale Telescope, 122
Mount Wilson Observatory, 122, 143
Mt. Fuji, 11
multiverse, 217–219
myths about the universe, 18–19

N

Nambu, Yoichiro, 171
National Astronomical Observatory of Japan Hawaii Observatory Subaru Telescope, 122
nebulas, 120, 145
negative curvature of space, 221–222
Neptune, 64, 77, 89
neutrinos, 109
Newton, Isaac, 224
Newtonian mechanics, 58
Newtonian representation of gravity, 224
nuclei, 163, 172
Nut (Egyptian Sun god), 18

O

Occam's razor, 55
Olympus Mons, on Mars, 85

On the Revolutions of the Celestial Spheres (Copernicus), 71
Oort cloud, 127
open universe, 223, 228–230
Opportunity rover, 85
orbits of planets
    according to Galileo, 72
    according to geocentric model, 70
    elliptical, 71–76
    and Kepler's Third Law, 77
Original Theory or New Hypothesis of the Universe, An (Wright), 120
Otsukimi (moon-viewing) festivals, 12–13, 22

P

pair production, 170
Paleozoic Era, 91
parallel universes. *See* multiverse
parsec, 126, 207
particles, elementary, 167–168
period-luminosity relationship, 188–189
photons, 167–168, 170, 229
Planck epoch, 163–165, 230
planetary system, 140
planets. *See also* geocentric model; heliocentric model; *names of specific planets*; orbits of planets; solar system
    distance between, 64
    Greek meaning of word for, 50
    motion of, 52, 58, 72
Pluto, 82, 90
positive curvature of space, 219–221
prism, corner cube, 67–68
Proxima Centauri, 109
Ptolemy, Claudius, 39, 51–54, 70
pulsating variable stars, 188
Pythagorean theorem, 66

Q

quarks, 163, 168, 170–171

R

radiation. *See* cosmic microwave background radiation (CMBR)
radiation pressure, 97
radio telescope, 124
radio waves, 124, 184–185
redshift, 146–152, 189

relativity, general theory of, 224–227
repulsive force, 224–225, 227
resolution of telescopes, 121
retrograde motion, 52
Russell, Henry Norris, 187–188

S

Sagan, Carl, 181
satellite measurements, 166
Saturn
    in diagram of solar system, 64
    facts about, 87
    orbit of, 77
    Titan, satellite of, 93, 183
scientific method, 58
Search for Extra-Terrestrial Intelligence (SETI) Institute, 184
Shapley, Harlow, 144, 188–189
sister hypothesis, 93
Slipher, Vesto, 146, 152
Sloan Great Wall, 141
soccer game example, for explaining galaxies, 133–139
solar system. *See also* geocentric model; heliocentric model; planets
    diagram of to fit map of Japan, 64
    formation of, 163
    size of, 127
spaceship travel, 185
specific gravity, 87
spiral galaxies, 106
Spirit rover, 85
Spitzer Space Telescope, 106
stars. *See also* Milky Way Galaxy
    absolute magnitude of, 187
    apparent magnitude of, 187
    beyond solar system, distance to, 126
    chemical composition of, 151–152, 166
    color of, 187
    distance of, measuring, 186–189
    formation of, 163
    and gravitational forces, 97
    luminosity of, 187
    number of in Milky Way, 106
    properties of, compared with Sun, 186–188
    variable, 188–189
    visibility of, 118
static universe, 224–225

stellar-mass black holes, 109
Subaru Telescope, 122
sublimation, 90
Sun. *See also* heliocentric model; solar system
    appearance of rotation around Earth, 34–39
    atmosphere of, 96
    core of, 96
    diameter of, 64
    distance from Earth, 41–45, 64, 125
    facts about, 95
    formation of, 96–97
    gravitational boundary of, 127
    internal structure of, 97
    photosphere of, 96
    size of compared to Moon and Earth, 45–49
superclusters of galaxies, 111, 141
supermassive black holes, 109
supernovas, 97, 189, 227
Swift Gamma-Ray Burst Mission, 232

T

Tale of Genji, The, 18
Tale of the Bamboo-Cutter, The, 10–11
Tanabata star festival, 206
tardigrades, 185–186
Tau Ceti (planetary system), 184
telescopes
    aperture of, 121
    early, 120–122
    famous, 122–123
    lenses of, 121
    mirrors in, 121–122
    radio telescope, 124
    resolution of, 121
    support of heliocentric model, 57
    visibility of stars using, 118
temperature of universe, 229
Theia (theoretical planet), 93
three-dimensional space, 209–210
tides, 94
time, 165
Titan (satellite of Saturn), 93, 183
Tombaugh, Clyde, 90
triangulation
    to figure distance to Moon, 68
    to figure distance to stars beyond solar system, 126
    to figure distance to Sun, 125

tube worm, 182
two-dimensional space, 220
Tycho Brahe, 70–72
Tychonic system, 70–71
Type Ia supernovas, 189, 227

U

United States, moon landing by, 13
universe. *See also* Big Bang; curvature of space; galaxies
   age of, 232–233
   ancient beliefs about, 18–19
   closed, 222–223, 228, 230, 232
   discovery that Milky Way is not whole universe, 144–145
   dynamic, 222–227
   early beliefs about, 116
   "edge" of, 177–178, 199–200, 209–210
   expansion of, 116, 146–165
      acceleration of, 227
      cone illustration of, 156–158
      ongoing, 226–227
      redshift as proof of, 146–152
      velocity of, 162
   flat, 220, 223, 228–232
   Friedmann models of, 222–223
   Friedmann-Lemaitre-Robertson-Walker (FLRW) model of, 227–229
   future of, 224–229
   large-scale structure of, 140–141
   multiple universes, 217–219
   open, 223, 228–230
   static, 224–225
Uranus, 64, 77, 88

V

variable stars, 188–189
Venus
   in diagram of solar system, 64
   facts about, 84
   orbit of, 77
   phases of, 57
   surface temperature of, 81
   visibility of in Western sky, 64–65
Very Large Array (VLA) observatory, 124
Virgo Cluster, 140
Virgo Supercluster, 207
VLA (Very Large Array) observatory, 124
voids, 207

W

water bears, 185–186
Weakly Interacting Massive Particle (WIMP), 226
Wilkinson Microwave Anisotropy Probe (WMAP), 166, 229–230
Wright, Thomas, 120

Z

Zwicky, Fritz, 226

ABOUT THE AUTHOR

Kenji Ishikawa is a scientific and technical journalist. He was born in Tokyo in 1958. After graduating from the College of Science at the Tokyo University of Science, he worked as a journalist for a weekly magazine and later became a freelance editor and writer. Besides writing novels and various columns, over the last 20 years, he has also written technical commentaries for general readers and conducted many interviews with leading engineers and researchers. His works cover scientific areas such as electricity, mechanics, aviation, astronomy, devices, materials, chemistry, computers, communication, robotics, and energy.

ABOUT THE SUPERVISING EDITOR

Kiyoshi Kawabata, PhD, ScD, is a professor emeritus in the Department of Physics, College of Science, at the Tokyo University of Science. Born in the Mie prefecture in 1940, Kawabata graduated from the School of Science, Division of Physics and Astronomy, at Kyoto University in 1964. While working on his doctorate, he studied abroad in the United States and received a PhD in astronomy from Penn State University in 1973. He was also awarded a ScD in astrophysics from Kyoto University. In 1981, he worked as a researcher at Columbia University and then worked for approximately eight years at NASA's Goddard Institute for Space Studies. In 1982, he began teaching as an assistant professor in the Department of Physics, College of Science, at the Tokyo University of Science, and he became a full professor there in 1990. He specializes in astrophysics, particularly observational cosmology and radiative transfer theory.

PRODUCTION TEAM FOR THE JAPANESE EDITION

Production: Verte Corp., Satoshi Arai, and Kenji Kawasaki
Illustration: Yutaka Hiiragi

MORE MANGA GUIDES

The *Manga Guide* series is a co-publication of No Starch Press and Ohmsha, Ltd. of Tokyo, Japan, one of Japan's oldest and most respected scientific and technical book publishers. Each title in the best-selling *Manga Guide* series is the product of the combined work of a manga illustrator, scenario writer, and expert scientist or mathematician. Once each title is translated into English, we rewrite and edit the translation as necessary and have an expert review each volume. The result is the English version you hold in your hands.

Find more *Manga Guides* at your favorite bookstore, and learn more about the series at *http://www.nostarch.com/manga/*.

THE MANGA GUIDE TO
STATISTICS

THE MANGA GUIDE TO
DATABASES

THE MANGA GUIDE TO
ELECTRICITY

THE MANGA GUIDE TO
PHYSICS

THE MANGA GUIDE TO
CALCULUS

THE MANGA GUIDE TO
MOLECULAR BIOLOGY

THE MANGA GUIDE TO
RELATIVITY

THE MANGA GUIDE TO
BIOCHEMISTRY

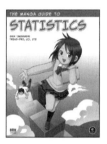

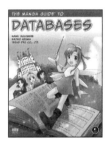

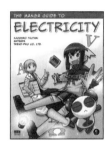

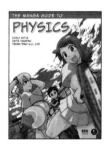

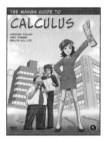

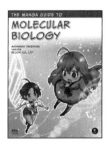

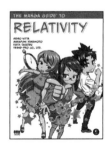

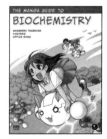

UPDATES

Visit *http://www.nostarch.com/mg_universe.htm* for updates, errata, and other information.

To view more photographs from NASA and other astronomers, check out these websites:

- *http://apod.nasa.gov/*
- *http://photojournal.jpl.nasa.gov/*
- *http://grin.hq.nasa.gov/*

If you'd like to help researchers categorize galaxies, discover exoplanets, analyze data, and participate in space exploration, visit *http://www.spacehack.org/*.

The Manga Guide to the Universe is set in CCMeanwhile and Chevin. The book was printed and bound by Friesens in Altona, Manitoba in Canada. The paper is Domtar Husky 60# Smooth, which is certified by the Forest Stewardship Council (FSC). The book uses a layflat binding, which allows it to lie flat when open.

GALLERY OF ASTRONOMICAL MARVELS

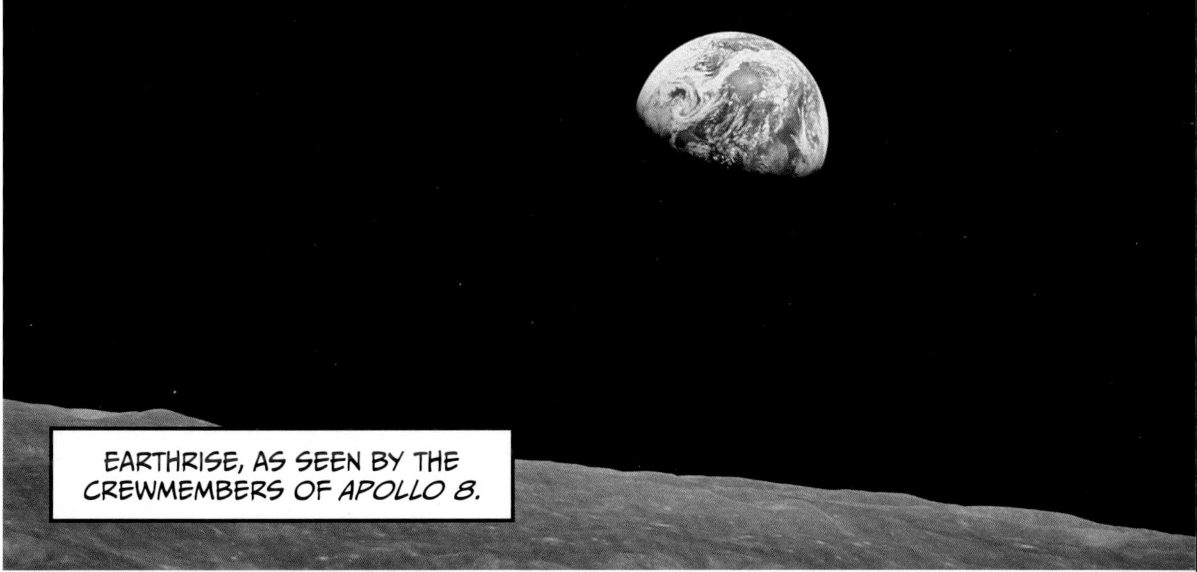

EARTHRISE, AS SEEN BY THE CREWMEMBERS OF *APOLLO 8.*

CREDIT: NASA/William Anders

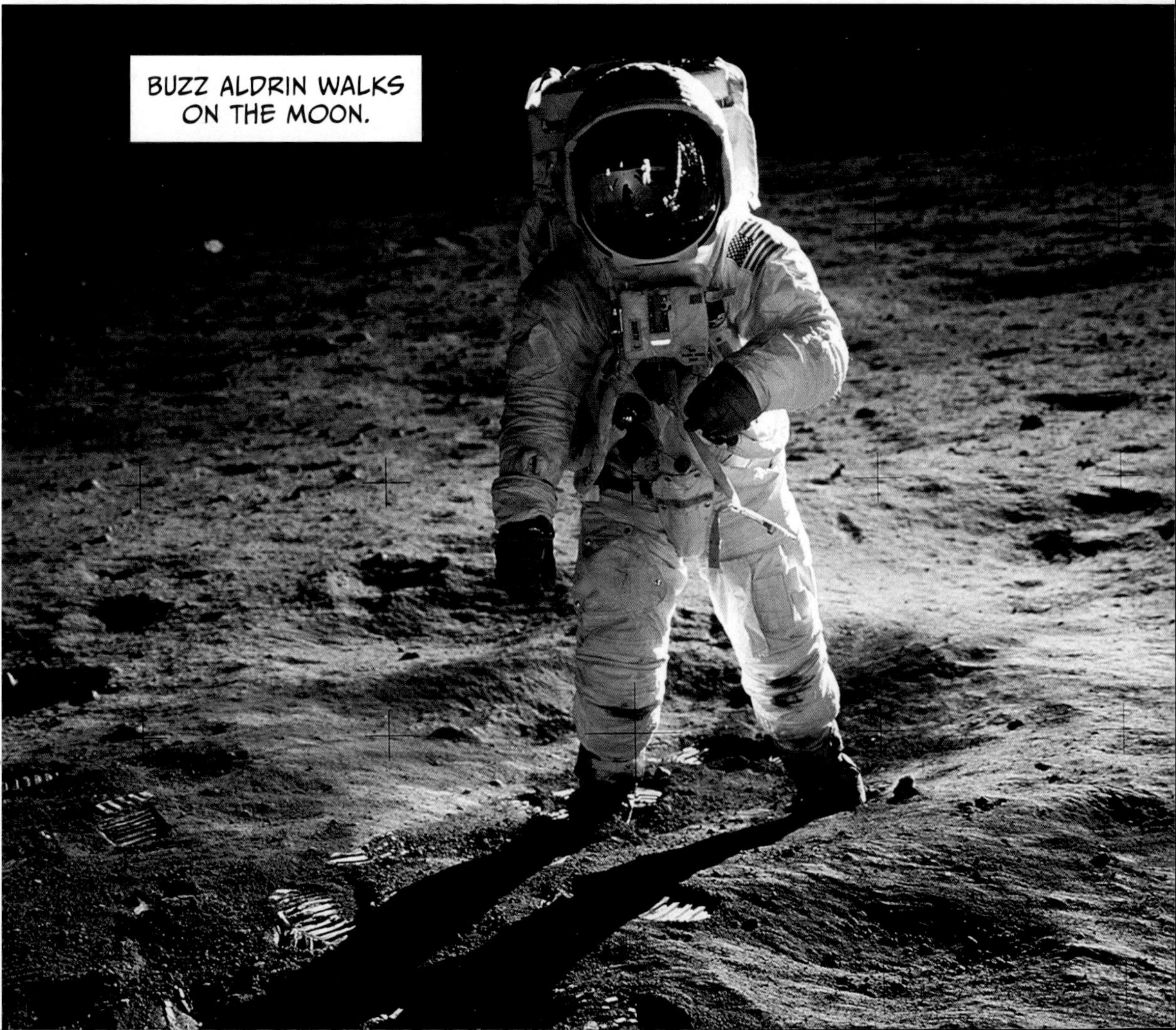

BUZZ ALDRIN WALKS ON THE MOON.

CREDIT: NASA/Neil Armstrong

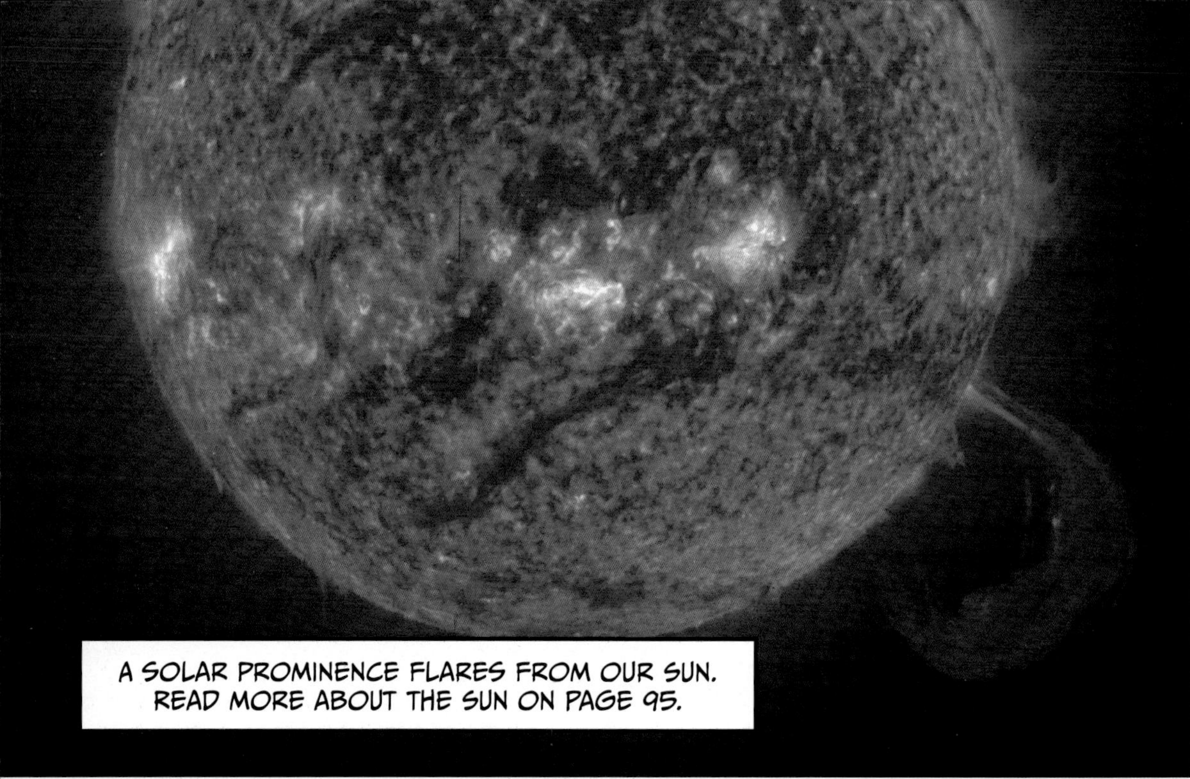

A SOLAR PROMINENCE FLARES FROM OUR SUN.
READ MORE ABOUT THE SUN ON PAGE 95.

CREDIT: ESA/NASA/SOHO

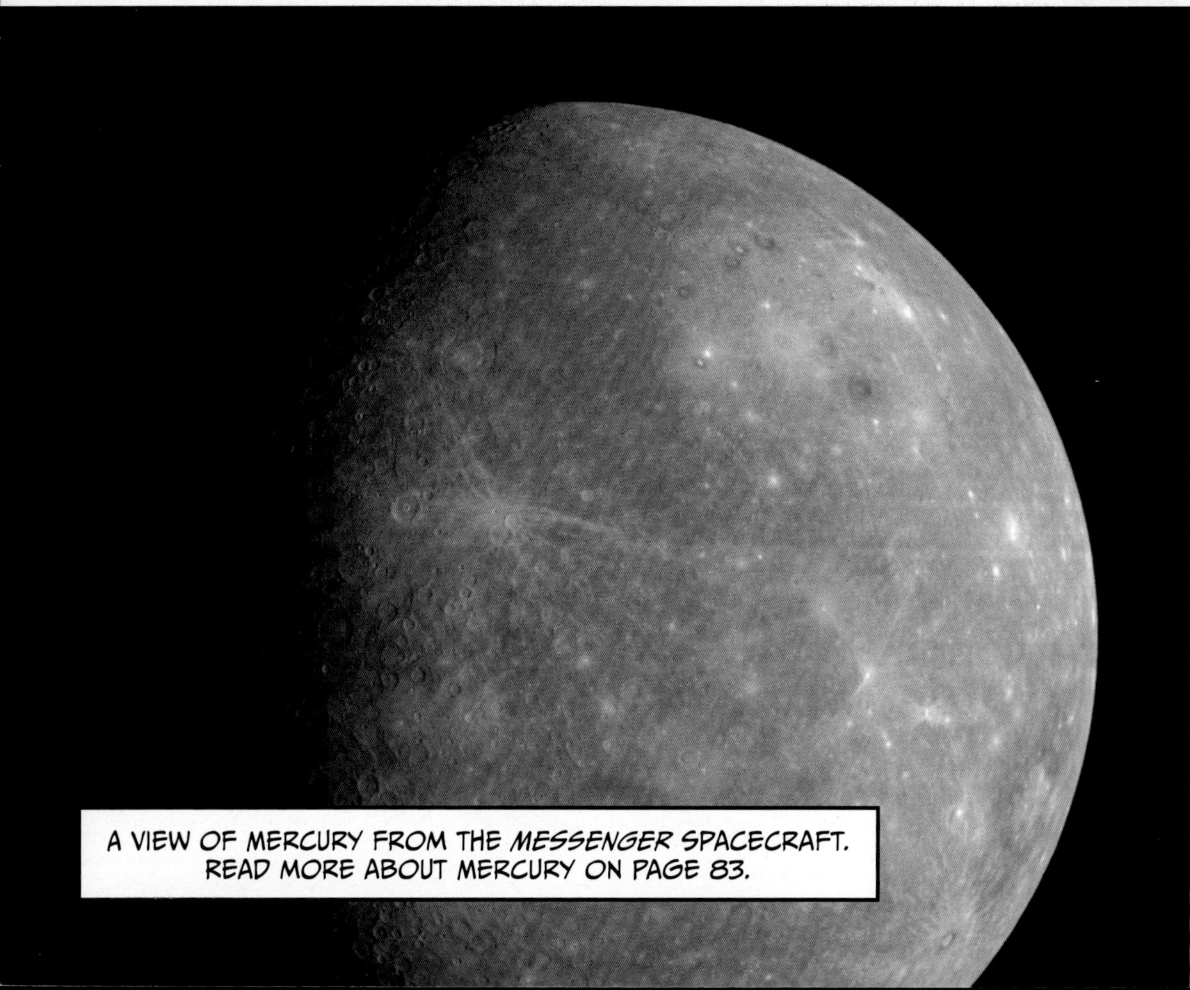

A VIEW OF MERCURY FROM THE *MESSENGER* SPACECRAFT.
READ MORE ABOUT MERCURY ON PAGE 83.

CREDIT: NASA

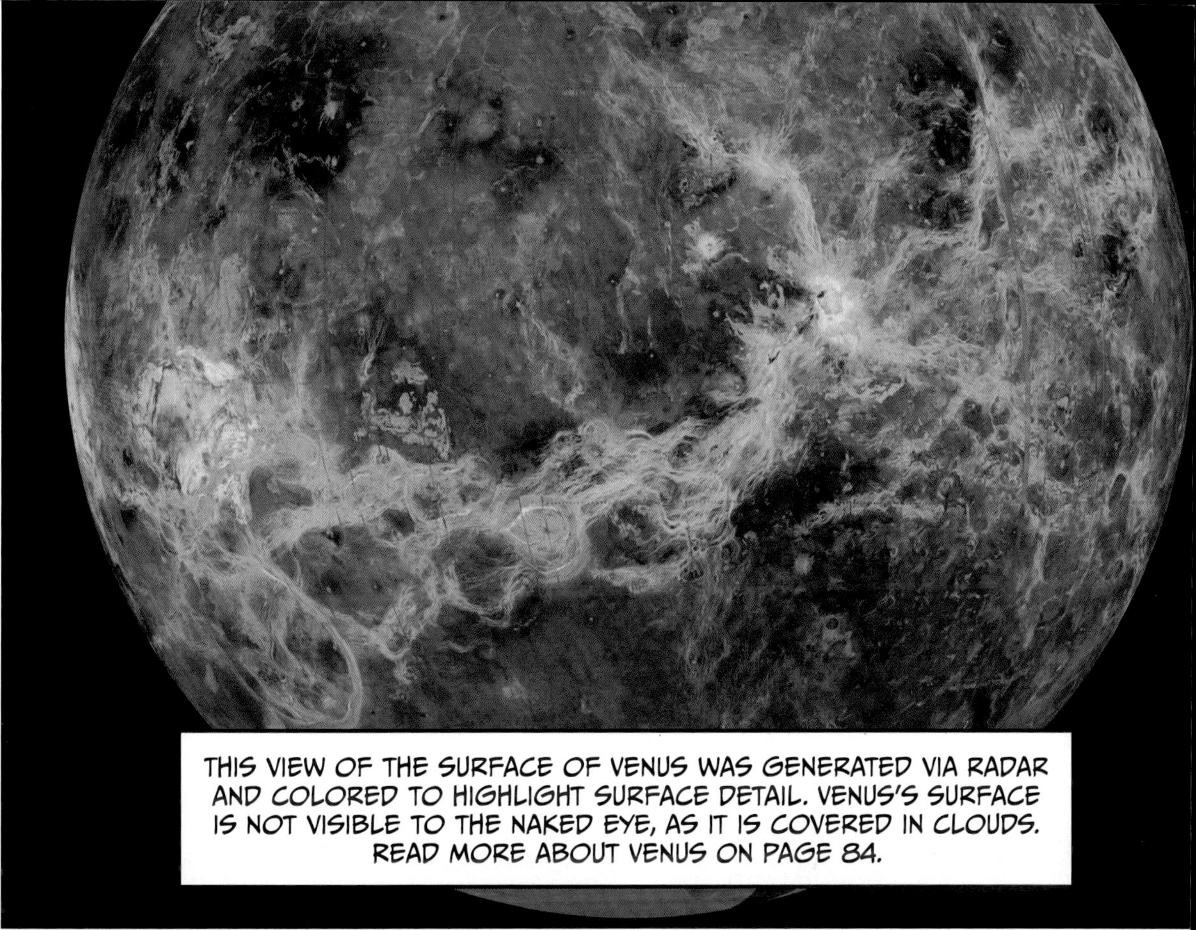

THIS VIEW OF THE SURFACE OF VENUS WAS GENERATED VIA RADAR AND COLORED TO HIGHLIGHT SURFACE DETAIL. VENUS'S SURFACE IS NOT VISIBLE TO THE NAKED EYE, AS IT IS COVERED IN CLOUDS. READ MORE ABOUT VENUS ON PAGE 84.

CREDIT: NASA/JPL

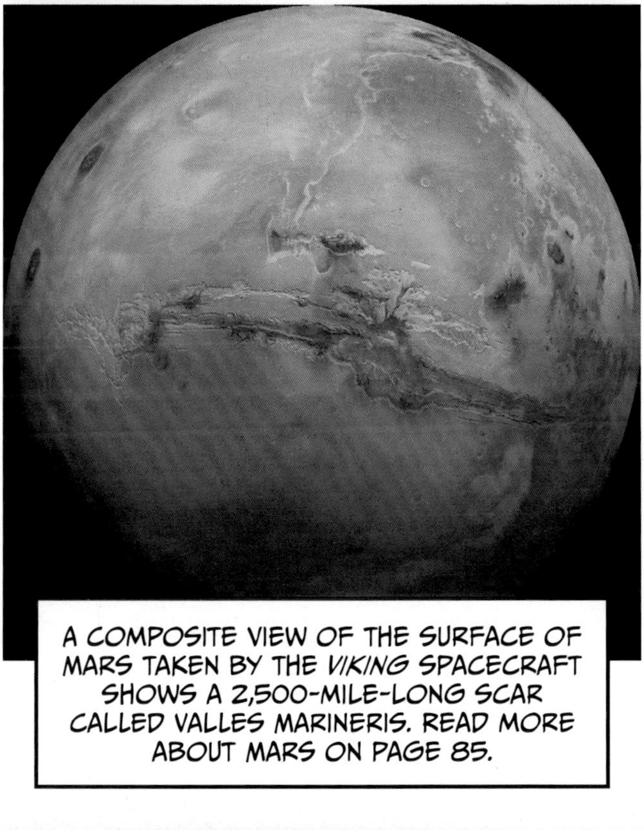

A COMPOSITE VIEW OF THE SURFACE OF MARS TAKEN BY THE *VIKING* SPACECRAFT SHOWS A 2,500-MILE-LONG SCAR CALLED VALLES MARINERIS. READ MORE ABOUT MARS ON PAGE 85.

CREDIT: NASA

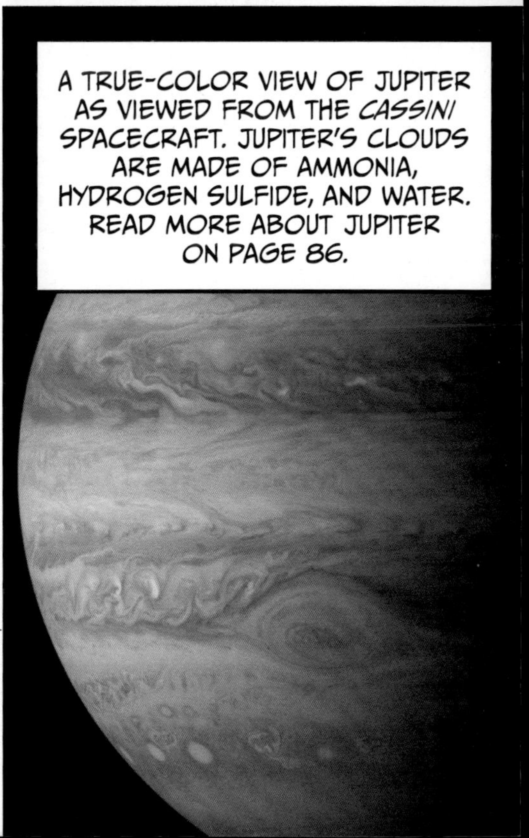

A TRUE-COLOR VIEW OF JUPITER AS VIEWED FROM THE *CASSINI* SPACECRAFT. JUPITER'S CLOUDS ARE MADE OF AMMONIA, HYDROGEN SULFIDE, AND WATER. READ MORE ABOUT JUPITER ON PAGE 86.

CREDIT: NASA/JPL/*Space Science Institute*

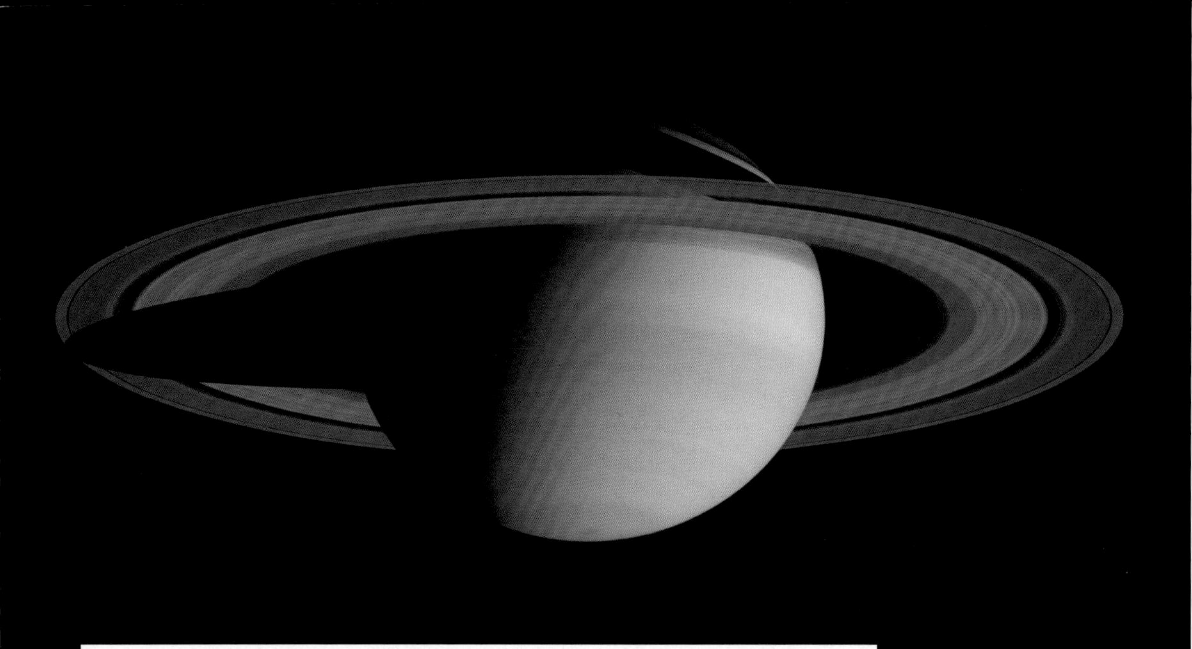

SATURN'S RINGS ARE VISIBLE IN THIS TRUE-COLOR IMAGE FROM *CASSINI*. READ MORE ABOUT SATURN ON PAGE 87.

CREDIT: NASA/JPL/Space Science Institute

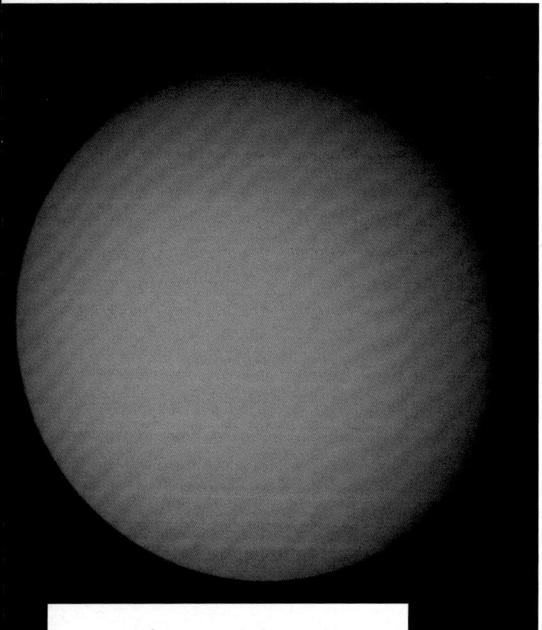

THE *VOYAGER 2* SPACECRAFT SHOWED THAT THE SURFACE OF URANUS IS HAZY AND BLUE. READ MORE ABOUT URANUS ON PAGE 88.

CREDIT: NASA/JPL

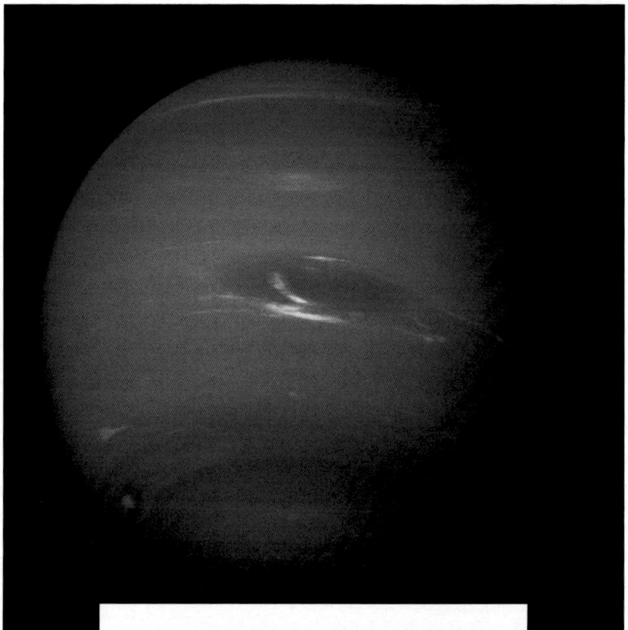

VOYAGER 2 ALSO CAPTURED A VIEW OF THE SURFACE OF NEPTUNE, INCLUDING A LARGE STORM THAT APPEARS AS A GREAT DARK SPOT. READ MORE ABOUT NEPTUNE ON PAGE 89.

CREDIT: NASA/JPL

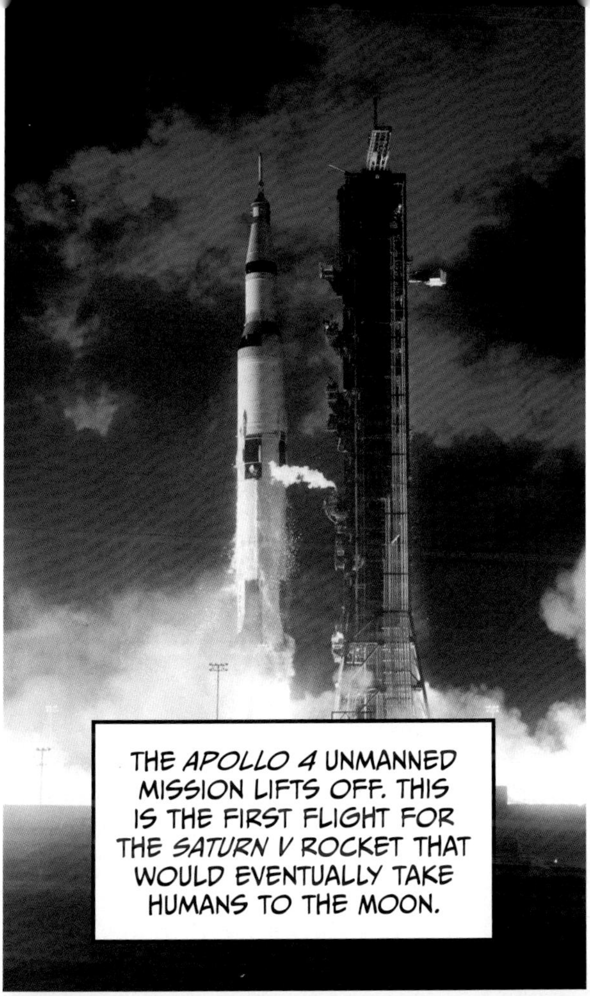

THE *APOLLO 4* UNMANNED MISSION LIFTS OFF. THIS IS THE FIRST FLIGHT FOR THE *SATURN V* ROCKET THAT WOULD EVENTUALLY TAKE HUMANS TO THE MOON.

CREDIT: NASA

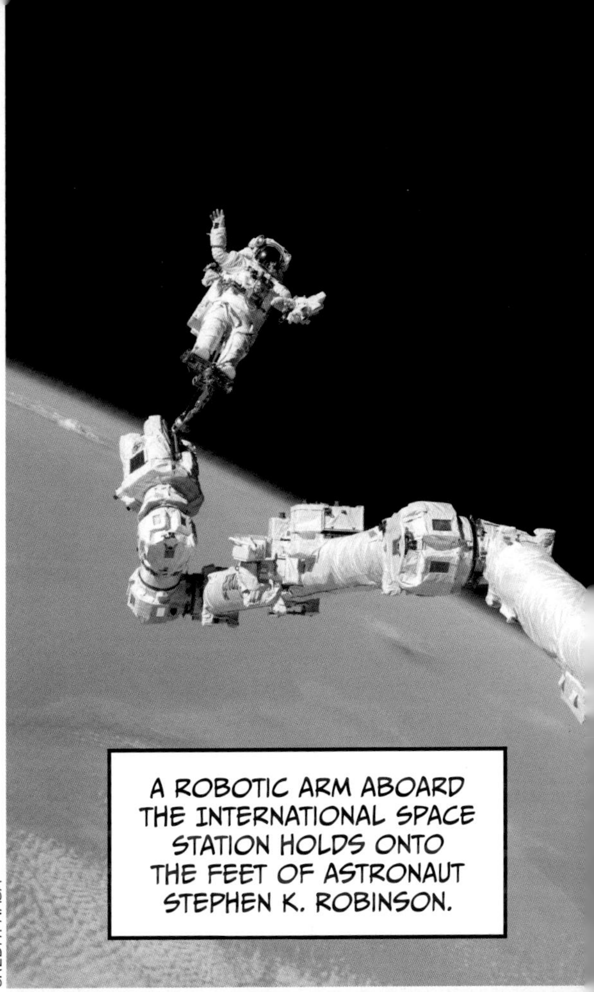

A ROBOTIC ARM ABOARD THE INTERNATIONAL SPACE STATION HOLDS ONTO THE FEET OF ASTRONAUT STEPHEN K. ROBINSON.

CREDIT: NASA

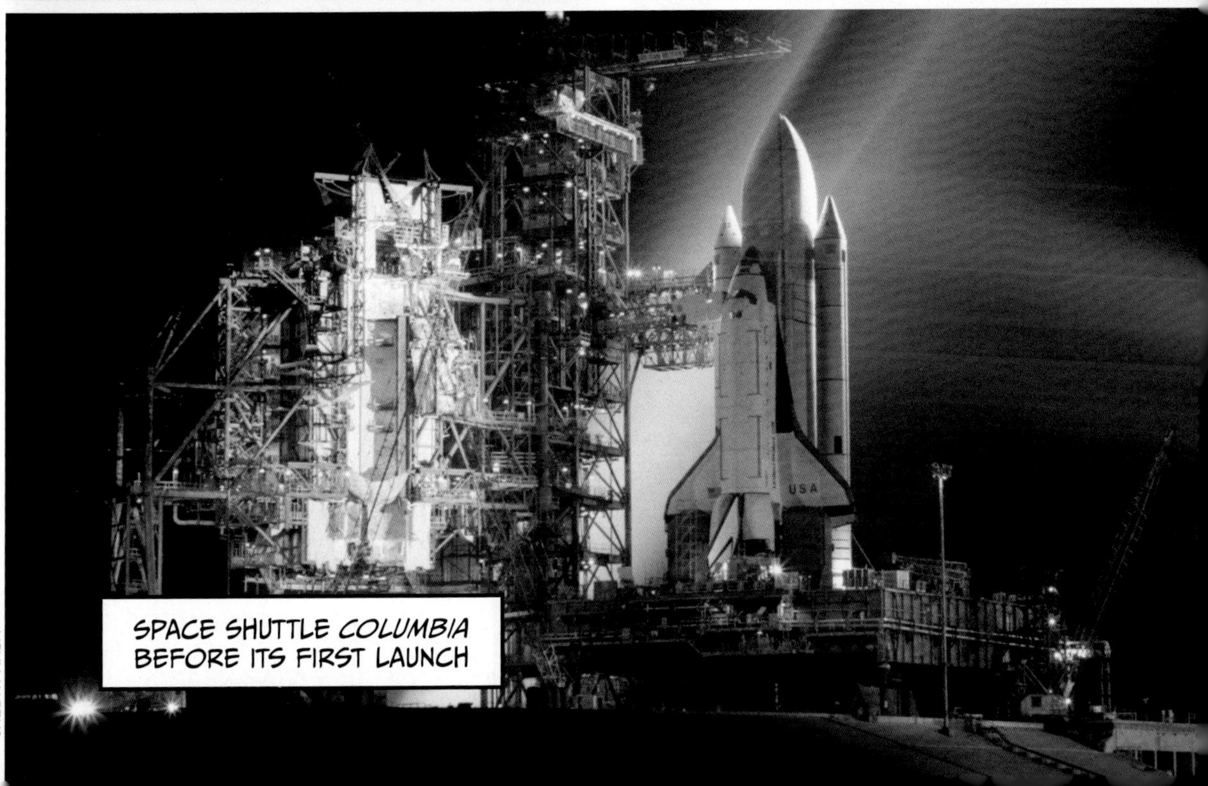

SPACE SHUTTLE *COLUMBIA* BEFORE ITS FIRST LAUNCH

CREDIT: NASA

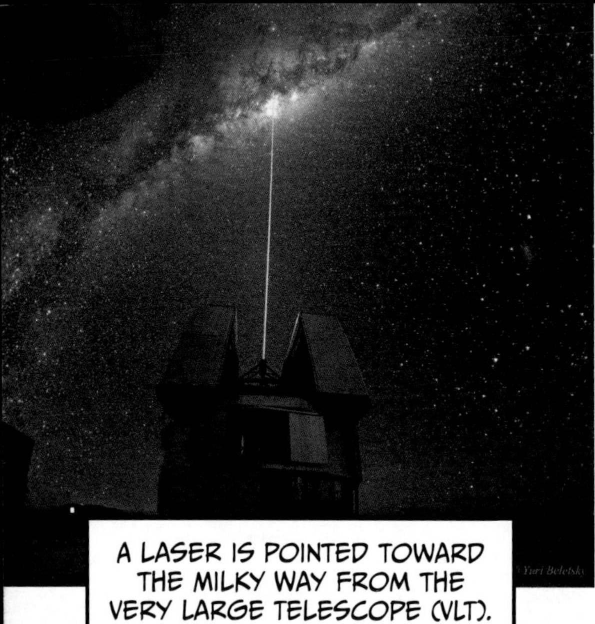

© Yuri Beletsky
A LASER IS POINTED TOWARD THE MILKY WAY FROM THE VERY LARGE TELESCOPE (VLT).

CREDIT: ESO/Yuri Beletsky (CC by 3.0)

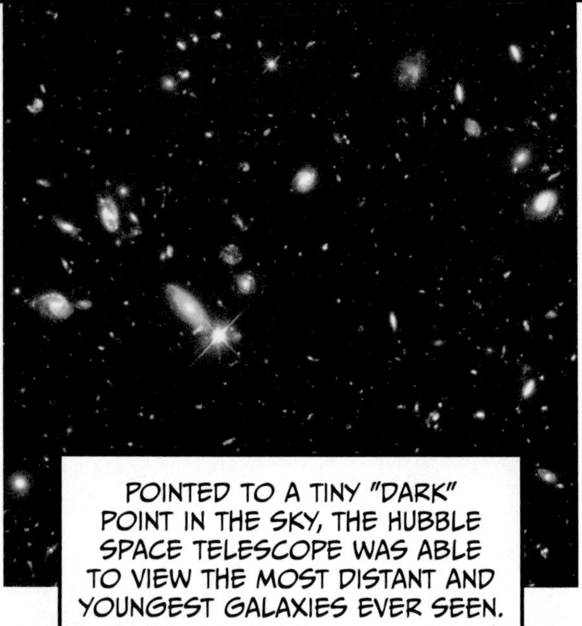

POINTED TO A TINY "DARK" POINT IN THE SKY, THE HUBBLE SPACE TELESCOPE WAS ABLE TO VIEW THE MOST DISTANT AND YOUNGEST GALAXIES EVER SEEN.

CREDIT: R. Williams (STScI), the Hubble Deep Field Team, NASA

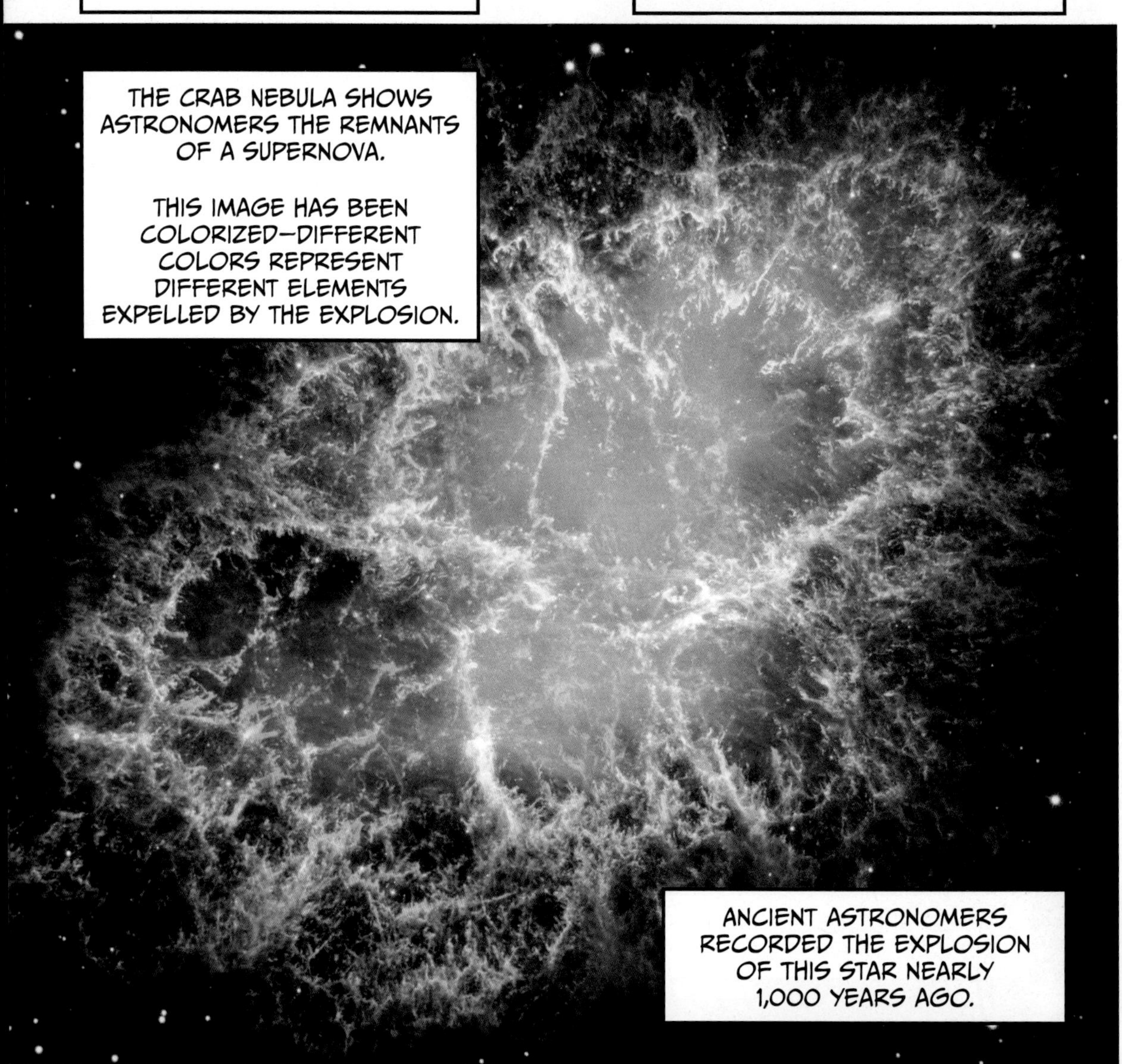

THE CRAB NEBULA SHOWS ASTRONOMERS THE REMNANTS OF A SUPERNOVA.
THIS IMAGE HAS BEEN COLORIZED—DIFFERENT COLORS REPRESENT DIFFERENT ELEMENTS EXPELLED BY THE EXPLOSION.
ANCIENT ASTRONOMERS RECORDED THE EXPLOSION OF THIS STAR NEARLY 1,000 YEARS AGO.

CREDIT: NASA/STScI

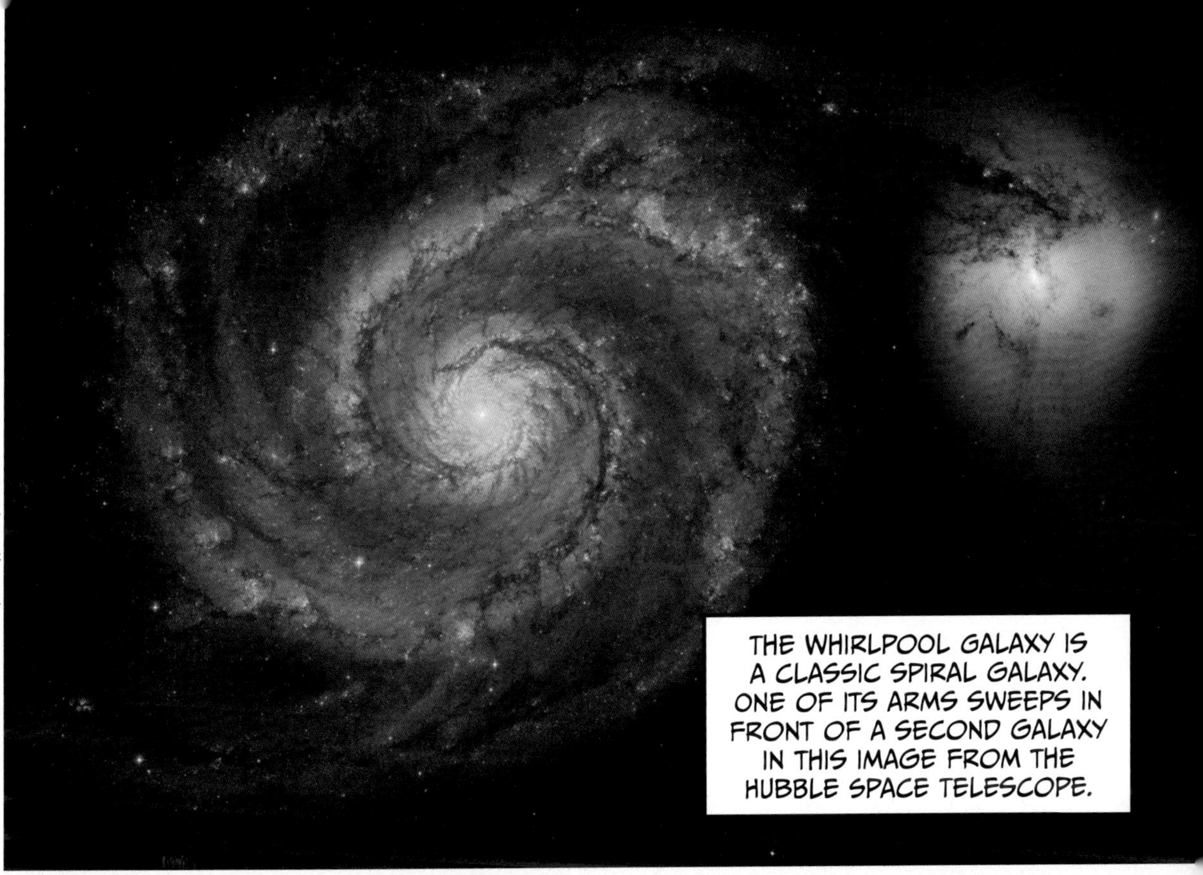

THE WHIRLPOOL GALAXY IS A CLASSIC SPIRAL GALAXY. ONE OF ITS ARMS SWEEPS IN FRONT OF A SECOND GALAXY IN THIS IMAGE FROM THE HUBBLE SPACE TELESCOPE.

CREDIT: S. Beckwith (STScI), ESA, NASA

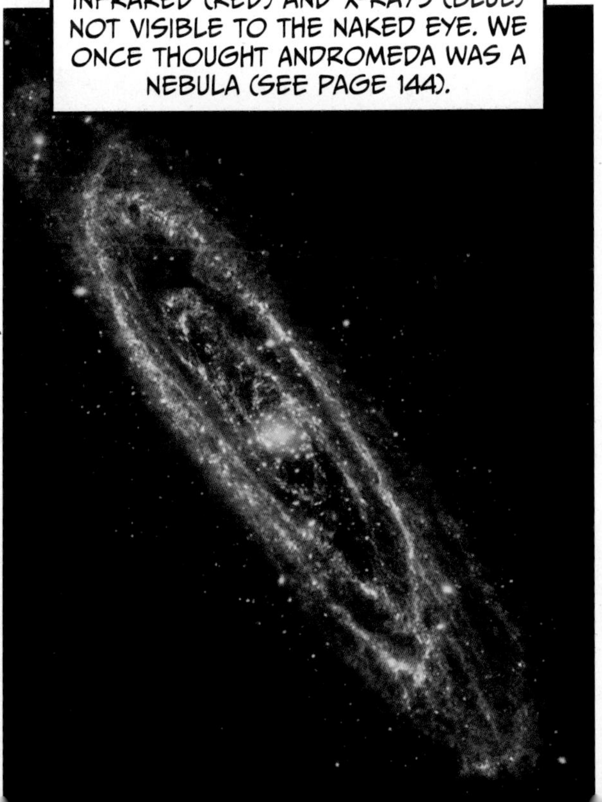

THIS IMAGE OF ANDROMEDA, OUR NEIGHBORING GALAXY, SHOWS INFRARED (RED) AND X-RAYS (BLUE) NOT VISIBLE TO THE NAKED EYE. WE ONCE THOUGHT ANDROMEDA WAS A NEBULA (SEE PAGE 144).

CREDIT: ESA/Herschel/PACS/SPIRE/J. Fritz, U. Gent; X-ray: ESA/XMM Newton/EPIC/W. Pietsch, MPE

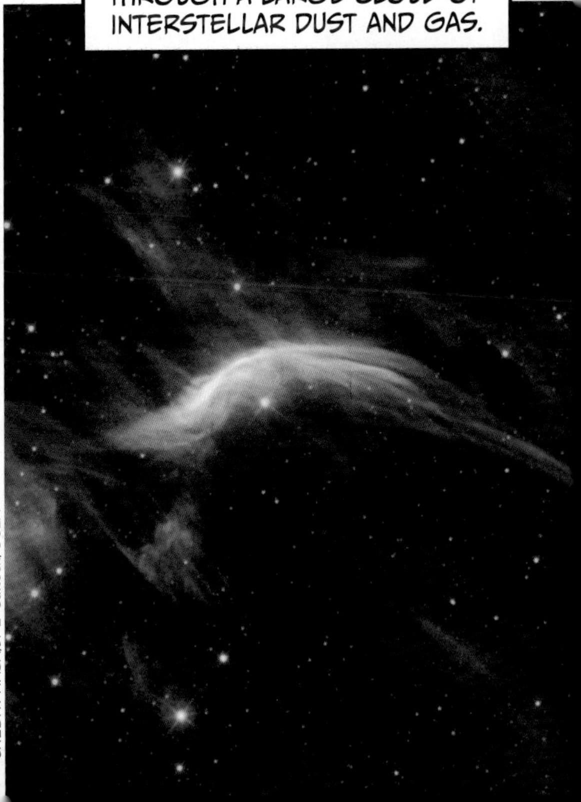

THIS INFRARED IMAGE SHOWS A BRIGHT BLUE STAR HURTLING THROUGH A LARGE CLOUD OF INTERSTELLAR DUST AND GAS.

CREDIT: NASA/JPL-Caltech/UCLA

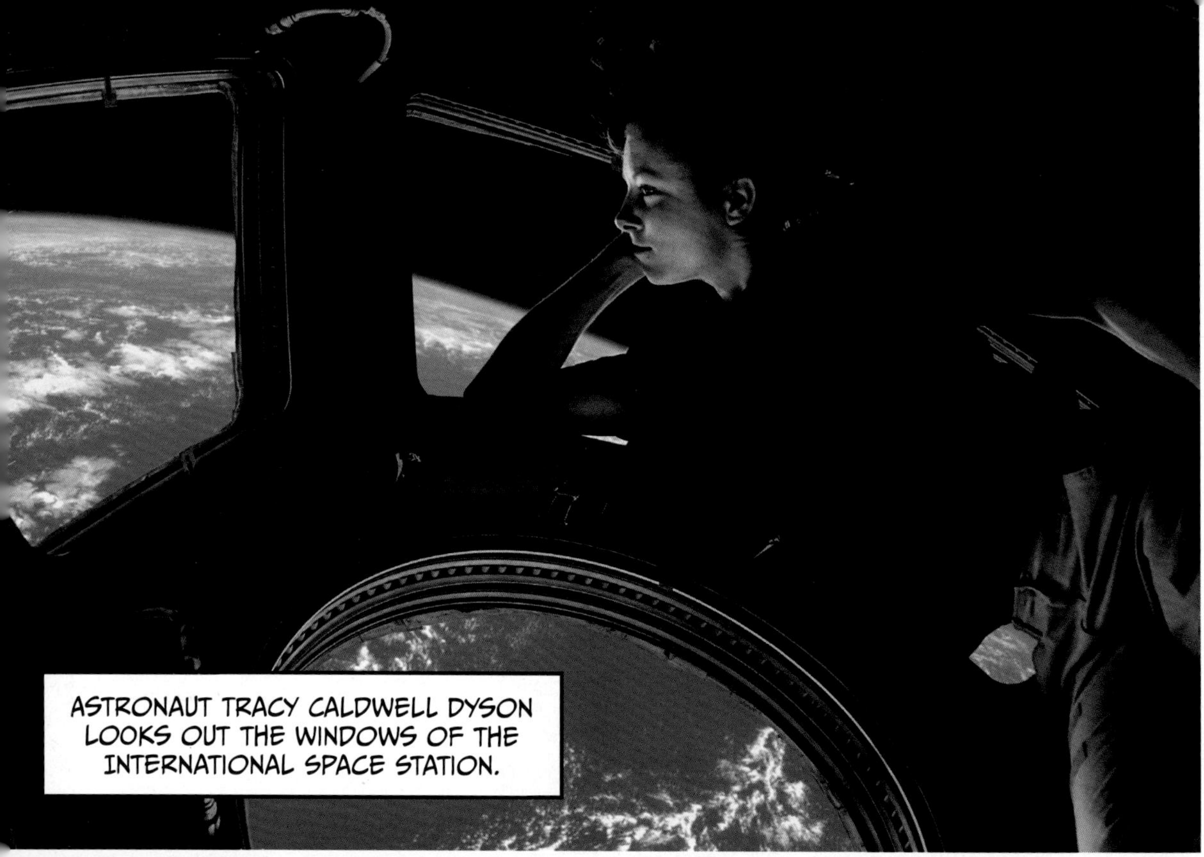
ASTRONAUT TRACY CALDWELL DYSON LOOKS OUT THE WINDOWS OF THE INTERNATIONAL SPACE STATION.

CREDIT: NASA/Douglas Wheelock

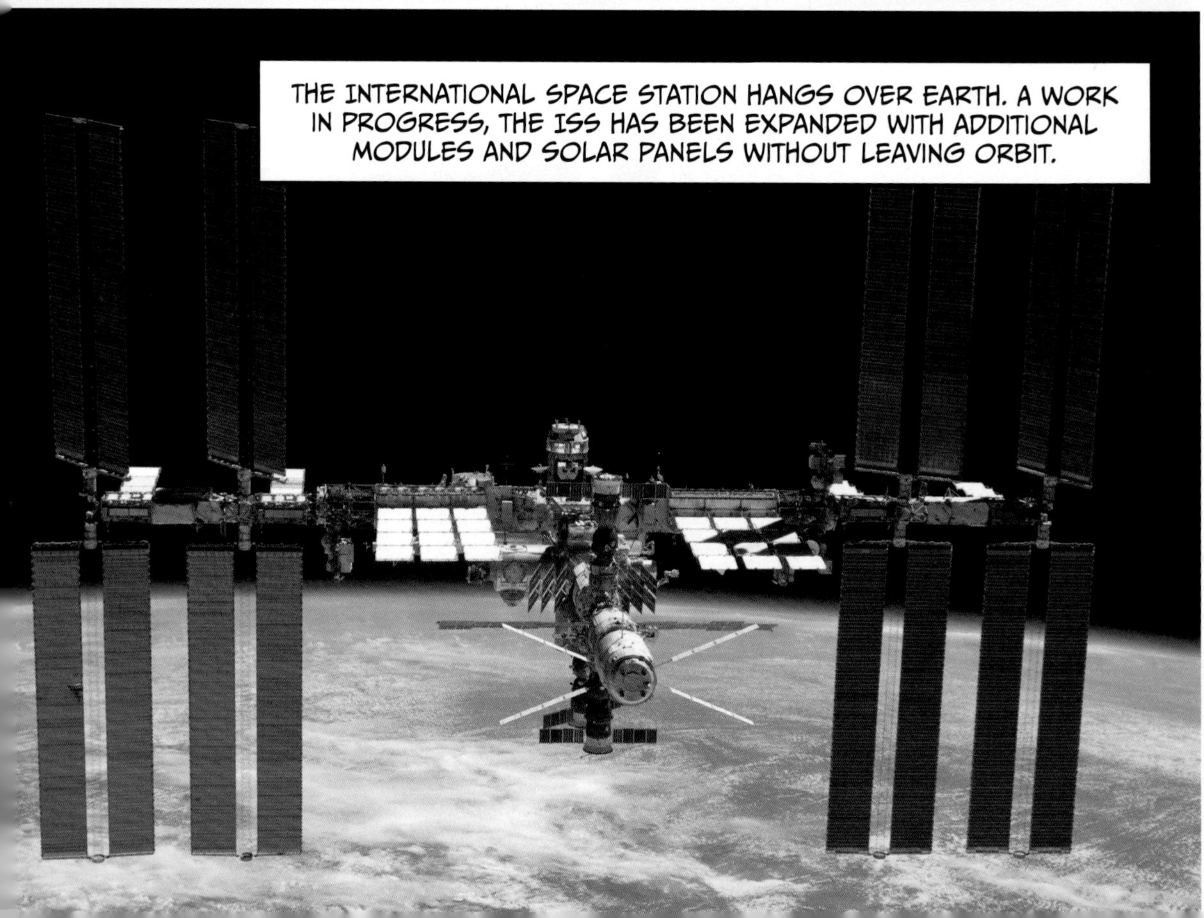
THE INTERNATIONAL SPACE STATION HANGS OVER EARTH. A WORK IN PROGRESS, THE ISS HAS BEEN EXPANDED WITH ADDITIONAL MODULES AND SOLAR PANELS WITHOUT LEAVING ORBIT.

CREDIT: STS-133 Shuttle Crew, NASA